Managing Water on China's Farms

Managing Water on China's Farms

Institutions, Policies and the Transformation of Irrigation Under Scarcity

Jinxia Wang
Center for Chinese Agricultural Policy, Chinese Academy of Sciences, and School of Advanced Agricultural Sciences, Peking University, Beijing, People's Republic of China

Qiuqiong Huang
Department of Agricultural Economics & Agribusiness, University of Arkansas, Fayetteville, AR, USA

Jikun Huang
Center for Chinese Agricultural Policy, Chinese Academy of Sciences, and School of Advanced Agricultural Sciences, Peking University, Beijing, People's Republic of China

Scott Rozelle
Freeman Spogli Institute for International Studies, Stanford University, Stanford, CA, USA

AMSTERDAM • BOSTON • HEIDELBERG • LONDON
NEW YORK • OXFORD • PARIS • SAN DIEGO
SAN FRANCISCO • SINGAPORE • SYDNEY • TOKYO

Academic Press is an imprint of Elsevier

Academic Press is an imprint of Elsevier
125 London Wall, London EC2Y 5AS, UK
525 B Street, Suite 1800, San Diego, CA 92101-4495, USA
50 Hampshire Street, 5th Floor, Cambridge, MA 02139, USA
The Boulevard, Langford Lane, Kidlington, Oxford OX5 1GB, UK

Notices

Knowledge and best practice in this field are constantly changing. As new research and experience broaden our understanding, changes in research methods, professional practices, or medical treatment may become necessary.

Practitioners and researchers must always rely on their own experience and knowledge in evaluating and using any information, methods, compounds, or experiments described herein. In using such information or methods they should be mindful of their own safety and the safety of others, including parties for whom they have a professional responsibility.

To the fullest extent of the law, neither the Publisher nor the authors, contributors, or editors, assume any liability for any injury and/or damage to persons or property as a matter of products liability, negligence or otherwise, or from any use or operation of any methods, products, instructions, or ideas contained in the material herein.

British Library Cataloguing-in-Publication Data
A catalogue record for this book is available from the British Library

Library of Congress Cataloging-in-Publication Data
A catalog record for this book is available from the Library of Congress

ISBN: 978-0-12-805164-1

For information on all Academic Press publications
visit our website at https://www.elsevier.com/

Publisher: Nikki Levy
Acquisition Editor: Nancy Maragioglio
Editorial Project Manager: Billie Jean Fernandez
Production Project Manager: Caroline Johnson
Designer: Greg Harris

Typeset by TNQ Books and Journals
www.tnq.co.in

Contents

About the Authors

Dr. Jinxia Wang is Deputy Director and Professor at the Center for Chinese Agricultural Policy in Chinese Academy of Sciences, Professor at Institute of Geographical Sciences and Natural Resources Research and School of Advanced Agricultural Sciences in Peking University. Her research focuses on water management, institution and policy, impact evaluation and adaptation strategies of climate change, and rural environmental policy. She received her PhD in Agricultural Economics (2000) at Chinese Academy of Agricultural Sciences. She has published more than 160 papers (more than 60 are in English) and five books. In 2009, she received the Outstanding Young Scientist Award from the National Natural Science Foundation in China.

Dr. Qiuqiong Huang is an Associate Professor at the University of Arkansas. Her research focuses on natural resource economics (with an emphasis on water and land) and development issues including education, migration, and labor market outcomes. She received her PhD in Agricultural and Resource Economics (2006) at the University of California, Davis. Her publications have appeared in *Agricultural Water Management*, *American Journal of Agricultural Economics*, *Australian Journal of Agricultural and Resource Economics*, *Ecological Economics*, *Environment and Development Economics*, *Food Policy*, *Journal of Productivity Analysis*, and *Water Resources Research*.

Dr. Jikun Huang is the Founder and Director of the Center for Chinese Agricultural Policy in Chinese Academy of Sciences, Professor at Institute of Geographical Sciences and Natural Resources Research and School of Advanced Agricultural Sciences in Peking University. Fellow of the World Academy of Sciences (TWAS) for the advancement of science in developing countries, and Honorary Life Member of the International Association of Agricultural Economists. He is also vice-president of the Chinese Association of Agricultural Economics, vice-president of the Chinese Association of Agri-technology Economics, and the elected president of the Asian Association of Agricultural Economists. He received his BS degree from Nanjing Agricultural University in 1984 and his PhD in Agricultural Economics from the University of the Philippines at Los Banos in 1990. His research covers a wide range of issues on China's agricultural and rural development, including work on agricultural R&D policy, water and land policy, agricultural price and trade policy, food demand and supply projection, and economics of climate change. He received Award for China's 10 outstanding youth scientists in 2002, Outstanding Achievement Award for Overseas Returning Chinese in 2003, Outstanding Contribution Award on Management Science in 2008, and IRRI's Outstanding Alumni Award in 2010. He has published about 460 journal papers, of which about 240 were published in international journals, including *Science* and *Nature*. He is the coauthor of 19 books.

Dr. Scott Rozelle is the Helen F. Farnsworth Senior Fellow and the co-director of the Rural Education Action Program in the Freeman Spogli Institute for International Studies at Stanford University. His research focuses almost exclusively on China and is concerned with agricultural policy, including the supply, demand, and trade in agricultural projects; the emergence and evolution of markets and other economic institutions in the transition process and their implications for equity and efficiency; and the economics of poverty and inequality, with an emphasis on rural education, health, and nutrition. Rozelle's papers have been published in top academic journals, including *Science*, *Nature*, *American Economic Review*, and the *Journal of Economic Literature*. He received his PhD (1990) from Cornell University.

Preface

China has emerged as a major player in the world economy. Water is the most critical factor that limits China's further growth, and water issues are among the top priorities of China's leaders. For example, the number one document of 2011 (the most important policy statement issued by China's central government annually) completely focused on water issues. China's growing water problems also have a large impact worldwide. If China were to rely heavily on food produced outside of China, the massive volumes of food imports by China would raise food prices internationally.

This book examines a series of water issues. We begin with describing the water shortage problems China is facing, particularly in the northern part of the country. We then look at what the government and farmers are doing and whether past policies have been effective in resolving the water problems. We document the change of existing and new water management institutional forms over time and across provinces throughout northern China. We also assess the impacts of these changes in the rural sector. Finally, we examine potential solutions that our research has uncovered, answering the question: Who can build the bridge over China's troubled waters?

One distinguishing feature of this book is that all analyses come from information we collected firsthand in China's rural villages. The series of surveys we have conducted over the years cover diverse geographic regions that are representative of northern China and collect data spanning a long period. Using structured questionnaires, the surveys also obtained perspectives from multiple stakeholders such as village leaders, water managers, and farmers. Our data collection efforts have made the book informative and the analysis quantitative and rigorous. For example, we were able to track the changes in water management institutions at the village level over time. The detailed information on agricultural inputs including water use allows us to study water use behavior at the household level. The detailed microlevel data also enables estimating strategies such as household fixed effects or instrumental variables to avoid biased estimates from unobservable heterogeneity or simultaneity. The policy-oriented research and rich analysis in this book make it of interest to both policy makers and researchers with a focus on China water problems. This book can also be used in a master's- or PhD-level resource economics course.

Jinxia Wang, Qiuqiong Huang, Jikun Huang, and Scott Rozelle
Beijing, China

Acknowledgments

We acknowledge the support of the following institutes and grants that funded our research: National Natural Science Foundation in China (70925001, 70733004), Ministry of Science and Technology in China (2012CB955700), Chinese Academy of Sciences (KZZD-EW-08-04; KSZD-EW-Z-021-1), International Water Management Institute, World Bank, Australian Center for International Agricultural Research, University of Minnesota Grant-in-Aid, University of Minnesota Global Programs and Strategy Alliance, and University of Minnesota Center for International Food and Agricultural Policy. We would like to thank our collaborators and colleagues for their contributions to the research; they include: Lijuan Zhang, Zhigang Xu, Amelia Blanke, Richard Howitt, Siwa Msangi, Bryan Lohmar, David Dawe, Anna Heaney, Ahmed Hafi, Stephen Beare, and Yuliu.

We also would like to express our appreciation to the following journals' publishers who have generously allowed us to include some peer-reviewed papers in this book: *Food Policy, Australian Journal of Agricultural and Resource Economics, Hydrogeology Journal, World Development, Agricultural Water Management, Environment and Development Economics, Agricultural Economics*, and *Journal of the American Water Resources Association.*

Introduction and Conclusions

INTRODUCTION

China is not particularly well endowed with water, yet over the past several decades, water has been treated as a cheap resource and used to expand agricultural and industrial production. While China's water resources are substantial compared with those of some other countries, its population is also much larger. Furthermore, its water is not evenly distributed across the country or across major agricultural regions. The nation's water resources are highly concentrated in southern China, while northern China, the area north of the Yangtze River Basin, has only one-fourth the per-capita water endowment of the South and one-tenth the world average (Ministry of Water Resources, 2012). The lower levels of rainfall in northern China are also much more seasonal than in the south, with more than 70% of the rain falling between June and September. Yet despite having only 24% of the nation's water resources, northern China contains more than 65% of China's cultivated land, and remains an important agricultural region and the site of much of China's industrial production. It accounts for roughly half of China's grain production and over 45% of national GDP (Ministry of Water Resources, 2012; SSB, 2012).

Increasing industrial output, expanding agricultural production, and rising domestic incomes have all contributed to higher demand for water resources in China. From 1949 to 2011, per capita use increased 136%, and total water use in China increased 492% (Ministry of Water Resources, 2012; SSB, 2012). Water use has increased across all sectors. Although China's irrigated area has continued to expand and agriculture still accounts for the majority of water consumption, industry's share has been growing rapidly, rising from 2 to 24% from 1949 to 2011. These changes have been putting a serious strain on China's water supply, raising the question of whether such growth rates are sustainable.

Indeed, water-related problems have been springing up across northern China, affecting the availability of both surface water and groundwater. The river systems that supply water to many irrigation districts sometimes do not provide sufficient water because upstream users withdraw more water than they are allocated by law. Because of excessive withdrawals, the Yellow River has run dry before reaching the ocean for at least some period during most years since the mid 1970s (Wang, 1999). Withdrawals from the Fuyang River, in the upper part of the Hai River Basin, have severely depleted the main river (Wang et al., 2005). In 1977–1997, almost no flow was recorded at the Aixingzhuang hydrological monitoring station, near the middle of the river basin. Cangzhou Prefecture, downstream from the Fuyang River, now receives only 10% of the surface water that it received in the 1970s.

China has also seen a steep fall in the level of the water table and related environmental issues. In several parts of the Fuyang River Basin, the shallow-water table has fallen at an accelerating rate over the past 20 years (Wang et al., 2005), while the deep-water table has been declining at an even faster rate in some areas. Excessive rates of groundwater withdrawal have generated large cones of depression under urban areas in six Hebei Province prefectures, and have also caused land subsidence in some predominantly rural counties (Hebei Hydrological Bureau and Water Environmental Monitor Center, 1999). Large groundwater extractions and the subsequent fall in the water table are also affecting the quality of groundwater, particularly through the intrusion of seawater. A survey carried out in the coastal provinces of northern China in the early 1990s found that more than 2000 km^2 of formerly freshwater table had fallen below sea level (Nickum, 1998). Farmers, industrialists, and city water managers were forced to abandon more than 8000 tubewells, and irrigated area declined by 40,000 ha. While these losses admittedly represent only a small part of overall agricultural production in northern China, they do significantly affect local residents and some observers predict that unless groundwater sources are allowed to replenish, the problems will increase at an accelerating rate.

China's government has identified the nation's rising water scarcity as one of the key problems that must be solved if the nation is to meet its national development plan in the coming years (Zhang, 2001). Shortages of water are attenuating efforts to alleviate poverty and are becoming a major source of environmental problems (World Bank, 1998; Zhang, 2000). In many regions of the country, rapidly growing industry and an expanding, increasingly wealthy urban population regularly out-compete the nation's farmers for limited water resources, threatening to curtail growth in food production.

In facing increasing water shortages, leaders have debated about which approach they should use to address water scarcity problems, although there has been very little success so far (Lohmar et al., 2003). Historically, developing more water resources to increase the water supply has been given the highest priority in resolving water shortages. Since the 1950s, China's government has invested more than 100 billion US dollars into constructing infrastructure for developing new water resources (Wang, 2000). The most well-known effort is the South–North Water Transfer Project that moves water from the Yangtze River Valley to northern China. Despite such ambitious goals, the high cost of developing new sources of water will make it so that the volume of water that can be added to northern China's water equation will still be marginal.

It has become clear that a supply-side approach alone is not sufficient to deal with the growing water scarcity. Gradually China's leaders have started to recognize the need to stem the rising demand for water from all sectors (Boxer, 2001). Leaders have promoted water-saving technologies and have even considered the implementation of a water pricing policy (Chen, 2002; Huang et al., 2010). Unfortunately, most of the efforts to encourage the use of sophisticated water saving technologies, such as drip and sprinkler irrigation, have failed and the Ministry of Water Resources has distanced itself from a water policy based on water-saving technology (Zai, 2002). Moreover, political considerations will most likely keep leaders from moving too aggressively on raising prices, at least in the agricultural sector (Huang et al., 2010).

With the failure and infeasibility of several of their policy options, leaders in recent years have begun to consider water management reform as a key part of their strategy to combat China's water problems (Wang et al., 2005). Despite water shortages, users in all sectors of the economy—but especially those in agriculture, by far the nation's largest consumer of water—do not efficiently use the water they are allocated. Xu (2001), for example, estimated that due to poor management of the nation's canal network, only 50% of the water from primary canals is actually delivered to the field. Local irrigation managers and farmers do not efficiently use the water that reaches the village's fields either, wasting between 20 and 30% of their water. Hence, overall, only about 40% of the water in China's surface water system that is allocated to agricultural production is actually used by farmers on their crops. Others have estimated even greater inefficiencies (Fang, 2000). In response, it has been proposed that local leaders reform the institutions that manage water in China's communities (Nian, 2001; Reidinger, 2002).

China has begun its own water management reform, creating new policies and institutions designed to more effectively manage the nation's water resources. Nevertheless, many difficulties remain. China's existing water management system was built haphazardly over the course of many years, with the goal not of conserving a valuable resource, but of exploiting that resource to maximize economic growth. Now that the objective is shifting, the system has been revealed to be a confusing mass of independent stakeholders. The challenge China's leadership now faces is to reorganize the national water management system in such a way as to align the stakeholders' incentives with that of the state. In this book, we describe the specifics of China's water shortage, the steps that have already been taken to make more effective use of a scarce resource, and the effects that both the water shortage itself and the subsequent policy responses are having on national production and development.

China's agricultural sector is not only the nation's largest consumer of water, but also employs 53% of China's population. When a national water shortage threatens China's agricultural sector, it has serious implications for rural development. It may even influence trade patterns, since China is the world's largest and most populated agricultural producer. The threat of losing irrigation in important temperate regions in northern China could cause China to import a much larger share of its grain needs than it has in the past. In addition, the need to make more economical use of water could also facilitate movement into high-value cash crops using sophisticated water-saving irrigation technologies. High-value crop production is also often labor intensive and thus matches China's comparative advantage on international markets. Thus, the need to increase the value of water used in agriculture may also induce a more general structural change in China's agriculture that many observers foresee as China's likely role in international markets: an importer of land-intensive grains and an exporter of high-value and labor-intensive crops.

A better understanding of water flows in China will help clarify how water allocation policies will affect aggregate agricultural production. Groundwater recharge rates, and the sources of these recharges, will help identify areas where water tables are particularly threatened and the interaction between surface-water use and falling groundwater tables. Finally, understanding the hydrology of river basins will help determine how policies to encourage water saving upstream will affect downstream users.

Perhaps the most important factor determining the effect of water scarcity on agricultural production is the response of individual farmers and local water managers. Because they have the autonomy to make such decisions as the volume of water applied to crops, the timing of water allocation, and which irrigation technologies and practices to use, the incentives they face will determine how effectively water conservation policies are carried out at the ground level. A wide variety of institutional responses have been established to encourage farmers and local leaders to adopt water-saving practices, including reforming irrigation management, raising water prices and reforming water fee collection, and investing in water-saving irrigation technology. We ask how these institutions work, which types are the most effective, what the determinants of adopting such measures are, and how they affect agricultural production. It is also important to empirically examine how the incentives faced by local water managers induce different water allocation policies, thus allowing for a better understanding of how policy changes affect water deliveries, and thus, agricultural production.

This book is organized into four sections. The first section is to set the stage for the research, and the second and third sections are to examine groundwater and surface water management separately. The fourth section is to discuss the future options for water management in China. Within the first section, four chapters are included. Chapter Water Scarcity in the Northern China examines the water scarcity situation in the major study region, the Northern China; chapter Irrigation, Agricultural Production and Rural Income analyzes the contribution of irrigation to China's agricultural production and rural income; and chapter China's Agricultural Water Policy Reforms: Increasing Investment, Resolving Conflicts, and Revising Incentives identifies the major issues related with water management institution and policies in China. Chapter Water Survey Data summarizes water survey data used for all analysis in the book. For the second section, chapters Evolution, Determinants, and Impacts of Tubewell Ownership/Management, Development of Groundwater Markets in China, and Impacts of Groundwater Markets on Agricultural Production in China include studies of the evolution, determinants, and impacts of tubewell ownership/management, and development and impacts of Groundwater markets. The third section covers four chapters. Chapter Water User Associations and Contracts: Evolution and Determinants analyzes the evolution and determinants of WUAs and Contracts; chapter Determinants of Contractual Form analyzes the determinants of contractual form; chapter Impacts of Surface Water Management Reforms looks at the impacts of surface water management reforms; and chapter Evaluation of Water User Associations is about the evaluation of WUAs. The final section (section four) discusses the future options from three aspects, the irrigation water pricing policy (chapter: Irrigation Water-Pricing Policy), water allocation through water rights institution (chapter: Water Allocation Through Water Rights Institution), and adoption of water saving technology (chapter: Adoption of Water-Saving Technology). Throughout the book, the focus is on water used for irrigation in agricultural production in rural China.

CONCLUSIONS

This book mainly includes the following findings.

First, groundwater is progressively used in China, and this is provoking a water table falling, especially in northern China. Increasing evidence indicates that China is facing a serious water scarcity, especially in northern China. This scarcity is due not only to a falling water supply, but also to rising water demand. With the decline of surface-water resources, farmers in northern China have begun to turn more toward groundwater resources. In 2004, most irrigation in northern China came from groundwater resources; the share of land irrigated by groundwater was nearly 70%. Unfortunately, this reliance on groundwater may have resulted in a number of adverse environmental effects. In the late 1990s the annual rate of overdraft exceeded 9 billion m^3. More than one-third of the overdraft was from deep wells, many of which are nonrenewable (at least in the short to medium run). The drop in the deep groundwater table in some areas has exceeded 2 m per year.

Second, irrigation has a positive impact on agricultural production. Because a larger share of their income comes from cropping, farmers in poor areas increase their incomes relatively more than farmers in richer areas. Irrigation also helps alleviate income inequality in rural areas. In addition, even after accounting for the increased capital costs and production costs associated with irrigation, returns from investments in irrigation are positive in the majority of villages that have invested in new irrigation systems. Although investments should not be made to increase irrigated area in all villages in all of China, in poor areas, irrigation projects should still be given extra weight as a development strategy, especially if benefits outweigh costs. A national water shortage will require some form of water rationing. Our cost-benefit analysis indicates that some communities benefit more from access to irrigation than do others. This finding offers some guidance on how water could potentially be allocated. Water should be shifted away from villages where the cost of irrigation outweighs the benefits, and redirected to those villages that are more reliant on irrigation.

Third, the evolution of tubewell ownership from collective to private has promoted the adjustment of cropping patterns and accelerated the fall in groundwater table in the North China Plain. Since the early 1980s collective ownership of tubewells has largely been replaced by private ownership. Most private tubewells are still owned jointly by several individuals as shareholding tubewells. Changes of natural resource endowments (falls in groundwater table and reduced deliveries of surface water resources) have been shown to lead to changes in the commonly observed forms of institutions. Fiscal measures have promoted the emergence of private tubewells; in contrast, targeted bank loan policies have slowed down tubewell privatization. The privatization of tubewells has promoted the adjustment of cropping patterns while having no adverse impact on crop yield. Specifically, after privatization, farmers have expanded sown area of water sensitive and high-value crops, such as maize, cotton, and noncotton cash crops (mainly horticulture crops). Consistent with the concerns of some observers, the privatization of tubewells has accelerated the fall in the groundwater table. Because of the positive effect privatization has had on income, however, we still believe that policy-makers should continue to support the privatization of tubewells in the North China Plain. Nevertheless, measures should be taken to address the falling groundwater table.

Fourth, tubewell ownership and resource endowment are major factors influencing the emergence of groundwater markets, and groundwater markets help the poor and increase the efficiency of water use. Groundwater markets in northern China have developed in terms of both their breadth (the share of villages in which there is groundwater market activity) and depth (the share of water which the average water-selling tubewell owner sells to others on a market basis). The form of ownership appears to be strongly correlated with the emergence of groundwater markets. Groundwater markets also appear in more villages, and tubewell owners sell a higher share of the water from their wells, when the groundwater table is deep (ie, water is scarce) and land is scarce. All of these findings suggest that when the factors that affect supply and demand for groundwater are present, there is a tendency for markets to emerge. Households that buy water from groundwater markets are poorer than water-selling households. Such a finding implies that groundwater markets have provided greater access to groundwater to poor farmers and possibly help reduce income inequalities in rural China. Many farmers on the North China Plain purchase water from private owners of tubewells. Many of these farmers pay more per cubic meter for their water than farmers who have their own tubewell or those with access to water from collectively owned wells. Our results suggest that farmers who buy water from local groundwater markets use less water than farmers who have their own tubewells or use collective tubewells. However, yields do not diminish. In addition, there is no measurable negative effect on income. Our findings imply that as water in China becomes scarcer, necessitating increased water efficiency, the emergence of markets for groundwater may be an effective way to provide irrigation services. Leaders should consider supporting privatization and encouraging the development of groundwater markets.

Fifth, China's surface water management has been reformed; and only those reforms that established incentive mechanism have played a significant role on improving water use efficiency. Since the 1990s, collective water management has been replaced by WUAs and contracting in many locations. In some villages, reform has been only nominally implemented, and the nominal reform has had little effect on water use. However, in villages that provided water managers with strong incentives, water use fell sharply. The incentives also must have improved the efficiency of the irrigation systems since the output of major crops, such as rice and maize, did not fall, and rural incomes and poverty remained statistically unchanged. Although the literature emphasizes the importance of participation for water management reform, we find little if any effect of participation by farmers on water use in our sample sites. When trying to explain the water management reform as a choice among different contractual forms including fixed-wage, fixed-rent, or profit-sharing, we find localities (such as conditions of canals, nature of the village's resources, and its economic environment) really matters. Therefore, officials that want the reforms to succeed should make an effort to ensure that more emphasis be put on the effective implementation and also need to take into account the features of the area where the reform is going to take place.

Sixth, the Bank WUAs have excelled in many dimensions, particularly in the implementation of the Five Principles of WUA management. In the case of Principle 1, Bank WUA villages had a number of characteristics that showed that they were endowed with more reliable water supply. In the case of Principle 2, Bank WUAs have been set up and are operating with a relatively high degree of farmer participation. The leaders are more consultative. The procedures are clearer and the processes more formal. Bank WUA villages are also set up in a way that makes them consistent with Principle 3; WUAs are organized largely within their hydraulic boundaries. Finally, the Bank WUA villages are successful in implementing Principles 4 and 5. For example, most of the Bank WUA villages can deliver water volumetrically (Principle 4); and all Bank WUA villages collect water fees from farmers (Principle 5). Hence, from this analysis, WUAs in the Bank villages can really be thought of as operating according to the best practices in terms of the Five Principles. While the positive record is true in Bank villages—relative to all of the comparison cases, it is also true that there is evidence that the Bank's effort to promote WUAs extends beyond their own project villages. In addition, water is being used more efficiently in Bank WUA villages and other WUA villages in the same regions. While the Bank WUAs have brought nominal changes, they have not generated the more fundamental changes that can lead to improved economic welfare or structural change.

Seventh, the current cost of water is far below the true value of water in northern China. Since water is severely underpriced, water users are not likely to respond to small increases in water prices. Therefore, a necessary step in establishing an effective water pricing policy is to increase the price of water up to a point that it equals the value that water has to the household. Increases in water prices once they are set at the value of water can lead to significant water savings. Unlike past water research, our study shows that water-pricing policy, by directly giving users incentives, has the potential of resolving the water scarcity problem in China. Dealing with the negative production and income impacts of higher irrigation costs, however, will pose a number of challenges to policy makers, at least in the short run. If China's leaders plan to increase water prices to address the nation's water crisis, an integrated package of policies will be needed to achieve water savings without hurting rural incomes or national food security. For example, subsidies that are decoupled from production decisions can be used to offset the loss in income.

Eighth, considerable gains from water reallocation among different uses can be expected. When water is moved downstream to higher-value agricultural use under conditions of free trade, the economic benefit, in terms of the increased value of agricultural production, was around \$129 million per year. If farmers in water-exporting regions had the property rights to transferred water, income from water sales would more than offset the forgone income from reduced agricultural production. The income from water sales is estimated to be around \$65 million per year. In the absence of property rights, the lost value of agricultural production lowers farm household incomes substantially. Conversely, with revenue from the sale of water, farm household incomes in the exporting regions would rise substantially. Importantly, without compensation the regions with the lowest incomes are likely to be affected the most by water transfers. Water markets would provide a mechanism to transfer water to higher value uses on a large scale and to the other productive uses, such as industry, and the environment. However, due to many barriers (such as existing administrative framework, small scale of farming and the consequent transactions costs of implementing water property rights), the water property rights in China still has not been established.

Finally, our analysis has shown that adoption of water-saving technologies depends strongly on incentives. Farmers and villages alike appear to behave rationally in their decision to adopt water-saving technologies, making a calculated cost-benefit analysis. Unsurprisingly, factors that make adoption cheaper or water more valuable have a significant positive impact on technology adoption. This includes perceived water scarcity, governmental financial support and extension services, average income, and the production of highly valued cash crops. Conversely, factors that make water less valuable have a significant negative impact on technology adoption. The availability of nonagricultural employment is an attractive alternative to investing in better irrigation technology. Likewise, when a village has a high rate of arable land per capita, it is less urgent that every piece of land is used as efficiently as possible. Of the different types of water-saving technologies, household-based technologies have grown most rapidly, and traditional technologies have the highest rates of adoption. The most successful technologies have been highly divisible and low cost ones that can be implemented without collective action or large fixed investments. Technologies that do not fit this description are adopted on a more limited scale, at least in part due to the failure of policy makers to overcome the constraints to adoption. Farmers in many parts of the region have not adopted even rudimentary water-saving technologies. This suggests that the incentives are not in place to encourage efficient water use.

REFERENCES

Boxer, B., 2001. Contradictions and challenges in China's water policy development. Water International 26 (3), 335–341.

Chen, L., 2002. Revolutionary measures: water saving irrigation. In: The National Water Saving Workshop. Beijing, China (In Chinese).

Fang, S., 2000. Combined with allocating and controlling local water resources to save water. Journal of China Water Resources 439, 38–39.

Hebei Hydrological Bureau and Water Environmental Monitor Center, 1999. Heibei Water Resources Assessment.

Huang, Q., Rozelle, S., Howitt, R., Wang, J., Huang, J., 2010. Irrigation water demand and implications for water pricing policy in rural China. Environment and Development Economics 15, 293–319.

Lohmar, B., Wang, J., Rozelle, S., Dawe, D., Huang, J., 2003. China's Agricultural Water Policy Reforms: Increasing Investment, Resolving Conflicts, and Revising Incentive. United States Department of Agriculture, Economic Research Service, Washington, DC. Agriculture Information Bulletin Number 782.

Ministry of Water Resources, 2012. Water Resources Bulletin. Ministry of Water Resources, Beijing.

Nian, L., 2001. Participatory Irrigation Management: Innovation and Development of Irrigation System. China Water Resources and Hydropower Publishing House, Beijing, China (In Chinese).

Nickum, J., December 1998. Is China living on the water margin? The China Quarterly 156.

Reidinger, R., 2002. Participatory irrigation management: self-financing independent irrigation and drainage district in China. In: Paper Presented at the 6th International Forum of Participatory Irrigation Management, Held by the Ministry of Water Resources and the World Bank, Beijing, China, April 21–26, 2002.

SSB (State Statistical Bureau), 2012. China Statistical Yearbook. China Statistics Press, Beijing, China.

Wang, J., Xu, Z., Huang, J., Rozelle, S., 2005. Incentives in water management reform: assessing the effect on water use, production and poverty in the Yellow river basin. Environment and Development Economics 10, 769–799.

Wang, J., 2000. Innovation of Property Right, Technical Efficiency and Groundwater Irrigation Management (Ph.D. thesis). Chinese Academy of Agricultural Sciences.

Wang, J., 1999. Situation and strategy of the Yellow river cutting off. Journal of China Water Resources 4, 10–11 (In Chinese).

World Bank, 1998. Rural China: Transition and Development. East Asia and Pacific Region. World Bank, Washington, DC.

Xu, Z., 2001. Studying on increasing water use efficiency. Journal of China Water Resources 455, 25–26.

Zai, H., 2002. Speaking on the National Workshop of Water Saving Irrigation. Held by the Ministry of Water Resources, 16–19 October, Beijing, China.

Zhang, Y., 2000. Ten challenges faced with China water resources in the 21st century. Journal of China Water Resources 439, 7–8.

Zhang, Y., 2001. Carefully implement the fifth national conference, promoting the development of water resources into a new stage. Journal of China Water Resources 450, 9–10 (In Chinese).

SECTION

I

Setting the Stage

CHAPTER

Water Scarcity in Northern China

1

China's water resource availability is among the lowest worldwide. The most recent estimate of the annual renewable internal freshwater resources per capita is 2093 m^3 (FAO, AQUASTAT), which is far below the estimated world average of 8349 m^3 (ESCAP, 2010). In addition, water resources are not evenly distributed. Northern China has only 21% of the country's water endowment (Ministry of Water Resources of China, 2011). Northern China, however, remains an important region, with about 35% of the total population, 38% of the nation's GDP, and almost half of the grain production (NBSC, 2013). In China, 81% of water supply comes from surface water resources and groundwater supply only occupies 18.8% of total water supply (Table 1.1). In addition to a few provinces, most provinces rely on the supply of surface water resources. In 2012, groundwater supply in five provinces occupied more than 50% of total water supply; they are all located in northern China (Hebei, Henan, Beijing, Shanxi, and Inner Mongolia provinces). For other provinces, the percentages of groundwater supply are all lower than 50%; some are even less than 5% (such as Shanghai, Zhejiang, Chongqing, Guizhou, Hubei, Guangdong, Jiangsu, Jiangxi, Yunnan and Guanxi Provinces).

China's rapidly growing industrial sector and an increasingly wealthy urban population compete with farmers for limited water resources. Between 1949 and 2004 total water use in China increased by 430%, a level of growth that is similar to the world's average increase of 400% but greater than the average growth rate for developing countries (Wang et al., 2005). Industrialization and urbanization have caused China's water to be increasingly allocated to nonagricultural uses. From 1949 to 2012 the share of agricultural water use declined from 97% to 63% (NBSC, 2014). At the same time the share of industrial water use increased from 2% to 23% and the share of domestic water use increased from 1% to 12%. Before 2000, there was almost no water allocated for environmental use. In 2012, about 1.8% of water was allocated for environmental use. However, water allocation among sectors differed by provinces (Table 1.1). Among 31 provinces, 52 percentage of them (16 provinces) allocated more than 63% (national average level) of water to agricultural sectors, and most of them (except Guangxi, Yunnan, and Jiangxi provinces) locate in northern China. For other 48% of provinces (15 provinces, most locating in southern China), the percentages of water allocation to agricultural sector are lower than national average level, and near half of them (seven provinces) are even lower than 50%.

Managing Water on China's Farms. http://dx.doi.org/10.1016/B978-0-12-805164-1.00001-4

Table 1.1 Water Supply and Water Allocation Among Sectors in 2012 by Province in China

	Water Supply (10^8 m^3)				Water Allocation Among Sectors (%)			
		Among Total Water Supply (%)						
	Total	**Surface Water**	**Groundwater**	**Others**	**Agriculture**	**Industry**	**Domestic**	**Ecological**
China	6131	80.8	18.5	0.7	63.6	22.5	12.1	1.8
Xinjiang	590	81.0	18.8	0.2	95.2	2.1	2.0	0.7
Tibet	30	88.3	11.7	0.0	90.9	5.7	3.4	0.0
Ningxia	69	91.9	7.9	0.0	88.5	7.1	2.3	2.2
Heilongjiang	359	55	45	0.0	82.2	11.6	4.5	1.7
Qinghai	27	86.9	12.8	0.4	82.1	9.1	8	0.7
Gansu	123	77.9	20.9	1.2	77.3	12.8	7.6	2.4
Hainan	45	92.7	7.3	0.2	76.6	8.4	14.6	0.4
Inner Mongolia	184	48.6	50.4	0.9	73.4	12.7	5.6	8.2
Hebei	195	21.1	77.5	1.4	73.2	12.9	12	1.9
Guangxi	303	96.2	3.6	0.2	69.9	17	12.1	1.0
Shandong	222	56.9	40.3	2.9	69.5	12.7	14.8	3.0
Yunan	152	95.7	3.6	0.7	68.4	18.3	12.6	0.7
Shaanxi	88	61.4	38	0.7	66.1	15.1	16.8	1.9
Jilin	130	66.2	33.4	0.5	65.3	20.9	9.2	4.6

Liaoning	142	54.6	43.1	2.3	64.3	16.2	16.5	3.1
Jiangxi	243	96.2	3.9	0.0	64.2	24.2	10.8	0.9
Sichuan	246	90.6	7.6	1.9	59.3	22.2	17.4	1.0
Shanxi	73	43.3	52.9	3.8	58.2	21.1	16.1	4.5
Hunan	329	94.4	5.6	0.0	57.2	29.8	12.3	0.8
Henan	239	42.1	57.5	0.4	56.8	25.4	13.4	4.4
Jiangsu	552	98.2	1.8	0.0	55.3	35.0	9.1	0.6
Anhui	293	88	11.8	0.3	54.0	33.9	10.6	1.6
Tianjin	23	69.3	23.8	7.4	50.6	22.1	21.6	6.1
Guangdong	451	95.9	3.8	0.4	50.5	27.0	21.2	1.4
Hubei	299	96.3	3.4	0.3	48.9	40.6	10.3	0.1
Guizhou	101	97.3	1.1	1.7	47.3	39.4	13	0.3
Fujian	200	96.4	3.3	0.3	46.4	37.8	14.2	1.5
Zhejiang	198	97.9	1.7	0.4	46.1	30.6	21.0	2.3
Chongqing	83	97.9	1.9	0.1	30.4	47.5	21.1	1.0
Beijing	36	22.3	56.8	20.9	25.9	13.6	44.6	15.9
Shanghai	116	99.9	0.1	0.0	15.1	62.8	21.5	0.6

Data sources: National Bureau of Statistics in China, China statistical yearbook (2013).

Increasing evidence indicates that China is facing serious water scarcity, especially in northern China. This scarcity is due not only to a falling water supply but also to rising water demand. On the supply side, the total estimated water resources of China dropped by 16.5% from 1997 to 2011 (Ministry of Water Resources, 2012; Ministry of Water Resources of China, 2011). One of the most obvious pieces of evidence is that surface water resources in some parts of China are diminishing. From 1980 to 2000, the flows of rivers in several major river basins in northern China have declined significantly. For example, the runoff in the Hai River Basin decreased by 41% (Ministry of Water Resources, 2007). The runoff in other river basins—the Liao, Yellow, and Huai River Basins—also have fallen by 9% to 15%. Because of this decline, some river basins (eg, the Hai River Basin and the Yellow River Basin) have changed from open to closed ones in some years; in other words, throughout the year, water does not flow into the lower reaches of the basin (Wang and Huang, 2004). In the future, with socioeconomic development and the pressure of climate change, the supply of surface water resources for those river basins locating in northern China (Hai River Basin, Huai River Basin, Yellow River Basin, Songhuajiang River Basin, and Liao River Basin) will face obvious decline while water demand will increase, which will result in the increase of a water supply-and-demand gap (Wang et al., 2013).

With the decline of surface water resources, farmers in northern China have begun to turn toward groundwater resources. In fact, groundwater has become the dominant source of water for irrigation in northern China. In the early 1950s, there was almost no groundwater-based irrigation in northern China (Wang et al., 2007). However, by the 1970s groundwater irrigation had risen to account for 30% of irrigation. By the mid-1990s, a decade or so after the launching of the economic reforms, groundwater irrigation continued to expand, reaching 58% in 1995. In 2004, most irrigation in northern China came from groundwater resources; the share of land irrigated by groundwater was nearly 70%.

Unfortunately, this reliance on groundwater may have resulted in a number of adverse environmental effects. According to a report by the Ministry of Water Resources in 1996, the overdraft of groundwater is one of China's most serious resource problems (Ministry of Water Resources and Nanjing Water Institute, 2004). In the late 1990s the annual rate of overdraft exceeded 9 billion m^3. More than one-third of the overdraft was from deep wells, many of which are nonrenewable (at least in the short to medium run). The drop in the deep groundwater table in some areas has exceeded 2 m per year. Based on our own previous research in Hebei Province, the shallow groundwater table also is falling in some areas—about 1 m per year (Wang and Huang, 2004). Beyond the drop of the groundwater table, overdrafting groundwater can cause land subsidence, the intrusion of seawater into fresh water aquifers, and desertification (Wang et al., 2007).

In addition to declining water availability, an upsurge in demand is increasing the pressure on the supply of irrigation water. Between 1949 and 2004 total water use in China increased by 430%, a level of growth that is similar to the world's average increase of 400%, but greater than the average growth rate for developing countries (Wang et al., 2005). Industrialization and urbanization have caused China's water to be increasingly allocated to nonagricultural uses. From 1949 to 2012 the share of irrigation water use declined from 97% to 62% (Ministry of Water Resources, 2012). At the same time, the share of industrial water use increased from 2% to 24% and the share of domestic water use increased from 1% to 13%. Before 2000, there was almost no water allocated for environmental use. In 2011, about 1.8% of water was allocated for environmental use (Ministry of Water Resources of China, 2011).

Facing the simultaneous decline of water availability and increase of water demand, it is natural to believe that China is facing a severe water crisis. A reading of the literature and policy documents supports this notion. For example, in 1999, Chinese Prime Minister Wen Jiabao, then the vice premier in charge of agriculture, warned of the dire situation that China's water crisis was creating (McAlister, 2005). Senior officials from the Ministry of Water Resources in China frequently point out that China is fighting for every drop of water and that the water crisis is threatening national grain production (Wang, 2008). Brown (2000) predicts that the falling water table in China may soon reduce the nation's output by so much that it will raise international food prices, since China—a nation that can afford massive volumes of imports—will have to rely on food produced outside of the country. Nankivell (2004) believes that China has reached a point at which critical decisions must be made to resolve its water crisis. Although other observers have made more moderate predictions, they also forecast that many agricultural producers may have to forgo irrigation due to falling water supplies (eg, Crook and Diao, 2000).

Despite these dire predictions about the severity of the water situation in China, we believe that it is premature to claim that China is faced with such a dire crisis given the lack of large-scale empirical evidence. While conceding that many localities are facing water scarcities, we do not believe that the discussion on China's water problems is based on compelling big picture evidence. Most studies are anecdotal and rely on observations of producers or users in a single location. Few studies draw their data from large-scale field-level surveys. When empirical work relies on information from a single location, it is hard to judge the severity of the nationwide water shortage. Without systematic data across space, it is impossible to know definitively whether China really is facing a nationwide water crisis.

In this chapter, we attempt to fill in the gap in their literature by empirically evaluating the status of China's groundwater economy. To do so, we will pursue three specific objectives. First, we will characterize China's groundwater resources, briefly reviewing the main physical and geographic properties of northern China's groundwater resource development, describing the role of groundwater in the economy and examining the technology that producers are using to extract and utilize the resource. Second, we will examine the main problems that the sector is facing, including falling groundwater levels and deteriorating water quality. Third, we will examine perception of village leaders and farmers about water scarcities. Finally, changes in surface and groundwater supply reliability will also be analyzed.

Given that a water shortage has implications for virtually every aspect of the economy, we need to limit the scope of our study. We focus on northern China, and specifically the rural economy there. We use the level of groundwater in rural villages to illustrate one aspect of China's water challenge, acknowledging that there are many other dimensions of the water problem that we cannot examine here (eg, pollution of underground aquifers; land subsidence). In our study, northern China can be thought to include the following regions: North China, Northeast China, and Northwest China. The data for this chapter come from the China Water Institutions and Management (CWIM) survey and the North China Water Resource Survey (NCWRS).

CHINA'S GROUNDWATER RESOURCES

Groundwater resources in China are both unevenly distributed and unevenly used across regions. Only about 30% of groundwater resources are located in northern China, but users in northern China use over 70% of the nation's groundwater.

Despite this inequity, the groundwater resources in the north are still abundant and accessible, existing across wide expanses of northern China's river basins. Alluvial deposits consisting primarily of sand, loess silt, and clay extend to a depth of more than 500 m below the surface in some areas (Kendy et al., 2003), contributing to the aquifers that supply groundwater to regions in all major river basins in the North China Plain. The distribution of these aquifers, however, varies greatly across northern China. In the North China Plain, mountainous areas are often groundwater deficient, unlike the south, where villages in mountainous areas can tap groundwater resources. In north China, almost all provinces have both mountainous areas and flat plains; it is hard to describe which regions are mountainous and which are flat. In the plains, the aquifers are multilayered. In the Hai River Basin, aquifers typically have two to five layers, with the first and third layers being the most water abundant. The first layer is typically an unconfined aquifer made up of large grained homogeneous sand and gravel. The other layers are typically confined aquifers. In some areas, especially in the eastern parts of the Hai River Basin, there is a naturally occurring saline layer. Created during a previous Ice Age, saline water is often found in the confined second layer, and has a salt content high enough that it is typically unusable for agriculture without treatment.

THE RISE OF TUBEWELLS

According to national statistics, the installation of tubewells began in the late 1950s and, although the number of wells has grown continuously, the pace of increase has varied from decade to decade (Ministry of Water Resources and Nanjing Water Institute, 2004). During the 1950s, the first pumps were introduced to China's agricultural sector. Although still fairly limited, the growth rate was fast. During the time period of the Great Leap Forward (the late 1950s and early 1960s), however, statistical reporting was suspicious and many irrigation projects that were started during the period were badly engineered and often abandoned. After the recovery from the Great Leap Forward and the famine that followed, statistical agencies recovered and statistical series since the mid-1960s are relatively consistent.

Since the mid-1960s, the installation and expansion of tubewells across China has been nothing less than phenomenal. In 1965 it is reported that there were only 150,000 tubewells in all of China (Shi, 2000). Since then, the number has grown steadily. By the late 1970s, there were more than 2.3 million tubewells. After stagnating during the early 1980s, a period of time when irrigated area, especially that serviced by surface water, fell, the number of tubewells continued to rise. By 1997, there were more than 3.5 million tubewells; by 2003, the number rose to 4.7 million.

The path of tubewell expansion shown in the official data is largely supported by the information we have from the NCWRS. During the survey we asked the village leaders to tell us about the initial year in which someone (either the village leadership or an individual farmer) in his/her village sank a tubewell. According to the data, we find that by 1960, less than 6% of villages had sunk their first tubewell. Over the next 20 years, between the early 1960s and the onset of reform, the percentage of villages with tubewells rose to more than 50%. During the next 10 years, between 1982 and 1992, the number of villages with tubewells rose by only 7%. After the early 1990s, however, the pace of the expansion of groundwater accelerated, and by 2004, almost 75% of villages had wells and thus access to groundwater.

While the growth of tubewells reported by the official statistical system is impressive, we have reason to believe the numbers are far understated. According to the NCWRS, on average, each village in northern China contained 35 wells in 1995. When extrapolated regionally, this means that there were more than 3.5 million tubewells in the 14 provinces in northern China by 1995. As important, according to our data, the rate of new well installation has grown rapidly. By 2004, the average village in northern China contained 70 wells, suggesting that the rise in tubewell construction since the mid-1990s has been even faster than indicated by official statistics. In 2004, we estimate that there were more than 7.6 million tubewells in northern China. At least in our sample villages, the number of tubewells has grown by more than 12% annually between 1995 and 2004.

GROUNDWATER RESOURCES FROM FARMERS' PERSPECTIVE

To obtain an understanding of how farmers view their water resources, we asked village leaders to describe the nature of the aquifers that are under their villages. Because most village leaders have not participated in any hydrogeological surveys of their villages, they often were unable to answer questions concerning the existence, size, or other geological properties of the aquifers under their villages. Instead, village leaders know more precisely how many shallow or deep wells are in their village and the depths of the wells. Although there is not a complete correlation between the depth of the wells and the nature of the aquifer, in many cases, the existence of a shallow or deep well coincides with that of shallow or deep layers of a village's aquifers. Regardless of their exact hydrogeological properties, according to our data (and the perception of village leaders), "deep wells" are almost always wells that have a depth of at least 60 m. If a village needs to drill through an aquitard (a clay layer in most cases) to sink a well, the well is always defined as a deep well. Shallow wells, in contrast, are mostly less than 60 m and do not penetrate an aquitard.

Whether deep or shallow, groundwater resources are extensive across regions of northern China. We asked village leaders about the existence of groundwater resources in the village. Most village leaders responded that there are groundwater resources; the share of villages having groundwater resources was almost 95% in 2004. However, not all villages with access to groundwater use it for irrigation. In 2004, more than 15% of irrigated villages with groundwater did not use it to irrigate. According to the village leaders, there are two major reasons for this. One reason is the existence of cheap and sufficient surface water resources (51% of villages). The other reason is insufficient funding for the digging of tubewells (37% of villages). Such findings suggest that there are still many untapped groundwater resources in north China. With increasing water scarcity and rising water demands, more villages have begun to use local groundwater sources. From 1995 to 2004, the share of villages using groundwater resources for the first time has increased by almost 12%.

Relying on the observations of our NCWRS respondents, one of our most prominent findings is the great diversity of aquifer development in northern China. Our data show that in 33% of our sample villages, the groundwater supply from shallow aquifers is sufficient to support current local water demand for irrigation. In other villages (42%), perhaps due to exhausted or unusable shallow aquifers, farmers extract groundwater only from deep aquifers. Although we have not asked farmers why they only extract groundwater from deep aquifers, based on our experience in the field, it should be due to exhausted or unusable shallow aquifers. In the remaining 25% of sample villages, both shallow and deep aquifers are used. The groundwater supply from shallow unconfined aquifers is highly dependent upon precipitation, which supplies groundwater recharge. When rainfall is above average, as it was in 2004, water levels increase in shallow aquifers due to above average recharge. This may be the reason that we observe more villages extract groundwater from shallow aquifers in 2004 than in 1995. We need more investigation to explore the reason in the future.

According to our respondents, the depth to water also varied across northern China. Although the average depth to water in 2004 was 26 m, it varied sharply across our sample villages. In fact, in most villages depth to water was fairly shallow. In 2004, the average depth to water for the villages in the shallowest quartile was only 4 m and that for the second quartile was only 9 m. Villages in the third quartile were pumping from an average depth to water of more than 30 m. In only 4% of groundwater villages were villagers pumping from more than 100 m.

GROUNDWATER PROBLEMS AND CHALLENGES

There are many misperceptions about the nature of China's water problems, especially as they relate to the rural economy. Many of China's water-related problems, although serious regionally, are not national in scope. Other problems are confined to urban areas or rural areas, but not both. In this section, we provide a brief assessment of the main problems facing China's groundwater economy. Given the fact that most of the work in the past has had an urban focus, our work centers on those problems affecting the rural sector.

OVERDRAFTING CHINA'S GROUNDWATER RESOURCES

According to a comprehensive survey completed by the Ministry of Water Resources (MWR) in 1996, the overdraft of groundwater is one of China's most serious resource problems (Ministry of Water Resources and Nanjing Water Institute, 2004). The survey provides evidence that groundwater overdraft is a widespread problem and may be getting worse. According to the report, overdraft is occurring in more than 164 locations and affects more than 180,000 km^2. The areas of overdraft range from 10 to 20 km^2 to more than 10,000 km^2 and are in 24 of China's 31 provinces. Groundwater overdraft is affecting all types of aquifers: the shallow groundwater table (87,000 km^2), the deep groundwater table (74,000 km^2), and both the shallow and deep layers of multilevel aquifers (13,000 km^2). There are also several other minor types of aquifers being overdrafted, accounting for about 7000 km^2. Since the 1980s, the annual overdraft of groundwater has averaged about 7.1 billion m^3. In the late 1990s, the annual rate of overdraft exceeded 9 billion m^3. More than one-third of the volume of overdraft is from deep wells, many of which may be nonrenewable on a short time scale.

While the problem of overdraft is usually discussed in general, in fact, the problem appears to be particularly acute in cities. The Ministry of Land Resources recently finished an evaluation of groundwater resources in China. According to the final report, groundwater resources in most large- and medium-sized cities in northern China are either in overdraft (extractions exceed recharge) or in serious overdraft conditions (the fall of the groundwater table exceeds 1.5 m per year). For example, in many cities the volume of water extracted from the aquifer is nearly double the volume of average annual recharge. According to a comprehensive survey completed by the MWR in 1996, groundwater overdrafting is a widespread problem and may be getting worse.

The definition of overdraft here is from MWR in China (Ministry of Water Resources and Nanjing Water Institute, 2004). It is important to note, however, that there are other definitions. In a sustainable system, groundwater recharge should equal discharge over time. Extractions (groundwater pumping) are only a small part of total discharge from an aquifer. Other parts include natural discharge to rivers (which explains why rivers flow even long after rain and snow stop falling), and natural discharge to wetlands, lakes, and plants. If extraction exceeds recharge, then all those other components of groundwater discharge would cease. Overdraft is better defined by long-term water-level declines.

Our data—focusing on rural areas as opposed to urban ones—shows a slightly different picture. As described in chapter "Irrigation, Agricultural Production and Rural Income," we find that nearly half of the survey villages using groundwater experienced little or no fall in the groundwater table. Indeed, in 8.5–16% of villages, the groundwater level was actually higher in 2004 than in 1995. Of course, this still leaves a large number of villages in which the water table is falling. Nevertheless, only 10% of villages using groundwater from 1995 to 2004 have water tables that are falling at a rate greater than 1.5 m per year—meeting the MWR's definition of "serious overdraft."

In summary, then, although most places in northern China are not misusing their water resources, we do not want to minimize the problems that are occurring in some places. There are a large number of rural areas in which the water table appears to be falling at a dangerously fast pace. Where the resource is being misused, steps will be required to protect the long-run value and use of the resource; however, it is important to realize that there will be associated costs in implementing and adopting water-saving techniques and technologies. Because measures to counter overdraft are not needed in all villages, China's leaders should not take a one-size-fits-all approach, and thus avoid inflicting unnecessary costs on villages where overdraft conditions do not exist.

SUBSEQUENT EFFECTS OF OVERDRAFT

As the groundwater table falls, producers face a number of impacts. Above all, the cost of pumping rises. According to our data, for every meter by which the groundwater table falls, pump costs rise by US\$0.06 per 1000 m^3 (or about 2% of the mean level of pumping costs in 2004). In addition, wells may have to be replaced, although this problem is not as common as one might suppose. Although the typical groundwater-using village sank about 22 new wells (5 deep and 17 shallow) between 2002 and 2004, only two of these new wells replaced old wells that had been abandoned due to a falling water table.

Another major consequence of overdraft is land subsidence. In Hebei province alone, by 1995, more than 5000 km^2 had subsided more than 600 mm. In Tianjin Municipality, the total subsided land area exceeded 7000 km^2. Groundwater overdraft may also lead to the intrusion of seawater into fresh water aquifers (Ministry of Water Resources and Nanjing Water Institute, 2004). By the mid-1990s, overdrafting had allowed seawater to intrude and contaminate aquifers under more than 1500 km^2 of land, especially in the coastal provinces of northern China, such as Liaoning, Hebei, and Shandong. The MWR has also been concerned about the impact of groundwater overdraft on desertification and the depletion of stream flow that had previously been supplied by natural groundwater discharge.

Although the consequences of overdraft are widely discussed in the literature and are equated with China's water problems in general, interestingly, none of these problems appears to be in any way associated with rural areas. According to our survey, no village leader ever reported any land subsidence problem, nor did any village leader report that the village's groundwater was contaminated by seawater intrusion. Finally, there is no evidence that villages using groundwater—regardless of the level of the water table—experienced a fall in cultivated area due to desertification.

This lack of impact in the rural areas, while good for rural development in the short-run, is worrisome. Because agriculture is the largest user of water in all of China, behavioral changes in this sector will surely be necessary in order to curb national water use. But if farmers are not largely unaffected by their overuse of water, they will have no incentive to change. Policymakers will have to find a way to align farmers' incentives with their own if they hope to see the adoption of water-saving farming methods.

OTHER PROBLEMS WITH GROUNDWATER

GROUNDWATER POLLUTION

Both the literature on groundwater and our survey also report a number of other problems that are not directly related to groundwater overdraft. For example, it has been widely reported in the press and in academic journals (eg, Kendy et al., 2003) that pollution from municipal sewage has contaminated the groundwater of many villages in China. Part of the problem is created when farmers pump from effluent canals, using sewage-laced water on their fields. The recharge from irrigation with such water can affect the entire aquifer. Even when villages do not use the water for irrigation purposes, recharge from streams and riverbeds can contribute to groundwater pollution. According to the Ministry of Water Resources and Nanjing Water Resources Institute (2004), the groundwater resources of more than 60% of the 118 largest cities in China have contaminated groundwater.

Drawing on our survey (in which we ask leaders about their perception of pollution), we find that the groundwater is polluted in 5.4% of sample villages. However, unlike villages near cities, which are mainly affected by municipal sewage waste, our respondents identified industrial pollution and runoff from mining operations as the most common source of pollution. In fact, of all of the villages that reported contaminated groundwater, 95% (5.15% of all villages) said that the main source of pollution was from industrial and mining wastewater. Only 0.25% of villages (or less than 5% of villages that report contamination) said that their groundwater was polluted by agricultural chemicals, and none attributed it to urban sewage.

While the perceived extent of the rural groundwater pollution problem appears to be less serious than in urban and suburban areas, it is still serious. Extrapolating our results to all of northern China, we can estimate that more than 20 million rural residents living in 20,000 rural villages are using groundwater that has been contaminated by industrial runoff. Moreover, unlike their urban and suburban counterparts, most villages in China lack any type of drinking water processing facilities. In most cases, the pollution causing the problems in one rural village is being created by industrial and mining facilities in another village. There is no clear advocate to force upstream villages either to stop polluting or to compensate downstream villages for the damage. There also is little funding for rural groundwater pollution abatement. In short, there is no incentive or means to address and/or curtail the activities that are polluting the groundwater of millions of rural villages.

SOIL SALINIZATION

Across China, the appearance of salinized soil has been a widespread problem. According to the Ministry of Water Resources and the Nanjing Water Resources Institute (2004), more than 1 million km^2 of China's land has become salinized over the past several decades. The majority of the most serious problems have occurred in the Northeast, the Northwest, and in some places on the North China Plain. Despite the widespread nature of the problem, the area affected by salinization has fallen in recent years. Ironically, it may be that the same forces diverting surface water away from agriculture and forcing producers to rely increasingly on groundwater are also the primary causes of such improvements. Without access to cheap and abundant surface water, which led to the salinized soil problem, the problem has gradually disappeared as farmers have turned to groundwater and the water table has fallen (Nickum, 1988). Salinization is caused by different factors and responds to different solutions in different settings. Over time, continued groundwater use is likely to increase soil and water salinization. Each time groundwater is "recycled" through the pumping and reinfiltration process, it becomes more saline.

In our sample of villages, we find that the salinization of the soil is one of the most commonly reported problems, although, consistent with national statistics, it is improving over time. According to respondents, in 2004 16% of villages reported having some salinized soil. Since the process that caused the soil salinization does not affect all cultivated area in a village, only 3.4% of cultivated area was reported to be affected. Moreover, the scope of soil salinization is improving over time. In 1995, 20% of villages reported salinized soil and 4.4% of cultivated area was affected. Hence, between 1995 and 2004, there was nearly a 25% reduction in the severity of the nation's soil salinization problem.

PERCEPTION OF VILLAGE LEADERS AND FARMERS ABOUT WATER SCARCITY

During our field survey, enumerators asked village leaders and farmers to describe the nature of water resources in their villages in 1995 and 2004. Village leaders and farmers were asked to choose from one of three options: there is no water scarcity (at least currently); there is a water scarcity but the scarcity is not severe; and

Table 1.2 The Opinions of Village Leaders and Farmers Regarding the Nature of Water Scarcities in Northern China's Villages

	Share of Sample Villages (%)	
The Village's Water Environment	**1995**	**2004**
Not scarcity	35	30
Scarcity	65	70
When scarcity, severely short	14	16

Data sources: Authors' survey in 2004.

there is water scarcity and it frequently constrains agricultural production (or water scarcity is very severe). Based on the responses of village leaders and farmers, we can have at least some insight into the status and the patterns of water scarcities in northern China.

The responses of village leaders and farmers did not give an optimistic picture about the state of water resources in the future. Survey results show that respondents in 70% of the villages in our samples villages reported that they are facing water scarcity in 2004, or at the very least those in these villages felt they had serious water scarcity (Table 1.2). Among those villages that reported water scarcity, 16% of the villages were facing "severe" water scarcities in 2004 (row 3, column 2). Only 30% of the respondents reported that they were not having any water scarcity problems (row 1, column 2).

More importantly, water scarcities have become more serious over the past decade. Survey results show that between 1995 and 2004, the share of water short villages increased by 5% (Table 1.2, row 2). The share of villages that reported severe water scarcity problems also increased (row 3). It is clear that from the perception of the village leaders and farmers in those villages that it is believed that the locality's level of water scarcity is becoming a serious problem. Although not happening everywhere, most villages are facing such problems. In some villages, respondents replied that water scarcities were so serious that water had become the most limiting factor in agricultural production.

CHANGES IN SURFACE AND GROUNDWATER SUPPLY RELIABILITY

The reliability of water supply is another indicator that can measure the degree of water scarcity from an angle different from that discussed above. If surface water supplies are reliable, in most cases we can assume that water is not scarce. On the other hand, if the water supply is not reliable, water scarcities can frequently emerge even when, during most times of the year (or during other years), water resources are sufficient. To get a sense of whether or not water resources in a village were reliable or not we asked farmers the following question: "When considering the time period between 2001 and 2004, was there any part of any year (either one or more periods among a single year or during either one or more years) that you were not able to have access to irrigation water (either surface or groundwater)?" We asked farmers that same question about the time period between 1991 and 1995. If farmers said there was at least one part of a year that they were not able to access irrigation water, the village was said to have "unreliable access." If farmers never met such a situation, we defined their surface or groundwater supplies as reliable.

Survey results show that surface water is highly unreliable in northern China. Our data suggest that the farmers in many villages do not believe that they have a reliable supply of surface water (Table 1.3). During the period from 2001 to 2004, more than 60% of the villages did not have access to reliable surface water resources.

Over the past decade the extent of water reliability has declined. In the early period of the 1990s (from 1991 to 1995), surface water supplies in most villages were fairly reliable. About 64% of the villages reported that they had access to reliable supplies of water at that time (Table 1.3). However, in the early 2000s (from 2001 to 2004), the share of villages with reliable supplies of surface water resources declined to 39%, a drop of 25% points. According to our data (and our assumption about the relationship between water scarcity and reliability), as water resources have become less reliable in our sample villages, this is clearly exacerbating the already scarce water resources.

At the same time surface water supply became less reliable, the reliability of groundwater supply also declined, although to a much less extent than surface water (Table 1.2). Groundwater resources have always been valued for their reliability since groundwater can serve frequently as a buffer against the uncertainties associated with surface water resources (Tsur and Graham-Tomasi, 1991). Our data indicate that groundwater supply is in general much more reliable than the supply of surface water. More than 90% of the villages had reliable supplies of groundwater between 1991 and 1995, a share that was 27 percentage points higher than villages that relies on surface water (Table 1.3). However, over time, even groundwater has become an unreliable source of water for some farmers. For example, in the early 1990s (from 1991 to 1995), 9% of the villages reported that their groundwater

Table 1.3 The Reliability of the Water Supply in Northern China's Villages

Nature of the Water Supply in Terms of Reliability	Share of Sample Villages (%)			
	Surface Water Reliability		Groundwater Reliability	
	1991–1995	2001–2004	1991–1995	2001–2004
Reliable	64	39	91	85
Unreliable	36	61	9	15
Total	100	100	100	100

Data sources: Authors' survey in 2004.

supplies were unreliable. The share of villages without reliable groundwater resources rose to 15% in the early 2000s (from 2001 to 2004). Therefore, in both groundwater-reliant and surface water–reliant villages, the reliability of the supply of water has deteriorated. This will no doubt intensify water scarcity in northern China.

CONCLUSIONS AND POLICY IMPLICATIONS

Our findings show that there is indeed a water crisis in northern China, although the severity of this crisis depends on the indicator used to measure water scarcity. Based simply on the qualitative responses of village leaders and farmers asked about the existence of a water scarcity problem, about 70% of villages are facing increasingly tight water supplies, 16% of which are also experiencing severely constrained agricultural production due to an insufficient water supply. Farmers in these villages report that delivery of both surface water and groundwater is less reliable than in the past.

When using other measures of scarcity, a more nuanced picture emerges. According to data on the change in the groundwater table in villages over the past 10 years, approximately 11% of villages are experiencing a rapid fall in the groundwater table, consistent with the self-reported number of 16%. However, this metric also indicates that around half of rural villages reliant on groundwater in northern China have experienced little or no decline in groundwater levels over the last decade; indeed, in some villages the groundwater tables have actually risen.

We interpret these conflicting pieces of information to mean that there is a water crisis in northern China, but that the crisis is confined to certain areas. Half of northern China is suffering from falling groundwater tables, affecting between 50 and 100 million farming households. Hence, policies to address water scarcity problems should be implemented, but should also be carefully targeted.

Perhaps our more important conclusion is that it is going to take a more active role of the government—or an adoption of a new strategy—for China to begin to solve the problems in the groundwater sector. Although the government in China has begun to take a number of policy responses, implementation has not been very effective. Farmers, in contrast, have been shown to be much more responsive. However, since farmers are often on their own, the actions of farmers have both helped and hurt water resources. According to our data, at times the actions of farmers are helpful in alleviating the crisis (eg, when they adopt water-saving technologies and reduce water use as the cost of water rises). At other times, they only make the groundwater scarcity worse (eg, when they continually drill wells—even in areas where the groundwater table is falling rapidly).

Therefore, one of the main messages of our research is that it is possible to address China's water crisis. It is just going to take more effective actions by the government to channel farmer behavior in a productive, nonharmful way. Above all, it must be realized that nominal reform cannot realize the policy goals of increasing water-use efficiency and conserving water. The government must create the institutions and infrastructure through which they can build an environment that will provide the incentives to make farmers save water. If they can, we have shown that farmers will save water, make adjustments, and adopt new technologies that are helpful to overcoming the water crisis. We believe a sustainable environment needs to be built on effective water pricing and water rights policies. To make this work, a huge commitment is needed to set up the institutions and infrastructure to implement them. Although this is a challenge, we believe it will be more effective and much cheaper than other technological solutions (such as the South to North Water Transfer Project).

REFERENCES

Brown, L.R., 2000. Fall Water Tables in China May Soon Raise Prices Everywhere. http://www.qmw.ac.uk/~ugte133/courses/environs/cuttings/water/china.pdf.

Crook, F., Diao, X., January–February 2000. Water Pressure in China: Growth Strains Resources. Economic Research Service, United States Department of Agriculture, pp. 25–29. Agricultural Outlook.

Economic and Social Survey of Asia and the Pacific (ESCAP), 2010. Statistical Yearbook for Asia and the Pacific 2009. United Nations Publication, Bangkok.

Kendy, E., Gerard-Marchandt, P., Water, M.T., Zhang, Y.Q., Liu, C.M., Steenhuis, T.S., 2003. A soil-water-balance approach to quantify groundwater recharge from irrigated cropland in the North China Plain. Hydrological Processes 17, 2011–2031.

McAlister, J., 2005. China's Water Crisis. Deutsche Bank China Expert Series. http://www.cbiz.cn/download/aquabio.pdf.

Ministry of Water Resources, 2012. Water Resources Bulletin 2011. Ministry of Water Resources, Beijing.

Ministry of Water Resources of China, 2011. Water Resources Bulletin 2011. Ministry of Water Resources, Beijing.

Ministry of Water Resources, 2007. Water Resources Bulletin. Ministry of Water Resources, Beijing.

Ministry of Water Resources and Nanjing Water Institute, 2004. Groundwater Exploitation and Utilization in the Early 21st Century. China Water Resources and Hydropower Publishing House, Beijing.

Nankivell, N., 2004. China's mounting water crisis and the implications for the Chinese Communist Party. Paper Presented in the 12th Annual CANCAPS Conference Quebec City, December 3–5. http://www.cancaps.ca/conf2004/nankivell.pdf.

National Bureau of Statistics in China (NBSC), 2013. China Statistical Yearbook. National Bureau of Statistics in China, Beijing.

National Bureau of Statistics of China (NBSC). National Data. http://data.stats.gov.cn/workspace/index?m=hgnd. (accessed 16.02.14.).

Nickum, J., 1988. All is not wells in North China: irrigation in Yucheng County. In: O'Mara, G. (Ed.), Efficiency in Irrigation. World Bank, Washington, DC, pp. 87–94.

Shi, Y., 2000. Groundwater development in China. In: Paper for the Second World Forum, the Hague, 17–22 March (Not published).

Tsur, Y., Graham-Tomasi, T., 1991. The buffer value of groundwater with stochastic surface water supplies. Journal of Environmental Economics and Management 21 (3), 201–224.

Wang, J., Huang, J., Blanke, A., Huang, Q., Rozelle, S., 2007. The development, challenges and management of groundwater in rural China. In: M.Giordano, Villholth, K.G. (Eds.), The Agricultural Groundwater Revolution: Opportunities and Threats to Development, Compresenhensive Assessment of Water Management in Agriculture Series. Cromwell Press, Trowbridge, UK, pp. 37–62.

Wang, J., Huang, J., Rozelle, S., 2005. Evolution of tubewell ownership and production in the North China Plain. Australian Journal of Agricultural and Resource Economics 49 (2), 177–195.

Wang, J., Huang, J., Yan, T., 2013. Impacts of climate change on water and agricultural production in ten large river basins in China. Journal of Integrative Agriculture 12 (7), 1267–1278.

Wang, J., Huang, J., 2004. Water problems in the Fuyang River Basin. Natural Resources Transaction 19 (4), 424–429.

Wang, S., 2008. Resolve Water Problems to Ensure Food Security. http://news.qq.com/a/20050308/000441.htm.

CHAPTER

2 Irrigation, Agricultural Production, and Rural Income

We have established that China is indeed experiencing a water shortage. But in reality, how much of a problem is this? In this chapter, we seek to answer this question by looking at the economic impact of water in rural China. China's rural areas are dominated by irrigated farmland, so by studying the role of irrigation in China's rural economy, we aim to better understand the potential economic repercussions of a water shortage that would limit farmers' access to irrigation.

China's leaders have long recognized the important role of water in the rural economy, consistently using water control as a means of accelerating production growth and reducing poverty. China's success in achieving food self-sufficiency took place after massive government investments in irrigation infrastructure in the 1960s and the 1970s, lending evidence to the idea that irrigation plays a key role in rural development (Liao, 2003). The Chinese government is so convinced of this relationship that investment in water control dominates all other forms of government investment, including spending targeted specifically at poverty reduction ($10.74 billion vs. $2.90 billion). (Yuan is the currency used in China. Throughout the book, monetary terms in Yuan are first inflation adjusted to the year 2004 using rural CPIs in China provided by the Statistics Bureau of China, then converted into 2004 US dollars. The exchange rate in 2004 was $1 = 8.27 yuan.)

However, the relationship between irrigation and agricultural production or household income remains unclear. Inside China, Hu et al. (2000) find that irrigation (measured as the ratio of irrigated land to cultivated land) had no effect on total factor productivity (TFP) growth of rice in China between 1980 and 1995. Jin et al. (2002) extend the work of Hu et al. (2000) to other crops and cannot find a link between irrigation and TFP growth of any major grain crop (rice, wheat, or maize). Using a provincial level data set, Zhu (2004) finds that irrigation does not have any impact on the yield of wheat or maize between 1979 and 1997; using a county level data set, Travers and Ma (1994) demonstrate that returns from irrigation investments in the poor counties are lower than their costs.

Although China has made remarkable progress in increasing the standard of living in rural areas since the onset of economic reform, rural income growth has slowed in recent years and the distribution of income has skewed (Nyberg and Rozelle, 1999; World Bank, 2001). While the incidence of poverty declined from 30.7% to 3.4% between 1978 and 2000 according to the national poverty line (Wang et al., 2002), large discrepancies among households within regions and

Managing Water on China's Farms. http://dx.doi.org/10.1016/B978-0-12-805164-1.00002-6

among sets of households in different regions have begun to appear. Indeed, Rozelle (1996) shows that the interregional Gini coefficient increased from 0.28 in the 1980s to 0.42 in the 1990s in rural China, while Yang (2002) shows a similar rise in inter-household measures of inequality. This widening of income inequality has slowed the rate of China's poverty reduction (World Bank, 2001).

Because the expansion of the rural economy has driven a large part of China's past economic growth (Perkins, 1994), the recent sluggish rural income expansion and increasing inequality have begun to command national attention. One type of investment that China's leaders have traditionally relied on to increase rural livelihoods is investment in irrigation. In 2000, government spending on irrigation ($4.53 billion) exceeded the annual budget targeted specifically at poverty reduction ($2.90 billion) and was more than 10 times the spending on agricultural research ($439.92 million) (Huang and Hu, 2001; Ministry of Water Resource, 2001; National Statistical Bureau of China, 2001a).

However, the evidence of the impacts of irrigation on rural income has been mixed. Fan et al. (1999) suggest that government expenditures on irrigation have only a modest impact on agricultural production growth and even less on rural poverty and inequality, even after accounting for trickle-down benefits. Jin et al. (2002) cannot find a link between irrigation and TFP growth of any major grain crop (rice, wheat, or maize) in China between 1981 and 1995. Outside of China, Rosegrant and Evenson (1992) are unable to establish a positive link between irrigation investment and productivity in India. Hossain et al. (2000) is among the few to find a positive impact of irrigation on rural household income in Philippines and Bangladesh. However, household survey-based analyses use small samples that are not nationally representative. To our knowledge, no household study has been used to examine these questions in China. Despite the increasing seriousness of China's rural inequality problem, there has been little theoretical or empirical effort to study the effects of irrigation investment on inequality, perhaps due to the complicated nature of the linkages between irrigation and inequality.

This chapter aims to understand the impact that irrigation in China has had on agricultural production, rural incomes, and poverty alleviation. By doing so, we hope to gain a better understanding of the potential economic impacts a water shortage would have on China's rural areas. To achieve this overall goal, we divide the chapter into five sections. In the first section, we discuss and evaluate the impact of irrigation on crop yields and revenue. In the next section, we examine the relationship between irrigation and income at the village level, using both descriptive statistics and multivariate analysis. In the third section, we use a simulation approach to explore the impact of irrigation on the incidence of poverty. In the fourth section, in order to uncover the effect of irrigation on the income distribution, we decompose inequality by source of income, by group according to access to irrigation, and by estimated income flows due to specific household characteristics. Finally, we conduct a cost-benefit analysis of irrigation based on current prices. All analyses in this chapter use data from the 2000 China National Rural Survey (CNRS).

IRRIGATION AND AGRICULTURAL PRODUCTION

DESCRIPTIVE ANALYSIS

Compared to other countries in the world, the proportion of China's cultivated area that is irrigated is high (Table 2.1). Data from our survey show that 52% of cultivated land is irrigated (row 1). Of the area that is irrigated, farmers irrigate 61% with surface water and the rest with groundwater. Although the figure for the proportion of irrigated area is higher than that published by National Statistical Bureau of China (CNSB, 2001b), both are higher than most other countries in the world. For example,

Table 2.1 Proportion of Sown Area by Irrigation Type (%)

		Among Irrigated Area			
	(1) Irrigated Area[a]	(1a) Surface Water Area	(1b) Groundwater Area		(2) Nonirrigated Area
1	China	52	61	37	48
	Major grains: aggregate				
2	Rice	95	95	3	5
3	Wheat	61	34	63	39
4	Maize	45	31	65	55
	Major grains: by season				
5	Single season rice	94	94	4	6
6	Early season rice	99	99	0	1
7	Late season rice	99	99	0	1
8	Single season wheat	10	37	63	90
9	Wheat–rice rotation	98	96	2	2
10	Wheat–maize rotation	77	24	73	23
11	Wheat–other crop rotation	63	23	76	37
12	Single season maize	15	23	71	85
13	Maize–other crop rotation	49	72	27	51
14	Coarse grains[b]	28	26	71	72
15	Tubers[c]	40	88	10	60
	Cash crops				
16	Cotton	94	13	87	6
17	Peanut	69	8	92	31

[a] *Proportion of irrigated areas include areas irrigated by surface water, by groundwater and by both (conjunctively). Proportion of areas irrigated conjunctively is not reported here because it is less than 3%. Thus column (1a) and column (1b) does not sum up to 100%.*
[b] *Coarse grains include sorghum, millet, pearl millet, buckwheat, and others.*
[c] *Tubers includes white potatoes and sweet potatoes.*

the comparable statistic for India is 33%; for Brazil, 1%; and for the United States, 6% (Food and Agriculture Organization of the United Nations, 2002).

Our figure may be higher than that used by official statisticians for two reasons. First, in our sample, we do not choose those villages that are more than 4 h away from a township, so we miss the set of sample households from regions in which the proportion of irrigated cultivated area was lower than average, biasing our estimation upward. In addition, although almost a representative sample of China, our randomly selected sample did not choose some provinces that happen to be less irrigated than the average national level. For example, only 17% of cultivated land in Heilongjiang province is irrigated, only 27% in Inner Mongolia, and 19% in Gansu (National Statistical Bureau of China, 2001a,b). Importantly, China's major food grains are mostly irrigated (Table 2.1, rows 2 and 3). Around 95% of rice and 61percent of wheat are irrigated, levels that are above the national average. Henceforth, as shown in Huang et al. (1999), investment in irrigation has been central for China to maintain food security and will continue to be one investment that enables China to lift its future production of food and meet its food grain security goals of achieving 95% self-sufficiency for all major grains.

While around half of China's cultivated area is irrigated, the proportion of area that is irrigated varies sharply by crop. In contrast to the case of food grains, a majority of area for most feed grains and lower-valued staple crops is not irrigated (Table 2.1, rows 4, 14, and 15). Despite its growing importance in China's agricultural economy, only 45% of China's maize is irrigated and even a lower proportion of coarse grains and tubers (including white and sweet potatoes) are irrigated. Although the proportion of irrigated area in cash crops also varies by crop, most of the area of China's main cash crops is irrigated (eg, 94% of cotton area and 69% of peanut area).

Our descriptive statistics show that irrigation may contribute to the growth in crop production in at least two ways. First, irrigation helps increase crop yield. The positive and significant differences between yields of irrigated and nonirrigated plots indicate that for almost all crops (except for rice and tubers) the average yields of irrigated plots exceed significantly those of nonirrigated ones (Table 2.2, column 6). For example, wheat yields of irrigated plots are 70.9% higher than those of nonirrigated ones (row 2). Irrigated maize yields are 16.4% higher, and irrigated cotton yields are 177% higher (rows 3 and 20).

Second, irrigation improves crop production by increasing the cultivation intensity and, as a result, the *annual output* of a particular plot of land (Table 2.2). When two crops are planted in rotation with one another (rows 5–7, rows 9–15, and row 17), the annual output per plot rises steeply when compared to the yields of a single season crop (rows 4, 8, and 16). For example, the annual yields of wheat–rice (9266 kg/ha—with the yield of rice being 6327 and that of wheat being 2939) and wheat–maize (8263 kg/ha—with the yield of wheat being 3877 and the yield of maize being 4386) rotations far exceed those of single season wheat (1931), rice (6195), and maize (2876).

Table 2.2 Crop Yield by Irrigation Type (Unit: kg/ha)

		(1) Total Yield	(2) Irrigated Yield[a]	(3) Surface Water Yield	(4) Ground Water Yield	(5) Nonirrigated Yield	(6) Percentage Increase[b]
	Major grains: aggregate						
1	Rice	5947	5942	5919	6663	6002	−1.0
2	Wheat	3305	3853	3302	4518	2255	70.9***
3	Maize	4041	4378	4276	4522	3762	16.4***
	Major grains: by season[c]						
4	Single season rice	6195	6207	6202	6367	6087	2.0
5	Rice–rice rotation	9934	9949	9,943	11,250	9000	10.5
6	Early season rice	4516	4516	4513	5250	4500	0.4
7	Late season rice	5418	5433	5431	6000	4500	20.7***
8	Single season wheat	1931	3624	4025	3223	1698	113.4***
9	Wheat–rice rotation	9266	9284	9,251	11,357	7513	23.6
10	Wheat	2939	2949	2972	3000	1763	67.3***
11	Rice	6327	6334	6279	8357	5750	10.2***
12	Wheat–maize rotation	8263	9174	8309	9617	6271	46.3***
13	Wheat	3877	4439	3796	4746	2642	68.0***
14	Maize	4386	4735	4514	4872	3628	30.5***
15	Wheat–other crop rotation	3331	3926	3375	4212	2411	62.8***

Continued

Table 2.2 Crop Yield by Irrigation Type (Unit: kg/ha)—cont'd

		(1) Total Yield	(2) Irrigated Yield[a]	(3) Surface Water Yield	(4) Ground Water Yield	(5) Nonirrigated Yield	(6) Percentage Increase[b]
16	Single season maize[d]	2876	3720	3056	4309	2378	56.4***
17	Maize–other crop rotation	3941	3984	4181	2883	3893	2.3
18	Coarse grains	1457	1996	1836	2115	1119	78.3***
19	Tubers[e]	4631	3918	4072	2942	5141	−23.8***
	Cash crops						
20	Cotton	2357	2561	1190	2790	924	177.3***
21	Peanut	2538	2758	2731	2770	2143	28.7***

[a] *We did not include yield of the plots irrigated by surface water and groundwater conjunctively because there are few observations of them.*
[b] *Percentage increase means irrigated yield compared to nonirrigated yield. We also test whether the difference is statistically significant. *** indicates significant at 99% level.*
[c] *In this category, we divide rice into single season rice, double season rice (early season rice, late season rice). We divide wheat into single season wheat, wheat–rice rotation, wheat–maize rotation and wheat rotated with other crops than major grain. We divide maize into single season maize and wheat–maize rotation.*
[d] *We dropped Liao Ning province here because 80% are nonirrigated plots; 46% of the nonirrigated plots and 60% of the irrigated plots suffered from draught (lost of produce more than 50%).*
[e] *Tuber includes sweet potato and white potato.*

Although there are two crops (rice and tubers) that have lower yields in irrigated plots when compared to nonirrigated plots, closer inspection shows that even in these cases, irrigation increases yields or at least does not hurt them (Table 2.2, rows 1 and 19). If we divide rice into single season rice, rice grown in a rice–rice rotation (early season rice and late season rice), and rice grown in a wheat–rice rotation, we find for each subdivision, the differences between the yields of irrigated and nonirrigated plots are all positive and significantly different in several cases. The average yields of irrigated rice plots in the aggregate are lower because yields of single season rice (both those that are irrigated and nonirrigated) are 64% higher than those of other types of rice (rice grown in rice–rice or wheat–rice rotations). In the case of tubers, we find that the higher yields on nonirrigated plots can be accounted for by plots in the sample's three southern provinces (Zhejiang, Sichuan, and Hubei Provinces) since the main season for growing tubers coincides with the rainy season and tubers planted in irrigated areas that are typically more subjected to flooding do not do as well as those planted on nonirrigated plots.

In the course of increasing crop production, irrigation almost certainly also helped improve food security at the household level, especially for poor households. In our analysis we define poor households and rich households as those with household incomes in the bottom and top quintile of each province, respectively. Our data show that poor households rely more on grain production. Among poor households, 78% of the land is allocated to growing grain crops, a level that is nearly 10% higher than that among rich households (68%). By increasing the level of grain output, irrigation also contributed to better access to food for poor households.

Even larger differences appear when examining *differences between the level of revenue* (price times yields) earned by farmers on their irrigated and nonirrigated plots (Table 2.3). Overall revenue from irrigated plots is 79% higher than that of nonirrigated plots (row 1, column 6). While we cannot pinpoint the source of these changes, three factors account for the higher crop revenues of a plot when irrigation is introduced: (1) Higher yields (of same crop), (2) increasing intensity (producing more than one crop per season), and (3) shifts to higher valued crops that are possible after irrigation.

Finally, our results also provide evidence that new irrigation may help raise incomes in poor areas. Farmers in rich and poor areas earn higher revenue from their irrigated crops (rows 2 and 3). In rich areas crop revenue per hectare from irrigated plots is 89% higher than revenue from nonirrigated plots. In poor areas revenue from irrigated plots exceeds those of nonirrigated plots by 93%. To directly address the positive impact of irrigation on household income, we should have used crop income and analyzed the impact of irrigation on crop income, and then linked it to household income. Unfortunately, although information on yield, irrigation status, and other variables is collected at the plot level, information on cost of inputs including fertilizer, labor, machinery, and seeding is only collected at the household level. Hence, it is impossible for us to obtain the crop income at the plot level and carry out such analysis. However, in the cost-benefit analysis of irrigation, we show that, *in the majority of the villages that invested in new irrigation*, the benefit of increasing

Table 2.3 Gross Crop Revenue by Irrigation Type and China's Regions

	(1) Annual Income Per Capita (USD/person)	(2) Percentage of Crop Income in Total Income (%)	(3) Crop Revenue (USD/ha)	(4) Crop Revenue for Irrigated Plots (USD/ha)	(5) Crop Revenue for Nonirrigated Plots (USD/ha)	(6) Percentage Increases of Crop Revenue[a] (%)
China	256	20[b]	510	593	332	79[b]
By wealth level[c]						
Rich area	410	10	525	596	316	89
Poor area	152	34	429	567	293	93

[a] Percentage increase is calculated as (column 4 − column 5)/column 5.

[b] The national level is lower than both in rich and poor areas because we do not include middle-income area here that has 65% increases in crop revenue when plots are irrigated.

[c] Rich area includes households whose incomes rank the first 20th percentile in every province and all the households from Zhejiang province. Poor area means households whose incomes rank the last 20th percentile in every province.

irrigated area outweighs the cost of doing so. In summary, even given the limitations of our analysis, we believe the evidence is clear that increasing irrigation will increase total household income.

While the data show that irrigation is effective in both rich and poor areas, differences in the nature of rich and poor economies suggest that irrigation may have larger impacts on rural welfare in poor areas. Since people are poorer, and since we typically assume that utility functions are concave, if rich and poor areas enjoy equal income gains, the gains in the poorer areas will turn into larger increases in welfare. As seen above, crop revenues in the poorest areas (93%) increase slightly more than those in richer areas (89%). Moreover, crop revenues make up a much larger part of total household income in poor areas than in rich areas (only 10% in rich areas and more than 30% in poor areas; see column 2). If we multiply the percentage increase of crop revenue by the share of crop revenue in total income, increases in total income in rich areas will be lower than that in poor areas. Since China's poverty is typically characterized by the small gap between the income of the poor and the poverty line (World Bank, 2001), raising the income of the poor by more than one-third would almost certainly have the effect of pulling a vast majority of those in newly irrigated areas out of poverty.

MULTIVARIATE REGRESSION RESULTS

To measure the effect of irrigation on agricultural production, we regress the yield of a crop or the crop revenue on a plot's irrigation status. We control for plot-specific characteristics: *soil quality* (where soil quality is a subjective measure whereby if the farmer ranked his plot as "good," a dummy variable was set equal to 1 and was set equal to 0 if the farmer ranked his plot as "not very good" or "poor"); *topography–plain* (a dummy variable that was set equal to 1 if the plot was on a plain); *topography–hill* (a dummy variable that was set equal to 1 if the plot was on a hill); *plot size* (measured in mu, mu is the metric used to measure land area in China and 1 hectare equals 15 mu); *distance from home* (the distance of the plot from the farmer's house, measured in kilometers); *shock-severity of disaster* (a continuous variable based on the farmer's subjective opinion about the percentage reduction in yields on each plot caused by adverse weather shocks suffered during the survey year); *single season crop* (a dummy variable that was set equal to 1 if the crop is not grown in conjunction with other crops during the year and set equal to 0 otherwise).

When irrigation is used to explain crop yield or crop revenue, an endogeneity problem may exist in terms of the household-level choice of which crop to irrigate. In the case of rural China, the decision on which crop to irrigate is most likely to be exogenous to our model. First, whether a plot is irrigated or not largely depends on whether irrigation is available. In most parts of our sample, if irrigation water is available, it is used. It depends on whether the plot is located in the command area of a well in the case of groundwater irrigation, and on whether the plot is located within the reach of a canal in the case of surface water irrigation. Second, the choice of which crop to cultivate depends on the cropping cycle and the characteristics of the plot. To look at this issue more closely, we use the part of our data that collected information on the types of crops grown on the plot one year before the survey year. We find that households grow the same summer crop on 95% of plots in each of two consecutive years. Moreover, to examine this issue more rigorously, we performed a Hausman–Wu test. The result of the test failed to reject the null hypothesis that the variable irrigation is exogenous to our model.

Most importantly, household fixed effects are used to control for all nonplot factors including both observable variables (eg, household land holdings and the distance of a village to the county seat) and unobservable variables (eg, the household's off-farm employment opportunities and management ability). Although prices are not explicitly included in the regressions, it should be noted that we are holding the effect of price constant and are estimating a supply function. When using cross-sectional data, it is common *not* to include price explicitly in the analysis since there often is no variation in prices within the unit for which fixed effects are controlled (Lau and Yotopoulos, 1971; Udry, 1996; Yotopoulos and Lau, 1973). Hence, our regression can be seen as a way to examine the economic efficiency gains that farmers realize when their plots are irrigated. This economic efficiency can be thought of as a shift up of the supply curve caused by increased irrigation (everything else held constant). Because we use a supply–function approach, we do not include measures of other variable inputs.

Our approach to explore the relationship between irrigation and agricultural performance addresses the shortcomings of previous work. First, due to lack of data, most studies have only used rough proxies for irrigation, such as government expenditure on irrigation. These proxies, however, may not provide an accurate measure of irrigation because there is no guarantee that the allocation of funds to water control is ever turned into an effective increase in irrigation stock. Moreover, the addition of irrigated area (and the subsequent rise in yields) through public investment likely will occur only at a lag of a year or more after the investment is made. Therefore, to analyze the impact of public investment, a dynamic framework or times series data is needed. Most studies that look at public investment have only used a static framework and so do not account for such an investment lag. When we use the stock of irrigation itself (ie, whether the plot is actually irrigated during each season being analyzed), we avoid the need to use a proxy for irrigation.

Second, and most importantly for this type of study, most other researchers do not control for the unobserved heterogeneity that may be obscuring the relationship between irrigation and crop yields and crop revenue. For example, the inability to control for the household's off-farm employment opportunities could lead to an underestimation of the impact of irrigation on yield in rich areas (eg, Zhejiang province). Although the proportion of irrigated land might be higher in Zhejiang than in poorer provinces, households in richer areas have more opportunity to work off-farm and, *ceteris paribus*, they will almost certainly allocate less family labor to farming activities than households in poorer provinces that do not have as convenient access to off-farm jobs. An omitted variable problem, in this case the omission of off-farm employment opportunities, would make the estimated relationship between irrigation and agricultural performance unreliable (Kennedy, 1998). In our study we have collected information on approximately four plots for each sample household. Such data allow us to control for all of the nonplot varying factors that could be affecting yields by using a fixed effects framework.

Finally, most analyses have been highly aggregated, both across provinces and across crops (Travers and Ma, 1994; Zhu, 2004). Using aggregate data fails to account for variation in plot-specific factors that may affect crop yields. For example, failure to account for the variation in soil quality will cause a downward bias in the estimation of the coefficient on irrigation. The reason is that the plots of highest qualities are more likely to be irrigated first. In later years, when opening up newly irrigated area, the land that is brought into cultivation is of lower quality. Because of this, it is possible that the *average* yield could fall when it is evaluated at an aggregate level. By using a rich set of plot-level data, we can hold constant many of the plot-specific factors that could be affecting yields and which could be potentially correlated with a plot's irrigation status (such as soil quality).

In the multivariate analysis, we first examine the effect of irrigation on *yields for individual crops*. While interesting by itself, such a regression does not capture all of the dimensions of the effects of irrigation. In the next step, we estimate a second model to explain *agricultural revenues*. If irrigation allows farmers to cultivate two crops per year and/or if it allows shifting into cash crops that generate higher revenues per hectare, the aggregate household agricultural revenue will capture the higher output from irrigation. Finally, we explain yields separately for rich and poorer areas in order to gauge the differences in the effects of irrigation in rich and poor areas. Note all the dependent variables are in log form so the coefficients represent percentage changes in yields or revenues.

The findings from multivariate analysis support the hypothesis that irrigation raises yields for most crops (Table 2.4). For example, irrigation increases the yields of wheat by 17.7%, those of maize by 29.4%, and those of cotton by 28.4% (row 1). The multivariate analysis results of crop-specific yields do differ from the descriptive results when examining the magnitude of the differences. With the exception of maize, the magnitude of the impact of irrigation is lower in the regression results than in the descriptive statistics. Most likely this is because in the regressions the irrigation impacts are being conditioned on the level of other variables, such as soil quality, and these other variables account for part of the irrigation effect (eg, since most irrigated land is "good").

The impact of irrigation becomes even greater when we look at household crop revenue (Table 2.5). Overall, irrigation increases revenue by 76.1% (column 1), a figure that is only slightly less than the unconditional difference observed in the descriptive statistics (Table 2.3, row 1). In other words, according to these results, most of the differences between revenues on irrigated and nonirrigated plots are due to the addition of irrigation and not other plot characteristics. The magnitude of the coefficient drops to 42.9% when household dummies are replaced with four household-level variables and a set of village dummies (column 2). Using village dummies instead of household dummies moves the coefficient of interest in the

Table 2.4 The Impact of Irrigation on Crop Yield With Household Fixed Effects

		Dependent Variables: Log Crop Yield		
		(1) Wheat	**(2) Maize**	**(3) Cotton**
	Irrigation status			
1	Irrigated (by surface water or groundwater)	0.177 (2.81)***	0.294 (4.17)***	0.284 (5.28)***
	Land characteristics			
2	Good soil quality	0.174 (5.41)***	0.130 (3.50)***	0.008 (0.24)
3	Topography-plain	0.070 (0.65)	0.302 (1.38)	−0.001 (0.02)
4	Topography-hill	0.132 (2.53)**	0.181 (0.90)	0.083 (0.92)
5	Plot size	0.041 (0.39)	0.204 (1.42)	0.010 (0.65)
6	Distance from home	0.003 (0.21)	−0.005 (0.16)	0.008 (0.24)
7	*Shock*: severity of disaster[a]	−0.009 (6.22)***	−0.016 (12.78)***	−0.001 (1.65)
8	*Single season crop*[b]	−0.040 (0.87)	−0.106 (2.06)**	0.054 (2.43)**
9	Number of plots	1027	1116	141
10	Number of households with multiple plots	297	329	38
11	R-square	0.15	0.47	0.39

*Absolute value of t statistics in parentheses. ** significant at 5%; *** significant at 1%.*

[a] *Severity of disaster means percentage reduction of production.*

[b] *A dummy variable that equals 1 if the crop is not grown in conjunction with other crops during the year and is 0 otherwise.*

Table 2.5 The Impact of Irrigation on Household Crop Revenue Per Plot

		Dependent Variable: Log Annual Household Crop Revenue		
	Level of Fixed Effects	**Household**		**Village[a]**
	Irrigation status			
1	Irrigated (by surface water or groundwater)	0.761 (15.98)***		0.429 (13.83)***
2	Irrigated by surface water		0.681 (13.23)***	
3	Irrigated by groundwater		1.019 (11.70)***	
	Land characteristics			
4	Good quality	0.286 (7.09)***	0.281 (7.19)***	0.219 (7.83)***
5	Topography-plain	0.098 (0.94)	0.082 (0.80)	−0.004 (0.07)
6	Topography-hill	−0.009 (0.11)	−0.069 (0.89)	−0.104 (2.02)**
7	Plot size	0.095 (1.02)	0.053 (0.60)	
8	Distance from home	0.020 (1.12)	0.021 (1.44)	0.022 (1.58)
9	*Shock*: severity of disaster[b]	−0.009 (9.50)***	−0.008 (8.74)***	−0.009 (11.93)***
10	*Single season crop*[c]	0.755 (26.96)***	0.736 (26.86)***	0.716 (28.48)***
11	Number of plots	5352	5614	5347
12	Number of household with multiple plots	1043	1070	
13	Number of villages			60
14	R-square	0.23	0.23	0.20

*Absolute value of t statistics in parentheses. ** significant at 5%; *** significant at 1%.*

[a] *In the village fixed effects model, we use four household characteristic variables that are not reported here: household size, average education level, total wealth, and total household land.*

[b] *Severity of disaster means percentage reduction of production.*

[c] *A dummy variable that equals 1 if the crop is not grown in conjunction with other crops during the year and is 0 otherwise.*

same direction as was observed in the yield equations when no fixed effects are used at all. Apparently, the use of village dummies and four household-level variables absorbs some, but not all, of the unobserved heterogeneity in crop revenue function.

Decomposing revenue differences by crop illustrates differences among crops in the earnings potential that arises with irrigation (Table 2.6). When a plot is irrigated,

Table 2.6 Decomposed Impact of Irrigation on Household Crop Revenue With Household Fixed Effects

	Dependent Variable: Log Annual Household Crop Revenue		
	Interaction dummies		
1	Rice* irrigation	1.156 (24.41)***	
2	Wheat* irrigation	0.573 (10.34)***	
3	Maize* irrigation	0.619 (10.85)***	
4	Single season rice* irrigation		1.004 (18.36)***
5	Single season wheat* irrigation		0.206 (1.83)*
6	Single season maize* irrigation		0.912 (4.00)***
7	Rice–rice* irrigation		1.473 (15.46)***
8	Wheat–rice rotation* irrigation		0.106 (1.58)
9	Wheat–maize rotation* irrigation		0.989 (12.32)***
10	Wheat–other crop rotation* irrigation		0.863 (9.02)***
11	Maize–other crop rotation* irrigation		0.832 (9.18)***
12	*Coarse grains** irrigation	0.317 (3.78)***	0.532 (5.67)***
13	*Cash crops*: cotton* irrigation	1.365 (15.14)***	1.541 (14.79)***
14	*Cash crops*: peanut* irrigation	0.887 (9.45)***	1.135 (10.78)***
15	*Tubers** irrigation	−1.226 (17.74)***	−1.120 (14.82)***
	Land characteristics		
16	Good quality	0.217 (6.00)***	0.212 (5.29)***
17	Topography-plain	0.065 (0.69)	−0.046 (0.44)
18	Topography-hill	−0.027 (0.36)	−0.145 (1.68)*
19	Plot size	0.011 (0.12)	−0.033 (0.33)
20	Distance from home	0.009 (0.43)	0.028 (1.13)
21	Shock: severity of disaster[a]	−0.009 (10.35)***	−0.010 (9.69)***
22	Single season crop[b]	0.231 (6.39)***	0.400 (8.55)***
23	Number of plots	4858	4166
24	Number of households with multiple plots	978	953
25	R-square	0.45	0.48

*Absolute value of t statistics in parentheses. * significant at 10%; *** significant at 1%.*
[a] *Severity of disaster means percentage reduction of production.*
[b] *A dummy variable that equals 1 if the crop is not grown in conjunction with other crops during the year and is 0 otherwise.*

rising yields and the ability to shift into new crops, such as rice and cash crops, facilitates the largest rises in revenue (115.6% higher for rice, 136.5% for cotton, and 88.7% for peanuts; see column 1). Although somewhat lower, when plots are irrigated rising yields also help increase revenues on wheat (57.3%), maize (61.9%), and coarse grains (31.7%). Of all major crops in the sample, tubers are the only ones that do not enjoy increased revenue. Additionally, when the major grain crops such as rice, wheat, and maize are disaggregated by rotation, the impact of increasing intensity also emerges (columns 2). For example, when using household fixed effects, irrigated double-cropped rice increases yields by 147.3%, higher than single season rice (100.4%). When irrigation facilitates the shift to a wheat–maize rotation, revenues generated on a plot rise by 98.9%, higher than either the rise that accompanies single season wheat (20.6%) or single season maize (91.2%). Significantly, the wheat–rice rotation does not show any statistical difference between irrigated and nonirrigated areas. Most likely this is because in the case of only four households does a single household have both irrigated and nonirrigated plots (the requirement that needs to be met for the observations to be used).

In trying to understand such a result, it would be interesting if we could estimate the effect of irrigation on switching to another crop or by going to a more intensive rotation. To do so, we could add a set of variables that interact major crops with irrigation or types of rotation with irrigation. Unfortunately, the coefficients in such an equation are likely to be subject to an endogeneity bias, because we would expect crop choices to be affected by the same (unobserved) variables that also affect revenues. Since we do not have any valid instruments to control for the endogeneity, we report the results in Appendix and use them to check the robustness of our results.

When dividing the sample into rich and poorer areas, we find similar results (Table 2.7). In both rich and poor areas, irrigation has a significantly positive effect on crop revenue, increasing it by 132.8% in rich areas and 43.9% in poorer ones (columns 1 and 3). While the higher marginal effects of irrigation on crop revenue in rich areas may explain why more of the past investment in irrigation has gone into favorable areas, it does not mean that the poor do not benefit. In fact, in terms of welfare effects, the poor may benefit more. Results in Table 2.3 show that the share of crop revenue in total income is three times as high in poor areas (34%) as in richer areas (10%). Taking this into account, irrigation benefits farmers in poorest areas one and a half times more that it does farmers in richer areas (15% in poor areas vs. 13% in richer areas). Certainly, as discussed in the descriptive analysis section, this means that irrigation also will have a positive effect on household food security of the poor.

Our results show that the magnitude of the impacts of surface water and groundwater irrigation differs in rich areas and poor areas (Table 2.7). In rich areas, the percentage increases in crop revenues are higher when plots are irrigated by surface water than by groundwater (column 2). Two reasons may account for the lower return from surface water irrigation in poor areas. First, it could be that surface water is less reliable in poor areas. Poor areas are often located in areas that have relatively

Table 2.7 The Impact of Irrigation on Crop Revenue in Rich and Poor Areas in China With Household Fixed Effects

		Dependent Variables: Log Plot Crop Revenue			
		Rich Area		**Poor Area**	
		(1)	**(2)**	**(1)**	**(2)**
	Irrigation status				
1	Irrigated (by surface water or groundwater)	1.328 (14.11)***		0.439 (3.50)***	
2	Irrigated by surface water		1.470 (14.30)***		0.296 (2.02)**
3	Irrigated by groundwater		0.717 (3.54)***		0.793 (3.55)***
	Land characteristics				
4	Good soil quality	0.147 (1.90)*	0.167 (2.17)**	0.143 (1.50)	0.139 (1.47)
5	Topography-plain	0.155 (0.88)	0.131 (0.75)	−0.327 (0.85)	−0.309 (0.80)
6	Topography-hill	−0.006 (0.03)	0.100 (0.53)	−0.280 (1.52)	−0.277 (1.51)
7	Plot size	0.048 (0.25)	0.066 (0.34)	−0.220 (1.43)	−0.236 (1.54)
8	Distance from home	0.134 (2.96)***	0.111 (2.44)**	−0.302 (3.39)***	−0.284 (3.18)***
9	*Shock*: severity of disaster[a]	−0.010 (4.46)***	−0.011 (4.70)***	−0.011 (5.99)***	−0.011 (5.52)***
10	*Single season crop*[b]	0.624 (11.54)***	0.599 (11.05)***	1.086 (13.91)***	1.105 (14.06)***
11	Number of plots	1542	1542	959	959
12	Number of households with multiple plots	309	309	172	172
13	R-square	0.25	0.26	0.28	0.29

*Absolute value of t statistics in parentheses. * significant at 10%; ** significant at 5%; *** significant at 1%.*

[a] *Severity of disaster means percentage reduction of production.*

[b] *A dummy variable that equals 1 if the crop is not grown in conjunction with other crops during the year and is 0 otherwise.*

scarce water resources. In those areas, due to the nature of the water resources, surface water is often not delivered either at the time when irrigation is required or in the quantities that are needed. Sometimes it is not delivered at all. Second, the irrigation efficiency most likely is relatively lower in poor areas. Canals in most poor areas are often not lined or the linings of canals have deteriorated over time. Under such circumstances, the benefit from using surface water, although positive, may be reduced for a given quantity of water.

IRRIGATION AND INCOME

Table 2.8 shows that the amount of irrigated land per capita is strongly correlated with annual cropping income (column 3). Compared to households without irrigated land (row 2), annual per capita cropping income is 40% higher (296% vs. 211%) in households that have irrigated land holdings of up to 0.067 ha per capita (row 3). This is true despite the fact that the average household's cultivated land per capita is only 0.089 ha in the households with irrigated land, less than half that of households without irrigated land. Cropping income per capita continues to rise as irrigated land per capita increases (rows 4 and 5).

It is also observed that as irrigated land per capita increases, cropping income becomes a more important source of household income (Table 2.8, column 3; see figures in parentheses). For example, cropping income accounts for only 12% of total income for households without irrigated land. The share of income from cropping grows continuously as irrigated land per capita increases. For those with more than 0.2 ha irrigated land per capita, cropping contributes to 43% of total household income.

Total income per capita, however, does not show the same monotonically increasing relationship with irrigated land area (Table 2.8, column 2). Households with irrigated land between 0 and 0.067 ha per capita have a relatively high average annual income ($335 per capita). In contrast, those with irrigated land between 0.067 and 0.2 have total incomes that reach, on average, only $256 per capita.

Multivariate analysis is used to examine the relationship between irrigation and total income. We also examine the relationship between irrigation and the components in total income—that is, cropping income, off-farm income, or other income (in per capita terms). We follow the standard approach in the literature on permanent income analysis (Datt, 1998; Paxson, 1992) and on income inequality in rural China specifically (Benjamin et al., 2002; Morduch and Sicular, 2002). In these studies the determinants of income can be analyzed by making income a function of a set of household and village characteristics, including household irrigated area. Household characteristics also include household size, average age and education level of the household's labor force, degree of land fragmentation, proportion of good quality land, and proportion of land affected by negative shocks. Cultivated land per capita is included to control for land as a fixed input. We also have included household agricultural assets, self-employed business assets, livestock assets, and nonproductive

Table 2.8 Annual Household Income Per Capita and Are of Irrigated Land Per Capita

	Irrigated Land Per Capita (ha)	Number of Household	Total Household Income Per Capita[a] (USD)	Cropping Income Per Capita (USD)	Off-Farm Income Per Capita (USD)	Total Land Per Capita (ha)
1	China	1198[b]	203 (100%)[c]	60 (19%)	189 (66%)	0.148
2	0 (nonirrigated)	186	203 (100%)	27 (12%)	145 (71%)	0.235
3	Between 0 and 0.067	554	335 (100%)	38 (11%)	248 (75%)	0.089
4	Between 0.067 and 0.2	378	257 (100%)	85 (31%)	130 (52%)	0.148
5	Above 0.2	80	366 (100%)	162 (43%)	166 (46%)	0.407

[a] *Other income is not reported here since it is not the focus of this paper.*
[b] *Total number of households is 1198 instead of 1200 because information on two households is missing.*
[c] *The shares of different income sources in total household income are reported in parentheses.*

assets (in per capita terms) to control for factors including household access to credit markets or ability to adopt new technologies.

A set of village fixed effects is used to control for the observable village characteristics including a village's topography, its distance from the county seat, and all other village fixed effects that vary by village but are difficult to observe or measure (eg, the economic environment of the village, certain climatic and/or agronomic factors that affect village-wide yields and prices, etc.). If the village fixed effects are not used to control for unobserved heterogeneity, we could have an omitted variables problem that generate biased estimates of the impacts of irrigation on income. It should be noted that in the approach we use later to analyze the impact of irrigation on rural income distribution, a linear specification for the income equation is required to decompose the inequality by estimated income flows. Because household income is treated in the same way as firm profit, income should also, theoretically, be a function of output and input prices and other factors. However, since input and output prices are almost surely the same within each village, the effect of prices on income is also grouped with other village fixed effects and cannot be separated out.

We do not worry about the problem of the omitted variables at the household level in our analyses for three reasons. First, a priori there is not likely to be a high degree of correlation between unobserved household effects and irrigated area. Second, it is possible that, given our specification, we are missing important household-level determinants of income; in particular, measures of the ability of the household wage earners. To control for this, we add two measures of ability: the classroom grades of the household head during his/her last year in school and the educational attainment of the mother. Although these are not perfect measures, they have been used in other analyses. When we include these variables, the coefficient on the irrigation variable does not change substantially (results not shown in the text). Third, since the only part of a household's irrigated area that might be related to ability is irrigated land that he/she rents (the rest is allocated to the household by the village based mostly on demographic factors), we also have included an alternative measure of irrigated land per capita that includes only irrigated land allocated from the village to the household, which would make it exogenous. When we run our regressions with this alternative measure, the results are robust (results not shown in the text).

REGRESSION RESULTS

Our regression estimates of the effect of irrigation on income performed well (Table 2.9). The goodness of fit measure, R^2, is 0.18 for total income and 0.54 for the cropping income equation, which are sufficiently high for analyses that use cross-sectional household data. In addition, many of the coefficients associated with our control variables are statistically significant and of the expected sign. For example, we find that the level of education of the household's labor force positively

Table 2.9 Determinants of Income: OLS With Fixed Effects at the Village Level

		Dependent Variables (Dollars Per Capita)			
		(1) Total Income	**(2) Cropping Income**	**(3) Off-Farm Income**	**(4) Other Income[a]**
1	Area of irrigated land per capita (ha)	2628.459 (2.37)**	3082.936 (6.37)***	−28.624 (0.03)	−425.853 (1.52)
2	Household size (number of household members)	−67.308 (0.57)	−14.222 (1.26)	−69.120 (0.60)	16.034 (0.90)
3	Average age of household labor (year)	−1.748 (0.11)	−2.853 (1.61)	−13.664 (0.90)	14.769 (3.91)***
4	Level of education of household's labor force (attainment in years)	116.139 (2.60)***	−7.385 (1.09)	86.184 (2.06)**	37.340 (2.51)**
5	Degree of land fragmentation (number of plots per household)	−184.540 (2.96)***	13.531 (1.62)	−189.853 (3.09)***	−8.218 (0.64)
6	Proportion of good quality land (%)	−0.502 (0.13)	0.762 (1.82)*	−0.692 (0.19)	−0.572 (0.62)
7	Proportion of land affected by negative shock (%)	−0.811 (0.19)	−0.954 (2.32)**	−0.636 (0.15)	0.779 (0.99)
8	Cultivated land per capita (ha)	502.836 (0.89)	640.016 (1.74)*	−134.822 (0.27)	−2.357 (0.01)
9	Nonland agricultural assets per capita (USD)	−0.123 (0.44)	−0.004 (0.21)	−0.061 (0.22)	−0.058 (1.53)
10	Self-business assets per capita (USD)	0.036 (0.42)	−0.008 (2.08)**	0.036 (0.42)	0.008 (0.79)
11	Livestock assets per capita (USD)	0.351 (2.53)**	0.033 (1.28)	0.075 (0.59)	0.244 (4.10)***
12	Nonproductive assets per capita (USD)	0.077 (2.35)**	−0.000 (0.20)	0.071 (2.17)**	0.006 (1.18)
13	Constant	2115.979 (1.68)*	279.794 (2.33)**	2398.542 (1.93)*	−562.357 (2.61)***
	R-squared	0.18	0.54	0.18	0.17

*Robust t-statistic in parentheses. * significant at 10%; ** significant at 5%; *** significant at 1%.*
[a] *Other income includes livestock income, income from gifts (nonremittances), rental income, income from subsidies and pensions, income from interest, income from asset sales, net value of commercial agricultural (eg, vegetable and fruit), value of crop subsidiaries (eg, fodders), net value of processed crop products, and miscellaneous income.*

affects total income. Also, as expected, negative shocks significantly reduce cropping income, and cultivated land per capita significantly increases cropping income.

Most importantly, our results allow us to reject the null hypothesis that irrigated land area has no effect on cropping income and indicate that the descriptive results are largely consistent with multivariate analysis (Table 2.9, column 2). Increasing irrigated land per capita by 1 ha will lead to an increase of $398 in annual cropping income per capita, holding other household characteristics constant. Interestingly, unlike many of the studies using aggregate data (eg, Jin et al., 2002; Rosegrant and Evenson, 1992), when we use plot-level data, we find a strong relationship between irrigation and cropping income.

Our multivariate analysis also reveals that, when we hold other factors constant, irrigation has a positive impact on total household income, mainly through its impact on cropping income. An additional hectare of irrigated land per capita is associated with an increase of $340 in total household income per capita. Thus, an increase of irrigated land per capita of one standard deviation (0.097 ha) leads to an increase in household per capita income of $33, which is about 10% of average household per capita income. Since there is no significant effect of irrigation on off-farm income and other income (column 3 and 4), it can be concluded that the positive impact that irrigation has on total income per capita comes largely from the impact of irrigation on cropping income.

The finding that irrigation has a positive impact on cropping income, coupled with the structural characteristics of household income, suggests that irrigation may have an important role in poverty reduction. Cropping income accounts for a much larger share of total income in poorer households than in rich ones. Households in the two poorest deciles earn almost 60% of their income from cropping activities (Table 2.10). In contrast, households in the richest decile earn less than 10% from cropping. Given such an income structure it is not surprising that the

Table 2.10 Share of Cropping Income in Total Income by Percentiles of Total Income

Percentile of Total Income (%)	Share of Cropping Income in Total Income (%)
1–10	54.52
11–20	58.61
21–30	48.04
31–40	44.82
41–50	40.07
51–60	33.02
61–70	28.90
71–80	22.77
81–90	21.69
91–99	9.31

correlation coefficient between cropping income and total income (0.18) is much lower than that between off-farm income and total income (0.98). Hence, investment in irrigation, by increasing cropping income, would increase the total income of poor households and lead to poverty reduction.

IRRIGATION AND INEQUALITY

In addition to its positive impact on the incidence of poverty, given the structure of income in rural China, it also is possible that increased investment in irrigation could help lower income inequality. Since cropping income contributes heavily to the income of poor households, increases in irrigation should have a relatively larger impact on their income, which makes up the lower tail of the income distribution. Moreover, the regression results also showed that increases in irrigated land area do not contribute to higher off-farm incomes (Table 2.9, column 3), which has been shown by others to increase rural income inequality (Kung and Lee, 2001; Rozelle, 1996). In other words, irrigation might increase total income of households in the lower end of the income distribution while having a smaller impact on the income of those at the higher end, overall resulting in lower inequality of total income.

To analyze the impact of irrigation on inequality, we follow the methodologies developed in a series of previous studies (Stuart, 1954; Pyatt, 1976; Pyatt et al., 1980; Lerman and Yitzhaki, 1985; Taylor, 1997; Morduch and Sicular, 2002) and decompose inequality in three ways: by *source of income* (cropping income from irrigated plots, cropping income from nonirrigated plots, and off-farm income and other income); by *group according to irrigation access* (those with some irrigated land and those without any irrigated land); and by estimated *income flows due to specific household characteristics* (eg, irrigated land area per capita and the education level of the household's labor force). Our methodology is similar in all three cases. We decompose the Gini coefficient for total household income as a weighted sum of the inequality levels of incomes from different components, with the weights being functions of the importance of each component and the correlation of each component with total income. For example, if the income contributed by irrigated land accounts for a large share of total income and is itself highly unequally distributed, it is likely to increase the total income inequality. However, if income from a component is negatively correlated with total income (ie, this component is more concentrated in the hands of poor farmers), then larger shares of that factor might help equalize total income. Technical details of the methodology are documented in the Methodological Appendix to Chapter 2.

DECOMPOSITION RESULTS

The overall Gini coefficient of per capita income from our sample is 0.541 (Table 2.11, row 1). Compared to Gini coefficients of 0.28 in 1983 and 0.42 in 1992 as calculated by Rozelle (1996), inequality has continued to rise in the 1990s.

Table 2.11 Gini Decomposition by Income Sources

	Income Sources	(1)[a] S_k	(2)[b] G_k	(3)[c] R_k	(4)[d] $S_kG_kR_k$	(5)[e] $\partial G_0/\partial e_j$	(6)[f] $(\partial G_0/\partial e_j)/G_0$
1	Total income	1	0.5407	1	0.5407		
	Cropping income						
2	From irrigated land	0.1692	0.6121	0.3478	0.0360	−0.0555	−0.1026
3	From nonirrigated land	0.0398	0.8433	0.1045	0.0035	−0.0180	−0.0333
4	Off-farm income	0.6208	0.7324	0.9070	0.4124	0.0768	0.1420
5	Other income	0.1778	0.7767	0.6452	0.0891	−0.0070	−0.0130

[a] S_k: *share of income source* k *in total income.*
[b] G_k: *Gini coefficient of income source* k.
[c] R_k: *Gini correlation between income source* k *and the distribution of total income.*
[d] $S_kG_kR_k$: *contribution of income source* k *to the Gini coefficient of total income.* $S_kG_kR_k$ *of cropping income, off-farm, and other income sum to 0.5407.*
[e] $\partial G_0/\partial e_j$: *marginal effect on the Gini coefficient of total income due to a marginal percentage increase in income source* j.
[f] $(\partial G_0/\partial e_j)/G_0$: *relative effect of a marginal percentage increase in income source* j *upon the Gini coefficient of total income.*

The Gini coefficient in rural China, however, is well within the range recorded for rural areas in other developing countries, albeit on the high side. For instance, Adams (2001) shows the Gini coefficient in rural Egypt is 0.532 in 1997.

Decomposing the Gini coefficient by income source shows that irrigation may help to equalize income (Table 2.11). Cropping income from irrigated land is the most equally distributed with a Gini coefficient approximately 0.1−0.2 points lower than those of the other income sources (column 2). Cropping income is not concentrated in rich households since the Gini correlation between cropping income and total income, R_k, is 0.35, a figure much lower than that of off-farm income. More saliently, cropping income from irrigated land has the highest marginal effect on lowering inequality (column 6). A 1% increase in cropping income from irrigated land for all households would decrease the Gini coefficient for total income by 0.1%. Hence, just as Rozelle (1996) found that cropping income, in general, helped abate regional inequality, our results find that interhousehold inequality is attenuated by the presence of irrigation.

Our results from decomposing inequality by group show that while the income differences between the irrigated group and nonirrigated group have contributed to the overall income inequality, the between-group component is not dominant (Table 2.12, row 2). Only about 9% of the total inequality level appears to arise from the presence of barriers to irrigation, reflecting the positive but small unequalizing effect of irrigation. On the other hand, 77% of the total inequality comes from the inequality within each group (row 2, column 4). The large within-group effect indicates that there is substantial inequality among farmers that have irrigated plots and among nonirrigated farmers.

Results from decomposing inequality by income flows due to specific household characteristics further confirm irrigation's propensity to equalize income (Table 2.13). After controlling for other factors, a 1% increase of irrigated land per capita leads to a 0.05% decrease in the Gini coefficient for total income. The results also show, however, that irrigation is not the only factor that can decrease inequality. A 1% increase of education level of the labor force in the household will lead to a 0.23% decrease in inequality level of the total income. Hence, education, like irrigation, can reduce poverty and lower income inequality.

NEW IRRIGATION PROJECTS: BENEFITS VERSUS COSTS

Both the descriptive statistics and the multivariate analysis have shown that irrigation raises household crop revenue per hectare, raises household incomes, and decreases inequality. To complete our analysis of the impacts of irrigation on the welfare of rural households, a cost-benefit analysis of irrigation is conducted. In our analysis, we calculate the cost-benefit analysis by comparing the per hectare benefits of a switch from nonirrigated to irrigated cropping to the estimated per hectare costs that are associated with the new irrigation. The benefit from irrigation is

Table 2.12 Gini Decomposition by Subgroups (the Irrigated Group and the Nonirrigated Group)

	(1) Total Gini Ratio	(2) Between-Group[a]	(3) Overlap[b]	(4) Within-Group[c]
1	0.537[d]	0.048	0.077	0.412
2	(100)	(8.89)[e]	(14.31)[e]	(76.8)[e]
		Share of Population (%)	**Share of Income (%)**	**Gini Coefficients by Groups**
3	Irrigated group	84.81	89.34	0.534
4	Nonirrigated group	15.19	10.66	0.533
5	Total population	100	100	

[a] *Between-group refers to inequality arising from income differences between irrigated group and the nonirrigated group.*
[b] *Overlap refers to inequality arising from the fact the income ranges in different groups overlap. Some households could be relatively poor in the group with higher average income but rich in another group.*
[c] *Within-group refers to inequality arising from income differences among households in the same group.*
[d] *Gini coefficient of total population differs slightly from that in Table 2.7 because we did not include households without information on irrigation.*
[e] *Share of each source of inequality in total inequality level is in parentheses.*

measured as the increase in household annual crop revenue due to irrigation. In our regression, the coefficient on the irrigation dummy variable (either irrigated by surface water or groundwater) estimates by how much irrigating a plot increases crop revenue, holding other things constant (Table 2.5).

At the same time, of course, there could be increased costs. The costs associated with increasing irrigation mainly include two components. The first component is the increase in the input costs that a farmer incurs when a plot is irrigated. The most obvious added direct cost is that associated with the payment of a water fee (especially in the case of surface water) or the pumping cost (in the case of groundwater and some surface water systems). In addition, since irrigation allows the farmer to make more intense use of the land, farmers working on irrigated plot might expect to use higher levels of variable inputs, such as fertilizer, pesticide, and labor. Unfortunately, we did not collect all inputs by plots; we only collected inputs on a per household basis. To obtain estimates of the added costs associated with newly irrigated plots, we had to rely on two subsets of households from our overall sample: (1) households that only have irrigated plots, and (2) households that only have nonirrigated plots. The increase in the input costs is then measured as the difference in the total input costs between these two groups.

The second component of the additional cost of new irrigation is that associated with the cost of constructing and operating an irrigation system. For example, in the case of a new surface water irrigation system, such a cost would include the depreciation of the canal system, the expected maintenance of canal system and the

Table 2.13 Gini Decomposition by Income Flows due to Specific Household Characteristics

	Income Sources	(1)[a] S_k	(2)[b] G_k	(3)[c] R_k	(4)[d] $S_kG_kR_k$	(5)[e] $\partial G_0/\partial e_j$	(6)[f] $(\partial G_0/\partial e_j)/G_0$
1	Total income per capita (USD)	1	0.5407	1	0.5407	0	
2	Area of irrigated land per capita (ha)	0.0649	0.5199	0.2122	0.0072	−0.0279	−0.0517
3	Level of education of household's labor force (attainment in years)	0.2597	0.2347	0.2567	0.0157	−0.1248	−0.2308
4	Proportion of good quality land (%)	0.0199	0.2342	0.1155	0.0005	−0.0102	−0.0189
5	Cultivated land per capita (ha)	0.0402	0.4676	0.0470	0.0009	−0.0209	−0.0386

This table uses results from Table 2.2. Not all variables are reported for the sake of brevity.

[a] S_k: *share of income flow contributed by factor k in total household income. Column (1) does not sum to 1 because we did not list all explanatory variables in the regression or the residual.*

[b] G_k: *Gini coefficient of income flow contributed by factor k.*

[c] R_k: *Gini correlation between income flow contributed by factor k and the distribution of total income.*

[d] $S_kG_kR_k$: *contribution of income flow contributed by factor k to Gini coefficient of total income. The sum of the five ($S_kG_kR_k$) do not sum to 0.5572 because we did not list all explanatory variables in the regression or the residual.*

[e] $\partial G_0/\partial e_j$: *marginal effect on Gini coefficient of total income due to a marginal percentage increase in income flow contributed by factor j.*

[f] $(\partial G_0/\partial e_j)/G_0$: *relative effect of a marginal percentage change in income flow contributed by factor j upon Gini coefficient of total income.*

opportunity cost of investment. Since this information was not collected during the 2000 CNRS, we had to rely on another set of data—the 2003 Public Investment in Rural Poverty and Development Survey. Fortunately, in this data set, there is detailed information on the amount of total investment on irrigation infrastructure and the area of land covered by the project.

The data used to estimate the ownership and operating costs per hectare of new irrigation system were collected by the Center for Chinese Agricultural Policy, Chinese Academy of Sciences. The data come from a randomly selected, nationally representative sample of 2376 villages in six provinces of rural China (Hebei, Jilin, Jiangsu, Sichuan, Shanxi, and Gansu). Six counties were selected from each province, two from each tercile of a list of counties arranged in descending order of per capita gross value of industrial output. Within each county, the survey team also chose six townships, following the same procedure as the county selection. On average, enumerators surveyed around 11 villages in each township.

The results of our cost-benefit analysis show that most of the investments in irrigation have positive returns (Table 2.14). In a surface water irrigation system, irrigation will increase household annual crop revenue by \$205 per hectare (column 1). The input costs of households that use surface water irrigation increase by \$58 per hectare on average. Depending on the type of the surface water irrigation system that is being installed, the construction and operating costs vary. In villages that only have unlined canals, the construction and operating costs may be as low as \$20 per hectare. In other villages that have lined canals and invest in power lifting irrigation stations, the costs may be as high as \$91 per hectare. If we take these costs as typical of those villages that have access to new irrigation projects, then, on average, in about 62% of the villages that used surface water irrigation, the benefits of adding irrigation are higher than costs. In about 52% of villages that use groundwater irrigation, the benefits are higher than the costs.

While we believe our results are fairly robust, caution should be taken in interpreting the results of our cost-benefit analysis. Specifically, due to lack of data, it is possible that some elements of the costs and benefits of irrigation have not been accounted for. For example, the environmental costs associated with irrigation are not included. Indirect benefits from increased income due to higher crop revenue (and the benefits of reduced poverty) are not included either.

CONCLUSIONS AND POLICY IMPLICATIONS

Analysis in this chapter provides evidence of irrigation's strong impact on agricultural production, rural income, and regional income inequality, both descriptively and in the multivariate analysis. We find that although the marginal impact of irrigation on revenue appears to be higher in richer areas, since incomes of those in poor areas rely more on cropping, farmers in poor areas increase their incomes relatively more than farmers in richer areas. In addition, even after accounting for the

Table 2.14 Average Cost of Increasing Irrigated Land by 1 ha (USD/ha/year)

	Benefit	Cost Associated with Irrigation			Comparison
Type of Irrigation System	**(1) Increase in Annual Crop Revenue**	**(2) Total**	**(3) Input Costs[a]**	**(4) Ownership and Operating Costs[b]**	**(5) Percentage of Sample Villages That Have Positive Returns From Investments in Irrigation**
Surface water	205[c]	79–149	58	21–91[e]	62.3%
Groundwater	339[d]	104	87	17[f]	51.9%

[a] *Input costs associated with irrigation include costs on purchasing water, energy, and fertilizer, etc. Labor expenditures are also included.*

[b] *Ownership cost includes depreciation of the irrigation system and opportunity costs of capital. It is determined by spreading the purchase and installation cost over its expected use period. Operating costs include operation, repairs, and maintenance of the equipment.*

[c] *The increase in the annual crop revenue is calculated using the crop revenue from nonirrigated plots (column 5, Table 2.3) and the percentage increase in annual household crop revenue due to surface water irrigation (column 2, row 2 of Table 2.6).*

[d] *The increase in the annual crop revenue is calculated using the crop revenue from nonirrigated plots (column 5, Table 2.3) and the percentage increase in annual household crop revenue due to groundwater irrigation (column 2, row 3 of Table 2.6).*

[e] *The component of a surface water irrigation system includes canals in most villages. In villages that use lifting irrigation, pumps and investment in power lift station (in some villages) are also included. In other villages, ponds, small weirs, or dams may also be part of a surface water irrigation system. Henceforth, the component of a surface water irrigation system varies across villages. The ownership and operating costs differ across villages correspondingly.*

[f] *In most villages, the component of a groundwater irrigation system includes wells, pumps, underground pipes, and other equipment such as transformers.*

Source of data: Data for benefit and input costs are from 2000 China National Rural Survey; data for operating costs and annual ownership cost are from 2003 Public Investment in Rural Poverty and Development Survey.

increased capital costs and production costs, returns from investments in irrigation are positive in the majority of villages that have invested in new irrigation systems.

If irrigation has such a great effect on agricultural performance and rural incomes, it stands to reason that the economic impact of a water shortage, which reduces the availability of irrigation, goes beyond simple crop yields to include the overall income distribution within a community. Given this, it is no surprise that large portions of many different country's budgets have been directed toward irrigation. Moreover, although the costs of the project must be considered, the malaise that pervades the international community in irrigation may need to be questioned (Byerlee et al., 1999). One implication from our work is that the positive impact on food production and potential poverty reducing effects should be factored in the benefit side of irrigation in evaluations of irrigation projects. Poverty alleviation programs, in particular, may want to consider increasing, or at least should not underestimate, the role of irrigation in rural development when designing their portfolio of activities.

A national water shortage will require some form of water rationing. Our cost-benefit analysis indicates that some communities benefit more from access to irrigation than do others. This finding offers some guidance on how water could potentially be allocated. Water should be shifted away from villages where the cost of irrigation outweighs the benefits, and redirected to those villages that are more reliant on irrigation. Our finding, however, does not suggest that investments should be made to increase irrigated area in all villages in all of China because costs of increasing irrigation may outweigh benefits in some areas. In addition, returns to investment in irrigation may be lower when other factors are taken into account (Jaglan and Qureshi, 1996). One such factor is the negative impacts of irrigation on the environment. In other cases, the expansion of irrigated area increases demand for water and may lead to depletion of the groundwater resource that is increasingly scarce. Under such circumstances, investments in irrigation should be directed toward improving irrigation efficiency by providing water saving technologies or improving the performance of existing irrigation infrastructure.

The importance of irrigation in rural communities serves to underscore the potential impact of a water shortage on China's rural economy. Analysis in this chapter shows that irrigation increases the welfare of poor farmers more than it does rich farmers. A water shortage affecting the availability of irrigation would therefore lead to more income inequality and a higher incidence of poverty, which disproportionately affect rural farming areas. Given the Chinese government's renewed focus on rural development and narrowing national income gap, devising effective and enforceable policy solutions to the looming water problem should be of high priority on the government's policy agenda. Other investments, such as investment in education, should be considered since these investments may have higher returns than investment in irrigation (Fan et al., 2004).

REFERENCES

Adams, R.H., 2001. Non-farm Income, Inequality and Poverty in Rural Egypt and Jordan. PRMPO, MSN MC4-415. World Bank.

Benjamin, D., Brandt, L., Glewwe, P., Guo, L., 2002. Markets, human capital, and inequality: evidence from rural China. In: Freeman, Richard (Ed.), Inequality Around the World. Palgrave, London, pp. 87–127.

Byerlee, D., Heisey, P., Pingali, P., 1999. Realizing yield gains for food staples in developing countries in the early 21st century: prospects and challenges. In: Paper Presented at Food Needs of the Developing World in the Early 21st Century, by the Vatican, January 27–30, 1999.

Datt, G., 1998. Computational Tools for Poverty Measurement and Analysis. Discussion paper. International Food Policy Research Institute.

Fan, S., Hazell, P., Throat, S., 1999. Linkages between Government Spending, Growth, and Poverty in Rural India. International Food Policy Research Institute.

Fan, S., Zhang, L., Zhang, X., 2004. Reforms, investment, and poverty in rural China. Economic Development and Cultural Change 52 (2), 395–421.

Food and Agriculture Organization of the United Nations, 2002. FAO Statistical Database. http://faostat.fao.org/faostat/collections?subset=agriculture.

Hossain, M., Gascon, F., Marciano, E.B., 2000. Income distribution and poverty in rural Philippines: insights from repeat village study. Economic Political Weekly 4650–4656.

Hu, R., Huang, J., Jin, S., Rozelle, S., 2000. Assessing the contribution of research system and CG genetic materials to the total factor productivity of rice in China. Journal of Rural Development 23 (Summer), 33–79.

Huang, J., Hu, R., 2001. Options for Agricultural Research in the People's Republic of China. Project Report. Asian Development Bank.

Huang, J., Rozelle, S., Rosegrant, M.W., 1999. China's food economy to the twenty-first century: supply, demand, and trade. Economic Development and Cultural Change 47 (4), 737–766.

Jaglan, M., Qureshi, M., 1996. Irrigation development and its environmental consequences in arid regions of India. Environmental Management 20 (3), 323–336.

Jin, S., Huang, J., Hu, R., Rozelle, S., 2002. The creation and spread of technology and total factor productivity in China's agriculture. American Journal of Agricultural Economics 84, 916–930.

Kennedy, P., 1998. A Guide to Econometrics. The MIT Press, Cambridge, MA.

Kung, J.K.S., Lee, Y.F., 2001. So what if there is income inequality? The distributive consequence of non-farm employment in rural China. Economic Development and Cultural Change 50, 19–46.

Lau, L.J., Yotopoulos, P.A., 1971. A test for relative efficiency and application to Indian agriculture. American Economic Review 61 (1), 94–109.

Lerman, R.I., Yitzhaki, S., 1985. Income inequality effects by income sources: a new approach and applications to the U.S. The Review of Economics and Statistics 67, 151–156.

Liao, Y., 2003. Irrigation Water Balance and Food Security in China (Ph.D. dissertation). The Chinese Academy of Agricultural Science.

Ministry of Water Resource, 2001. China Water Resource Yearbook. China Water-hydropower Publishing House, Beijing.

Morduch, J., Sicular, T., 2002. Rethinking inequality decomposition, with evidence from rural China. The Economic Journal 112, 93–106.

National Statistical Bureau of China, 2001a. China Rural Poverty Monitoring Report: 2001. China Statistics Press.

National Statistical Bureau of China, 2001b. Statistics Yearbook of China. China Statistics Press, Beijing, China.

Nyberg, A., Rozelle, S., 1999. Accelerating China's Rural Transformation. World Bank, Washington, DC.

Paxson, C.H., 1992. Using weather variability to estimate the response of savings to transitory income in Thailand. American Economic Review 82, 15–33.

Perkins, D., 1994. Completing China's move to the market. Journal of Economic Perspectives 8, 23–46.

Pyatt, G., 1976. On the interpretation and disaggregation of Gini coefficients. The Economic Journal 86, 243–255.

Pyatt, G., Chen, C., Fei, J., 1980. The distribution of income by factor components. Quarterly Journal of Economics 95, 451–473.

Rosegrant, M., Evenson, R., 1992. Agricultural productivity and sources of growth in South Asia. American Journal of Agricultural Economics 74, 757–761.

Rozelle, S., 1996. Stagnation without equity: patterns of growth and inequality in China's rural economy. The China Journal 35, 63–92.

Stuart, A., 1954. The correlation between variate-values and ranks in samples from a continuous distribution. British Journal of Statistical Psychology 12, 37–44.

Taylor, J.E., 1997. Remittances and inequality reconsidered: direct, indirect, and intertemporal effects. Journal of Policy Modeling 14, 187–208.

Travers, L., Ma, J., 1994. Agricultural productivity and rural poverty in China. China Economic Review 5 (1), 141–159.

Udry, C., 1996. Gender, agricultural production, and the theory of the household. Journal of Political Economy 104 (5), 1010–1046.

Wang, J., Xu, Z., Huang, J., Rozelle, S., 2002. Pro-poor Intervention Strategies in Irrigated Agriculture in China: Mid Term Progress Report. Research report. International Water Management Institute.

World Bank, 2001. China: Overcoming Rural Poverty. World Bank.

Yang, D.T., 2002. What has caused regional inequality in China. China Economic Review 13, 331–334.

Yotopoulos, P.A., Lau, L.J., 1973. A test for relative economic efficiency: some further results. American Economic Review 63 (1), 214–223.

Zhu, J., 2004. Public investment and China's long-term food security under WTO. Food Policy 29, 99–111.

CHAPTER

3

China's Agricultural Water Policy Reforms: Increasing Investment, Resolving Conflicts, and Revising Incentives

We have already provided evidence indicating that China's water supply is dwindling. Here, we examine the demand side of China's water crisis. Who are the principal consumers of water in China? How are water rights allocated and enforced? How will the rapid increases in demand and competition for China's limited water resources be met? What have been the policy responses to the decrease in the water supply, and how effective have they been?

Rapidly growing industry, increasingly productive farmers, and a large population with rising incomes all compete for China's water resources. The sustained high industrial growth rate over the last 20 years has caused a significantly higher proportion of China's water to be allocated to industrial production, while the proportion allocated to residential users is also increasing, particularly as the number of urban residents and incomes grow. On top of this growing demand from the nonagricultural sector, China continues to expand its irrigated area. These trends have led to an overall increase in the demand for water, though agriculture is still by far China's largest user of water.

Observers of China's current water situation agree that the crisis has not yet manifested itself in a substantial loss of irrigated area or industrial production. Even the most pessimistic observers characterize the crisis as a rapid decline in water availability that, if left unchecked, will lead to a decrease in food production in the coming 20 years. Economically, to argue that a true water crisis exists in China, one must show that water deliveries have been disrupted or prices have risen to an extent that actually threatens economic activity. Disruptions of water deliveries have occurred in some areas but so far are not severe enough to affect aggregate production, either in industry or in agriculture. But where water delivery disruptions have occurred, they have affected farmers, industry, and residential users. Overall, irrigated area has expanded in recent years, and China plans to continue expanding this area. Industrial production has also grown rapidly in the past several years, even in the regions where water is relatively scarce. In addition, water prices, while higher than in other parts of Asia (Valencia et al., 2001), are still well below the marginal value of water use in each sector and the percentage of wastewater treated after use,

Managing Water on China's Farms. http://dx.doi.org/10.1016/B978-0-12-805164-1.00003-8

while increasing, is not large. Thus, there appears to be ample room to further increase water productivity and avert a more drastic crisis in the future. The real debate over the future severity of China's water problem comes down to a question of how well policymakers can respond to the various water-related issues confronting them.

The overall goal of this chapter is to provide a timely analysis of how China has managed water in the past, the challenges that the nation is currently facing, and the measures that have been implemented or are at its disposal to combat water shortages in the face of rapid economic growth and rising demand for food. To meet this goal, we have three specific objectives. First, we briefly review the state of China's water resources and water management structure. Next, we examine some of the main responses to the water shortage—both in terms of government-issued policy reforms and also conservation efforts on the part of individual farmers—and analyze their effectiveness. Finally, we look at some of the conflicts that have sprung up over water allocation issues, and consider ways in which these disputes might be lessened.

China is big, and water policy is complex, so it is impractical to attempt to cover all water-related topics in one volume. We focus here on the water shortage in the north, which includes the three most water-stressed river basins: the Hai, Huai, and Huang (Yellow) River. In addition, we concentrate on problems that affect water availability for irrigation in agriculture. In doing so, we touch briefly on the role of domestic and industrial users, since their behavior affects the water supply available for agriculture. Plans are currently underway to transfer water from the relatively water-abundant Yangtze River Basin in the south to the water-poor Hai River Basin in the north, but these projects may well take over a decade to complete, and the water deliveries will be too expensive for use in agriculture. We therefore do not consider these transfers in this report.

CHINA'S WATER MANAGEMENT POLICIES AND INSTITUTIONS

Over the past 50 years, China has constructed a vast and complex bureaucracy to manage its water resources. This system is so sprawling, and involves so many different stakeholders, that the rules followed at the local level bear only a passing resemblance to the official policies enacted at the national level. Moreover, the system itself was not designed to make efficient use of a valuable resource; indeed, this has only become a concern in recent years. Instead, the system was originally designed to construct and manage systems to prevent the floods that have historically devastated the areas surrounding the major rivers, and to effectively divert and exploit water resources for agricultural and industrial development. China's success in accomplishing this latter goal is largely why the nation faces water shortage problems today.

Water policy is created and executed primarily by the Ministry of Water Resources (MWR). The MWR has run most aspects of water management since China's first comprehensive Water Law was enacted in 1988, taking over the duties from its predecessor, the Ministry of Water Resources and Electrical Power. The policy role of the MWR is to create and implement national water price and allocation policy, and to oversee water conservancy investments by providing technical guidance and issuing laws and regulations to the subnational agencies. The national government invests in developing the water resources from all large rivers and lakes and projects that cover more than one province. Local governments are in charge of projects that are within their administrative districts. Historically, investment from national funding sources has been heavily biased toward new investments, while cash-strapped local governments have been responsible for maintenance funds.

Although much of China's water is still used by farmers in agriculture, the nation's water policy is becoming increasingly biased toward other users. Acting at the direction of the Water Law, the MWR gives priority to domestic, primarily urban, users over agricultural and industrial users in the allocation of all water. Provincial governments also have the power to allocate water based on local priorities. Since officials' incomes are directly tied to local economic performance, and industry contributes more to the local economy, this provision has led many provinces to favor industry at the expense of agriculture. These policy biases apply mostly to new sources of water; water is actually taken away from agriculture only in isolated cases. Although in some upstream regions that draw more than their allocated amount of water, enforcing the legal allocations results in agricultural users "losing" water.

Under the 1988 Water Law, the MWR is not solely responsible for all water-related policies; other ministries in China also influence water policy for both rural and urban areas. The diverse uses of water and divergent objectives and interests of water management agencies often result in conflicts and inefficient water use. In the use of agricultural water, the MWR shares its duties with the Ministry of Agriculture, particularly in developing local delivery plans and extending water-saving technology. In urban areas, Urban Construction Commissions (UCCs) (or Bureaus) are charged with managing the delivery of water to urban industrial and domestic users. UCCs also have taken responsibility for managing groundwater resources that lie beneath municipalities' land area. Groundwater levels, both urban and rural, are monitored jointly by the Ministry of Geology and Mining (MGM) and its local associates. In theory, the MGM's information about the groundwater level is used when deciding whether to grant groundwater pumping permits, though local water bureaus do not always use the information. China's State Environmental Protection Agency has the responsibility for managing industrial wastewater and municipal sewage treatment. Last, in the area of price setting, the MWR, in conjunction with the State Price Bureau and acting with the approval of the State Council, sets guidelines at the provincial level. Subnational Water Resources Bureaus (WRBs) and Price Bureaus (at the direction of the leaders in the localities) set the final price levels according to local supply and demand as well as other economic and political factors.

Outside of the central government, many subnational water management institutions also influence water policy. Provincial, prefectural, and county governments all have WRBs linked vertically to the MWR in Beijing. Formally, the subnational offices are charged with implementing the rules and policies advanced by the national authorities. In reality, however, the heads of local WRBs are appointed by, and report to, leaders of their own jurisdictions (such as provincial governors or county magistrates). These horizontal ties frequently dominate the vertical ones. Consequently, WRBs create and execute water policy and regulations based on the needs of their own jurisdiction, causing a considerable degree of heterogeneity in water policies across regions.

Since rivers, lakes, and aquifers do not always follow administrative boundaries, there are institutions that manage water across administrative boundaries. Each of China's seven major river basins has a National River Basin Commission (NRBC) to manage the basin's water resources. The NRBCs are directly under the MWR, and when they were set up, they were given the authority (at the direction of the MWR leadership) to approve or reject the provincial WRB's plans to withdraw water from the main stream of the river basin under their charge. Importantly, the NRBCs do not regulate water withdrawals from the tributaries of the main river under their charge—these are regulated by the local WRBs. Some observers believe that the commissions were not very effective in the immediate years after they were set up (Nickum, 1998). Provinces were able to implement their own plans, often to the detriment of other provinces and against the plans of the national commissions, which lacked adequate enforcement power.

Below the national level, irrigation districts (IDs) were developed to administer water resources that span lower level administrative boundaries. An ID reports to the WRB that encompasses its entire command area. For example, if an ID includes two or more prefectures, it is under the provincial WRB, but if it lies in two or more counties, all within the same prefecture, it is under the control of the prefecture's WRB.

RESPONSIBILITIES OF CHINA'S LOCAL WATER MANAGEMENT INSTITUTIONS

The ultimate duty of WRBs has always been to create and manage water allocation plans, to conserve limited water supplies in deficit areas, and to administer water infrastructure investment. The primary task of local water policy managers is to transform investment dollars into infrastructure, maintain the system once it is in place, and manage the water flows within and among IDs.

In the early 21st century, WRBs in most regions of northern China have been spending more of their time assisting with the development of and attempting to control groundwater resources, though control of these resources has been difficult. One approach has been to control the number and location of wells. Throughout the late 1980s, the monopolization of well-drilling activity gave local authorities fairly comprehensive control over access to groundwater since most deep wells (and many shallow wells) were sunk by well-drilling enterprises owned and operated by the WRB. In recent years, however, the rise of private well-drilling companies and competition among local collectively owned (by either a township or a village) well-drilling companies have reduced this avenue of control. In this new environment, local WRBs are still charged with controlling groundwater extraction by using their authority to issue all well-drilling permits for water extraction and management (Wang, 2000). There are, however, many exceptions to this process. For example, UCCs, are notoriously independent, and in many cases urban units operate on their own without the oversight of the WRBs.

The WRBs are also charged with overseeing a system of permit rights to draw groundwater in addition to well-drilling rights. This system is intended to allow them to operate a de facto groundwater allocation plan, but it has not always worked in practice. Because of the problems in monitoring groundwater extraction, there is little control over the quantity of groundwater extracted once the wells are in operation. Often, groundwater extraction fees from large government-owned wells are not charged by volume, but rather are based on a fixed negotiated amount per year. In general, except in cases where groundwater tables have fallen so much that they are causing an acute crisis, urban and rural localities are in charge of their own groundwater resources and little action is taken to restrict groundwater pumping.

FINANCING WATER MANAGEMENT AT THE SUBPROVINCIAL LEVEL

The financing of local WRB activities and the fiscal crisis facing many local water agencies have played a role in shaping the way that WRBs have developed and how they have set their priorities. Operations and investments of local water bureaus are financed by fees for water deliveries, water extraction, and well-drilling permits and by transfers from the administrative hierarchy above the local bureau. Limits on the pricing of water, however, frequently keep system officials from charging enough to cover their operation and maintenance costs (Nyberg and Rozelle, 1999). In addition, targeted budgetary allocations from upper-level governments often never arrive in full or are diverted for other matters (Park et al., 1996). The fiscal stress has led to distortions in the way investment funds are allocated among new and existing structures.

Shortages of operating funds have also led to innovative, although sometimes distracting, ways of meeting fiscal deficits. To make up the deficit between revenues and expenditures, local water agencies fulfill their financial obligations through a variety of means. Irrigation officials may tap funds intended for investment in infrastructure or hold back payroll expenditures to meet immediate operating expenses. Local bureaus also sometimes encourage employees to set up businesses around the use of water, such as fish farms or tourism assets in reservoirs, with the profits from the enterprise used to supplement the revenue side of the agency's balance sheet and provide wage payments, making it easier to meet payroll expenditures. A system that relies on individuals to use earnings from a quasiprivate business to subsidize a difficult-to-monitor policy task, such as the efficient delivery of water to farmers, is less likely to meet policy goals than a fully funded system.

Because of the recent signs of an impending water crisis, water management policies and institutions have changed at all levels. Nationally, China's leaders have increased investment in water delivery infrastructure, although perhaps more system-wide organizational reforms might be better suited to solve the problems of enforcement at the local levels. Vague policy statements have been issued underlining the need to curb inefficient water use and poor water management, but few concrete steps have been taken to improve policy implementation. In the next section we consider some of the policy responses to the rural water shortage.

INFRASTRUCTURE INVESTMENT AND POLICY REFORM

Faced with increasing water scarcities, China's government has responded in a number of ways. Local (township and county), regional (prefectural and provincial), and central governments have created and issued a number of laws, regulations, and policies, and have tried to encourage local officials to reform water management. In this section we focus on two major responses that government has made: the creation of new groundwater policies and the development and promotion of irrigation management reform for surface water resources. We also examine whether the government has succeeded in implementing these initiatives.

GROUNDWATER POLICIES AND REFORMS

An important determinant of China's overall water management capacity is the state of the water recovery, storage, and delivery infrastructure. While decollectivization in the late 1970s and early 1980s led to jumps in agricultural productivity and production, these same reforms led to ambiguous property rights over many local water delivery systems built during the collective period (1959–79) and a decline in local governments' ability to invest in large infrastructure projects. The ambiguity over ownership of these systems produced weak incentives to invest in and maintain them. Moreover, transfers of investment funds from the national to local governments fell, decreasing local governments' ability to invest in maintaining water storage and delivery infrastructure. With little incentive to invest in surface water delivery systems, and facing severe fiscal constraints, China's surface water systems experienced a decline in effectiveness, and farmers in northern China began to rely more heavily on groundwater.

Increasing numbers of wells drove the growth of agriculture in northern China during the 1980s and early 1990s (Stone, 1993). Farmers in China generally prefer groundwater, even in areas where surface water is inexpensive and villages are integrated into its canal network. They use surface water when it is available, but often complain that surface water is unreliable, and therefore maintain access to groundwater as well.

Despite the demand for groundwater deliveries, the fiscal reforms described earlier left many local governments strapped for funds. Many villages, particularly those without lucrative nonagricultural enterprises, eventually faced serious fiscal shortfalls and were unable to continue sinking wells, especially in areas where the groundwater table had fallen significantly and more expensive pumps were needed. In some areas, the water table fell below the reach of the village wells, and access to irrigation water ceased entirely.

Role of the Private Sector

As local governments' ability to invest declined, other investors stepped in to take their place. Individual entrepreneurs began investing in wells and delivery systems in the early 1990s and selling the water to farmers (Wang et al., 2000a,b). The lack of attention to this phenomenon by the national statistical service makes it impossible to observe what happened at the national level, but a survey of three Hebei counties shows the speed at which private well use expanded. From 1983 to 1998, privately owned and operated wells, and the corresponding water delivery systems, rose from 15% of the total wells to 69%. Across some parts of China, and particularly in the Hai River Basin, private entrepreneurs have raised the capital needed to sink deeper wells and to install underground, low-pressure piping networks to deliver water to farmers' fields. After making the investments, the entrepreneurs sell the water to local farmers.

Private wells are often more reliable than public ones, and more efficient. Econometric inquiries into the determinants of water supply suggest that the privately run systems deliver water in a more timely and less costly manner and that this water is used more efficiently on the farm (Wang and Huang, 2002). These results may arise because private enterprises have better incentives to lower costs, because volumetric pricing (more common with private IDs) gives farmers more incentive to use water efficiently, or because the more timely deliveries that come with the small, private groundwater districts allow farmers to use less water. The more reliable and timely deliveries provided by private groundwater IDs have also been linked to the cultivation of higher valued crops such as fruits and vegetables (Xiang and Huang, 2000).

The increased number of private wells, however, does not necessarily mean that China will be able to avert a more drastic water crisis in the future. This increased efficacy comes at a cost, however. Recent findings show that as the share of noncollective property rights in water delivery systems in three Hebei counties increased over time, the level of the ground water table fell (Wang, 2000). This may be because the private wells are being established in areas where the water table is already falling; alternatively, the falling water tables may be evidence of the tragedy of the commons: private entrepreneurs competing over a free, but limited, resource. Thus, although the establishment of private wells has allowed many regions to maintain irrigated agriculture in the face of falling water tables, the proliferation of private wells may also hasten the point at which pumping from the water table is either no longer profitable or no longer possible.

Role of Government

Although limited in scope, government officials have put some effort into issuing laws, regulations, and policies to manage groundwater resources (Wang et al., 2007a). For example, according to China's National Water Law, revised most recently in 2002, all property rights to groundwater resources belong to the state. This means that the right to use, sell, and/or charge for water ultimately rests with the government. The law also does not allow groundwater extraction if pumping is harmful to the long run sustainability of groundwater use.

Beyond the formal laws, officials also have launched a number of policy measures in order to rationally manage the use of the nation's groundwater resources. There have been policies promulgated to control the right to drill tubewells, to manage the spacing of tubewells, and to regulate the collection of water resource fees. Tellingly, while there initially appears to be many policy measures managing China's groundwater, when compared with the number of regulations concerning other issues, such as flood control and the construction of water-related infrastructure projects, in fact, there are few regulations relevant to groundwater management. Importantly at the national level, there is not a single water regulation specifically focused on groundwater management issues.

While these laws and policies may appear to be a step in the right direction, when the effort to establish a regulatory framework is weak, there has been an even greater deficiency in the implementation of official laws and policy measures on groundwater management (Wang et al., 2007a). According to our North China Water Resource Survey (NCWRS) survey data, less than 10% of well owners obtained a well-drilling permit before drilling, despite the nearly universal regulation requiring a permit. Only 5% of community leaders believed that well-drilling decisions required consideration of well spacing. Even more tellingly, water extraction charges were not imposed in any community and there were no quantity limits put on well owners. In fact, in most communities in China, groundwater resources are almost completely unregulated in practice.

Why is this so? Certainly, part of the problem is historic neglect. At the ministerial level, the size (in terms of employees) of the division of groundwater management is still relatively small. There are far fewer officials working in this division than in other divisions, such as flood control, surface water system management, and water transfer. Moreover, unlike in the case of the surface water management system (Lohmar et al., 2003), the groundwater management system has no umbrella organization taking responsibility for aquifers that cross local or provincial boundaries, nor has there been any effort to bring together or improve communication among all the local governments and private entities that use water extracted from different parts of a single aquifer. Without a single body controlling the entire resource, it is difficult to implement policies that attempt to manage the resource in a manner that is sustainable in the long run.

SURFACE WATER POLICIES AND REFORM

Looming water scarcities have pushed China's leaders to consider community-level irrigation management reform a key part of their strategy to combat China's water problems. Investment in surface water management systems has boomed since the 1980s, and more importantly, China is beginning to shift its investment priorities from new projects to renovation and maintenance of existing systems (Nyberg and Rozelle, 1999). Although it is too early to tell the depth of commitment to this new direction of investment spending, there are signs that the investments are effectively targeted at repairing IDs in decay. For example, one ID in Baoding Prefecture that had seen its command area fall by 80% over the last 20 years was recently granted funding to completely renovate its deteriorating canal system. Many farmers found the system so unreliable that they switched to groundwater, contributing to the rapid lowering of the water table. Irrigation officials said that the new grant, the first funding they had ever received from Beijing for system repairs, would allow the system to deliver water to fully restore its former command area and to reduce conveyance losses to negligible levels.

More recently, an effort has begun to establish unambiguous property rights in many smaller systems, an important first step to improving many of the smaller surface water storage and delivery systems. According to a survey in Ningxia Province, since the early 1990s, and especially after 1995, reform has successively established Water User Associations (WUAs) or contracting systems in place of collective management (Wang et al., 2007b). Collective management implies that the community leadership directly takes responsibility for water allocation, canal operation, and maintenance and fee collection. WUAs are groups of farmers along a lateral canal that select a leadership and a set of rules to manage water deliveries that they purchase directly from the ID on a per-unit basis. Under a contracting system, an individual is selected to take over the management of lateral canals and is provided with incentives to deliver water more efficiently. Both types of reforms take the management of irrigation deliveries away from the village collective and are intended to bypass the traditional village-township-county route of fee payment to the ID, thus saving money by reducing fees that accrue to these levels of the administrative bureaucracy. In general, they are geared toward improving the management of irrigation deliveries to farmers and improving fee collection to bolster the irrigation district's fiscal situation.

In Ningxia province the share of villages that manage their water under systems of collective management declined from 91% in 1990 to 23% in 2004. Contracting has developed more rapidly than WUAs. By 2004, 57% of the communities managed their water under contracting and 19% through WUAs. Our NCWRS (see data appendix) also found a similar reform trend of irrigation management (Huang et al., 2008), with collective management declining from 90% to 73% from 1995 to 2004. At the same time, WUAs increased from 3% to 10% and contracting management systems increased from 5% to 13%. If these figures accurately reflect northern China's averages, this means that nearly 100,000 communities in northern China are managed, at least nominally, under new types of surface water management institutions.

Although institutional reform is certainly an important part of behavioral change, in order for these reforms to be meaningful, the incentives motivating participants must change as well (Wang et al., 2005). Unfortunately, our China Water Institutions and Management survey data suggests that this has not been the case with the new WUAs and contracting management systems, few of which have been created in such a way as to give their members an incentive to save water. When managers have partial or full claim on the earnings of the water management activities (eg, on the value of the water saved by reforming water management), we say that they face strong incentives (or that the manager is managing with incentives). If the income from their water management duties is not linked to water savings, they are said to manage without incentives. For example, in 2001, on average, leaders in only 41% of communities offered WUA and contracting managers monetary incentives that could induce them to consciously save water. In the remaining 59% of sample communities, WUA and contracting managers are operating similarly to leaders in collectively managed communities—without any incentive to save water.

In effect, then, this structural reform has been ineffective. Both descriptive statistics and econometric analysis show that there is no significant relationship between water use in a community and reform of its irrigation management system. However, we found that communities that provided water managers with strong incentives to conserve water saw water use fall sharply, while agricultural production and rural incomes remained steady. Results described in this paragraph are reported in details in Wang et al. (2005). The efficiency of the irrigation systems improved as well. These studies show that, when faced with the right incentives, farmers are willing and able to adjust their behavior in order to curb their water usage. Better enforcement of already existing water regulations may be all that is needed to solve China's water shortage problem.

FARMERS' INCENTIVES TO REDUCE WATER CONSUMPTION

Despite the government's attempts to improve the water management environment in China, the fact remains that in many parts of northern China ground- and surface water sources are being depleted and current water-use levels are not sustainable with the current water supply system. As noted earlier, agricultural users will not be given priority for any additional sources of water that become available. Thus, using water more efficiently is the only way to maintain current levels of agricultural production without increasing total agricultural water demand.

Even with what seems to be an impending water crisis, farmers have hardly begun to adopt water-saving technologies or practices, largely because they have no incentive to do so. Until the 1970s, water was considered abundant in most parts of China and was even free for agricultural users. Farmers generally had to volunteer labor, however, to construct and maintain water storage and delivery infrastructure during this period. Collectives had de facto rights over the ground and surface water in their communities. Facing low or free water prices, farmers naturally used as much water as they wanted. Even today, the price of water does not reflect its scarcity; most farmers save water only when their deliveries are curtailed, not because the price is too high.

WATER PRICING IN CHINA

Since 1978, a system of volumetric surface water pricing has been in effect, with prices set by the Price Bureau in Beijing and modified by provincial price bureaus according to local conditions. The current price structure exhibits substantial variation across the country, and takes into account both scarcity and the ability to pay.

Typically, for a specific end-use (agriculture, industry, domestic) in a specific province, prices are uniform, although local exceptions can be made. In terms of ability to pay, agricultural users pay lower prices than domestic users, who in turn pay less than industrial users. For example, in Hubei Province, the price for agricultural users is $0.0051 per cubic meter, while domestic and industrial users pay $0.01 and $0.015 per cubic meter, respectively. In terms of scarcity, different prices prevail in different provinces, with prices increasing substantially as water scarcity becomes more severe (generally, as one moves from south to north). For example, in the late 1990s, agricultural surface water was priced at about $0.0013 per cubic meter in the southern province of Guangdong, $0.0051 per cubic meter in the central provinces of Hubei and Henan, and $0.0096 to $0.013 per cubic meter in the northern province of Hebei, where water shortages are most acute (Ministry of Water Resources, 1998).

Despite increasing water prices, current pricing policies do not effectively encourage water saving, and in fact they contribute to China's water problems in other ways. Since China's farmers each farm several small plots, charging each farmer according to how much water they use (volumetric pricing) is very costly and difficult to monitor. Some observers argue that water prices are so low that demand is relatively inelastic, thus raising water prices would only raise revenues and not decrease the demand for water by a significant amount.

DEBATE OVER WATER PRICE

China is in the midst of embarking on water price reform to better reflect water's status as a scarce resource and valuable public good. There is widespread agreement that water prices are too low in China, and well below the marginal benefit. The focus of the reform will be prices paid by domestic and industrial users; whether water prices will be raised for agricultural users is hotly debated. Some policy-makers believe that raising water prices for agricultural users is the only effective way to get farmers to implement sound water-saving measures. Others claim, however, that raising agricultural water prices will only further burden poor farmers facing low grain prices and, in many cases, high local tax rates. It would also directly counter another important policy goal in China: raising rural incomes and reversing a widening rural–urban income gap.

Under the current system, water is measured for volumetric pricing at some point above the household level, or even above the village level. Usually this is either at the main canal level or at the level of an irrigation group. Irrigation groups can range in size from as small as 30 households to as large as an entire township. Fees charged to individual households are usually a prorated amount of the total fee paid at the point of delivery (plus additional costs to cover the collection effort of the water officers and other water managers). The prorated amount is generally based on the size of the household's irrigated land endowment. There is some true volumetric pricing for individual farmers, but this is relatively rare in surface systems and is restricted to farmers near the head of main canals who have intake pipes directly from the main canal into their fields (groundwater deliveries, however, are often priced volumetrically).

Under this surface water pricing system, farmers have little incentive to reduce their water use since they are charged a fixed amount regardless of how much they actually consume. Indeed, the classic free-rider problem arises: There is an incentive to use more than one's share of the water, especially in large irrigation groups that are more difficult to monitor. This pricing system can also contribute to the interregional problems described earlier, as upstream users have more opportunities to free ride by using more water than they pay for, to the detriment of downstream users. Downstream users who pay the same water fee per hectare as upstream users actually pay more per unit of water because their available water supply falls as the upstream farmers use more than their share. Interviews produced repeated stories of how upstream users, after opening channels to deliver water to their fields, have no incentive to close them. In extreme cases, users at the end of the lateral canals do not get any water and refuse to pay water fees.

One practical solution to the free-rider problem is to change the pricing system so that each farmer pays for what he actually uses, rather than simply a fixed share. While on paper this may sound like a feasible plan—simply measure water volume at the household level instead of at the ID level—in reality, the fragmented and small-scale nature of China's farms will pose a significant problem. The high transaction costs of measuring water intake on hundreds of millions of small land parcels throughout China and collecting fees on a farm-by-farm basis, however, is not a cost-effective solution. Accounting practices instituted to minimize the transaction costs involved in fee collection have further divorced the farmers' production decisions from the value and amount of water that they use; indeed, most farmers in China do not know exactly how much or when they are paying for water. In essence, the ID bills the village for water usage, and the village accountant collects the fees from individual farmers. Since the village accountant must also settle accounts with farmers on other transactions, including local taxes, education fees, and collectively provided services (such as plowing or spraying), water fees frequently are lumped together in a single bill for all services and taxes. Moreover, the clearing of accounts is often done only once or twice a year, so in many cases the water that a farmer pays for has already been delivered, perhaps as many as nine or 10 months previous. In 2012 more than 1200 farmers across China (conducted by two of the authors), fewer than 20% of the farmers could tell enumerators the price they paid for water, either per hectare of land or per cubic meter.

Given that pricing policy does not currently provide a direct incentive to save water (a situation that will probably not change in the near future), another approach to reduce water use in agriculture could be outright restrictions on water deliveries. When water deliveries to agriculture are cut, farmers do tend to use the remaining water more efficiently. For example, in the irrigation district described by Hong et al. (2001) where agricultural water supplies fell by more than 50% from 1985 to 1990, irrigated area and production only fell by 30%, indicating that farmers found ways to increase water productivity so that irrigated area and production did not fall so much. The rise in productivity is probably due to improved water management at both the farm and system level. It is important to note that these improvements were not nearly enough to completely avoid a drop in production, but this represented a sudden decline in deliveries, which gave farmers just a couple of years to adjust their behavior. Over a longer period, and with better management of agricultural water use, agricultural production could be maintained.

PROMOTION OF WATER-SAVING IRRIGATION TECHNOLOGY

In addition to adjusting farmers' incentives to encourage more effective use of water resources, policymakers could also provide farmers with water-saving technologies and education on water-saving practices. This component of the larger policy effort to reduce agricultural water use is being pursued in China, but hurdles remain. Information dissemination has been poor, so even when farmers face strong incentives to save water, they are often unaware of their options to do so. In addition, several of the water-saving options available to farmers, such as drip or sprinkler irrigation, are prohibitively expensive and may not be suitable to the cultivation of some grains.

The millions of farm households with small landholdings make it difficult to design an effective means of informing farmers about water-saving irrigation technologies. The primary method to date is to set up model villages with water-saving irrigation technology and have farmers visit to see how the technology works and how effectively it reduces water use or increases yields. In June 2000, the authors visited one of these model villages. The central, provincial, and county governments each contributed \$58.23 per ha (a total of \$174.68 per ha) to help defray the \$388.17 per ha investment cost of sprinklers (meaning the farmer himself had to invest \$213.50 per ha). But the county's water bureau could not find individual farmers willing to make such an investment. Instead, they found some villages (such as the one we visited) willing to collectively invest in the sprinklers for the entire village and manage the entire purchase and installation of the sprinkler system.

Although effective in getting technology into the field, there are several problems with this approach for promoting widespread adoption. One problem is that there is little village-to-village interaction, and the mechanisms for getting farmers or village leaders from other areas to visit the village and see the technology demonstrated are not clear. Another problem is that the villages that adopt are often so unusual (eg, the village we visited had more than $360,000 per year in total village revenues) that there is little basis for assessing the true feasibility of adopting the technology. Moreover, this method tends to develop and promote technologies born in the sterile laboratories of research institutes, rather than responding to the real-world needs of farmers. Perhaps because of this disconnect, there is an emphasis on promoting expensive water-saving technologies rather than on teaching farmers and village leaders affordable water-saving practices, such as careful timing of water application and monitoring of soil moisture, many of which require no investment at all.

SUCCESSES

While farmers have yet to adopt many water-saving practices in China, there are some success stories. One strategy to save water (or increase the value of water in agriculture) is the widespread establishment of greenhouse production over the last several years. Greenhouses are established primarily to grow vegetables in the winter when the price is as much as 10 times the summer price. But greenhouses are also efficient consumers of water and effectively raise the value of water delivered to agriculture. The greenhouses are covered with plastic to prevent evaporation and utilize other water-saving technologies, such as drip irrigation or micro-sprinkler systems. Although national statistics do not cover the rise of greenhouse agriculture, it is clear that greenhouses have become a common feature in rural China, particularly in areas near urban markets.

Farmers in rural China are beginning to adopt other water-saving technologies and practices. Plastic sheeting to cover crops after watering is much more commonly practiced than it was 10 years ago, and there is potential for further adoption. Plastic sheeting not only prevents evaporation but also raises soil temperature, promoting plant development at early growth stages. Field leveling is a longstanding practice in rural China, but the practice is expanding and increasingly combined with border irrigation. The percent of households using field leveling refers to traditional methods of field leveling, not laser leveling. Laser leveling would likely lead to much higher water-use efficiency than traditional leveling methods (Nyberg and Rozelle, 1999). These practices ensure that water delivered to the field is evenly distributed, rather than leaving some areas dry and others with excess water. In many rice-growing regions, alternating wet and dry irrigation is practiced, a water-saving practice based on timing and requiring little capital investment. Usually, an entire village will adopt this practice and farmers simply accept the alternating irrigation deliveries.

IRRIGATION DISTRICT MANAGEMENT REFORM

The timing and reliability of surface water deliveries greatly affect agricultural production. Often, untimely deliveries or the risk of no delivery are due to deteriorating surface water infrastructure, or poor incentives facing water managers. But these problems are exacerbated when communication is poor between irrigation district managers and farmers or when water managers lack an incentive to make deliveries more timely. Water that is delivered at times when the crop does not particularly need it is more or less wasted, whereas well-timed water delivery can greatly increase agricultural production and has a much higher value.

To improve water delivery services, fee collection services, and communication with farmers, many IDs have developed more flexible and responsive ways to deliver water. Although the institutional response varies from village to village, there are many examples of how irrigation district managers have tried to win back the confidence of farmers and more effectively deliver surface water. In one Henan village, the irrigation district hired teams of three people to be the liaison between the irrigation district and the farmers. Called an irrigation association, each team serves to provide better information to the irrigation district so deliveries can be more timely and farmers will not switch to groundwater. In these villages, the increasing use of groundwater has led to competition in the delivery of the village's water, forcing the surface system to improve its water delivery services.

Irrigation management reforms can also facilitate the promotion of water-saving irrigation technology and practices. Irrigation district management reforms being tried in rural China generally separate water fees from other local fees. Under such a system, farmers are more aware of their water costs than under the system in which water fees are collected along with other village fees and taxes. Moreover, the groups and WUAs can ensure that the gains from aggregate savings are passed on to member farmers. Meetings of user groups can also be used to introduce water-saving irrigation technology or teach water-saving irrigation practices such as measuring soil moisture and timing irrigation.

EFFECTS OF FARMERS' CROPPING DECISIONS ON WATER USAGE

No change is made in a vacuum, and policymakers have to be prepared for the reality that any new policies regulating water use or allocation in China's agricultural sector will be met with a behavioral response from farmers that could fundamentally change the face of China's agricultural production. In this section, we consider what that behavioral response might be, how it could affect crop yields and water usage, and the policy environment that might spur such a change.

One effect of tighter water regulations might be the end of the common practice of wheat/corn double cropping in northern China. Currently, farmers first plant winter wheat in November and harvest it in June, then plant corn in June and harvest it in September or October. During the corn-growing season, rainfall is sufficient and irrigation is not usually needed, but during the winter wheat season, rainfall is scarce and crops rely heavily on irrigation from surface water and groundwater systems. If water prices increase or deliveries are reduced, many farmers may cease irrigated wheat production, decreasing yields substantially.

As China's farmers move away from irrigated wheat production, production of other crops will likely increase. Farmers may choose to continue growing wheat but to forgo irrigation, resulting in lower wheat yields but not much change in other crop yields. Farmers could also choose to abandon wheat production entirely and concentrate on a single crop of corn, which, with a longer growing season, could show significantly higher yields. Alternatively, farmers may switch to water-saving crops such as millet; indeed, there is some qualitative evidence that this has been their response to limited water availability. On a trip in June 2000, the authors visited a village where the wells dried up and irrigation was lost in the early 1990s. Some farmers in this village switched to millet and sweet potato rather than wheat because of the loss of irrigation water. Ultimately, a consortium of private investors and the village collective invested in a water supply company that sank a powerful pump 165 m down to supply water for irrigation; and wheat production was restored. On the aggregate level, however, this switch would be limited by the demand for such alternative crops.

One surprising potential cropping change may actually be a shift toward more water-intensive crops. While this seems counterintuitive, several factors could contribute to this sort of cropping decision. Water can become much more productive in agriculture and the price of water less of a concern if farmers adopt better water conservation practices and take advantage of increased timeliness and reliability of water deliveries. Because wheat is so land intensive, however, it is not particularly suitable to many of the most effective water-saving technologies, such as drip irrigation, micro-sprinkler technology, or greenhouse production. Other crops, such as fruits and vegetables grown in greenhouses, are better suited to take advantage of modern water-saving irrigation technologies. These crops also tend to be labor intensive rather than land intensive and therefore a better match China's comparative advantage. Moreover, these crops are better suited for the world market, as China opens its agricultural sector to international competition.

There are many ways in which farmers can adjust their behavior to limit water usage. But in order for this to occur, they must first be given the incentive to adjust their behavior; in this case, in the form of higher water prices or more strictly enforced water usage restrictions. Moreover, certain institutional changes must also take place. First, the water delivery system must undergo the investment and institutional reform necessary to ensure timely and reliable deliveries of water to agricultural users. If a high level of uncertainty remains in the water delivery system, farmers will not invest in the water-saving irrigation technologies necessary to produce high-value crops. Econometric evidence supports the idea that the reliability of water delivery encourages the cultivation of high-value crops in China (Xiang and Huang, 2000). Second, China must relax its grain self-sufficiency policy so that farmers and local leaders are not under pressure to produce grain. Local leaders are often encouraged to promote grain production in rural China, and this may cause them to resist shifting away from grain into other, more water-efficient crops. Third, farmers must have access to inexpensive and appropriate water-saving irrigation technology, which in turn requires an effective educational outreach program, as described earlier. Without this, farmers may have the will to change but be unsure of what steps may be taken to reduce their water usage.

WATER ALLOCATION DECISIONS

Because of the limited nature of China's water resources and the very palpable externalities that stem from unregulated water use—polluted lakes and rivers, land subsidence, and droughts—conflicts between users frequently arise, disrupting water deliveries and affecting the water supply to the communities involved. These conflicts are generally sharpest within a given basin and in a water-constrained region. In this section, we examine two separate sets of conflicts: those between upstream and downstream users in different geographical parts of a water basin and those between industrial and agricultural users. We then consider the allocation decisions made by China's water managers and describe how they either alleviate or contribute to the problems. One could also examine conflicts between rural and urban users, but, because domestic water use is relatively small, we focus on the above-mentioned two sets of conflicts. Urban–rural conflict could potentially be more serious in the future.

INTERREGIONAL CONFLICTS

In China, as in the rest of the world, some of the most serious water conflicts stem from problems that arise when trying to allocate water among regions. The most common example occurs when excessive upstream water use deprives downstream users of their share of surface water resources. It is also a problem when common-property water resources, such as a lake or a bay, are adjacent to two jurisdictional units.

The most high-profile conflicts have arisen on the Yellow River. Upstream urban growth and newly constructed irrigation projects in Ningxia, Gansu, Shaanxi, and Inner Mongolia siphoned off increasingly larger uptakes to meet the needs of their industrial and agricultural users. Because of lax enforcement by the Yellow River Basin Commission, these withdrawals frequently exceeded the allocated water allotment for the region. As a result, downstream agricultural and industrial users were forced to switch to groundwater or to go without water at all.

Similar problems were observed in the Hai River Basin on a series of several visits to Hebei Province from 1998 to 2001. Two upstream counties in Shijiazhuang Prefecture had monopolized an entire reservoir system's capacity, and downstream counties in Cangzhou Prefecture had to rely on groundwater despite a clearly unsustainable rate of extraction and deteriorating water quality. Similarly, in the early

1990s, irrigation-intensive cropping systems were being developed in the counties in the Taihang Mountains upstream from Baoding Prefecture. As a result, Baoding's lakes were drying up, and municipality wells were pumping so much that the ground was in danger of slumping and damaging parts of the city and its infrastructure. Without an effective enforcement agency with the power and authority to enforce community withdrawal limits, upstream communities will continue to exceed their water allocation at the expense of the communities downstream.

AGRICULTURE–INDUSTRY CONFLICTS

The allocation of water between industry and agriculture has given rise to similar conflicts. Although agriculture is the largest user of water, water demand from industry is growing rapidly, leading to tense disputes over water rights. Because water used in industry has a much higher economic value, and China's local leaders have strong incentives to maximize growth and profits in their districts, industry often wins out when there is a decision to be made as to whether water should be sent to an industrial facility or kept for agriculture.

Giving China's rapidly growing industrial users' priority over water supplies has led to declining water supplies to agriculture in many areas. For example, Hong et al. (2001) describe a large irrigation district in Hubei where there has been a substantial reallocation of water from agriculture to hydropower generation and industrial and domestic uses over the past several decades, especially during the 1990s. From 1985 to 1990, agriculture received 64% of the water from the reservoir, but this share fell to 35% from 1993 to 2001. Between these two periods, total water supplies available for agriculture (including from sources other than the reservoir) declined by more than half. This sharp decline in water supplies led to a 31% decline in irrigated rice area and a nearly commensurate fall in rice production (since farmers cannot grow rice without irrigation).

In addition to decreasing the share of water available for agriculture, this preferential treatment for industry can also lead to an increase in water contaminants, thus increasing the unusable proportion of available water supplies. The release of polluting effluents into the river systems is a serious problem in China (World Bank, 1997). Although industrial wastewater treatment capacity has grown tremendously in the past several years, in most cities a large portion of industrial effluents are still discharged directly into rivers. Pollution in many areas is often so bad that surface water cannot be used for irrigation, or, if used, can lead to soil contamination (Smil, 1993). In many cases, releases from factories have harmed a region's aquaculture industry (World Bank, 1997). Again, local officials often sidestep legislation and regulations designed to curb such pollution in order to protect their own interests and keep local industries profitable.

Industry's growing need for water is so acute, however, that increasing amounts of surface water are insufficient and industrial users seek out groundwater as an additional source. On the North China Plain, prolonged extraction of groundwater for industry has greatly lowered the water table under many urban districts, allowing for the intrusion of contaminated water and possibly causing subsidence as well. This has forced farmers in and around these regions to draw their water from deeper and deeper wells. Industrial water managers have attempted to purchase agricultural water supplies, but have not always been successful. Upstream agricultural counties that have built their own reservoirs and canal systems have little incentive to provide water to industrial centers, since their own agricultural activities would be adversely affected.

Rather than helping to alleviate these conflicts, local authorities tend to exacerbate them, largely due to the divergent goals and interests of the various stakeholders involved. Sometimes industry is denied access to agricultural water despite the fact that industry use is generally valued higher. For example, even though several major cities in Henan Province have so little water that some industries have been shut down, agricultural officials who control the water from new reservoirs have expanded rice production and have plans to develop water-intensive horticulture cultivation. Industries in one Hebei city, visited by the authors in 2000, had to shut down production in many of their factories and could barely operate their power generation plant during the peak irrigation season because agricultural officials drew almost all of the surface water. In both cases, agricultural users could have feasibly shipped water to the cities to be reused. It is even likely that farmers could earn more money by selling water to industrial users than by using the water for agriculture.

These anecdotes highlight not only the irrational decision making of local officials, but also the degree of variation in allocation decisions across regions. In a sense, China's water market is a giant free-for-all, with whichever bureau or institution that has the best access to water sources using as much water as necessary to meet their own needs, with little or no regard for the needs of other users who happen to have less convenient access.

RESOLVING INTERREGIONAL CONFLICTS

Officials and policymakers at all levels are developing ways to manage and resolve problems generated by interregional conflicts over water. As of yet, however, most of these responses are new, experimental, and difficult to implement. The most common solution to date is to increase the authority of higher-level administrative units so that the unit of decision making is broad enough to internalize the conflict. More recently, a system of water rights is being considered as a potentially more effective means to resolve these conflicts, particularly interprovincial conflicts along a river system.

An example of how China has moved to resolve interregional conflicts is the move by the State Council through the MWR to increase the authority exercised by the NRBCs, particularly the Yellow River Basin Commission. In response to the decreased flow to downstream provinces, the Yellow River Basin Commission in 1998 was given more personnel, a higher budget, and, along with the other NRBCs, more power to resolve conflicts among the provinces that use the water in the river basin. By 1999, the newly empowered commission restricted the upstream provinces' access to water and increased deliveries to downstream ones. After 2000, despite a drought, the water in the Yellow River flowed all the way to the ocean for the whole year.

In some cases, upper-level jurisdictions have redrawn boundaries of water districts or taken control of reservoirs to make what they believe is a more rational allocation of water. For example, in Hebei Province, Shijiazhuang Prefecture had built a reservoir that served several counties under its jurisdiction. When a downstream prefecture, Cangzhou, began to suffer serious groundwater shortages because of falling water tables, the province took control of the reservoir, linked an irrigation canal that went to the downstream county, and allocated water away from Shijiazhuang to Cangzhou.

Another way of lessening these sorts of interregional conflicts is the adoption of better management systems at the local level. The WUAs that were so unsuccessful at reducing water usage (described in a previous section of this chapter) have been effective at increasing cooperation between local water managers. One example of a successful WUA is the Hong Miao WUA in Hubei Province, organized in 1995 as a response to poor irrigation service and frequent conflicts between upstream and downstream users. Predictably, the downstream users were sometimes unable to irrigate their crops. Since the formation of the WUA, however, conflicts have lessened and irrigation services have improved. Because of better coordination among the water users, farmers have been able to work together to streamline the water delivery system, and the entire area can now be irrigated in 4 days instead of 2 weeks.

Ultimately, a system of water rights may be implemented to better manage conflicts between regions. The efficacy and feasibility of such a system are presently being debated within the relevant departments in China's Government as well as at international development agencies such as the World Bank, the Asian Development Bank, and the United Nations Food and Agriculture Organization. More research is needed to help these agencies understand what preconditions are needed to implement such a system, and how it will affect water allocation and agricultural production. The obvious political costs and difficulties in establishing such a system will likely serve to put actual implementation of a system of water rights off for some time. But the fact that such a system has been proposed and is being seriously considered portends well for progress that will help solve future water allocation problems in China.

RESOLVING AGRICULTURE–INDUSTRY CONFLICTS

Policymakers are also responding to rationalize the allocation of water between industry and agriculture. Most regions have attempted to deal with emerging problems by defining more clearly the priorities of different users. Generally, throughout China, both industry and agriculture are encouraged to save water, but there is little oversight, and it is understood that industry gets priority over new supplies while agriculture has to make do with its existing allocation, or less.

When water shortages become serious and chronic, stronger and more permanent solutions to water disputes are necessary. To resolve the problem of officials from competing ministries working to divert as much water for their constituents as possible, many provinces and municipalities are promoting reforms to merge the functions of different water management units into a single authority. Although such units have different names in different places, most commonly they are called the Water Affairs Bureau (WAB). The WABs, at the extreme, merge the personnel, resources, and duties of the local WRBs, UCCs, and the water protection division of the local Environmental Protection Bureau into a single unit (Ministry of Water Resources, 1999).

Often cited as an example of the establishment of an effective WAB, Shenzhen Municipality was one of the first prefectures to create a unified water authority. Although water pollution and other environmental considerations instigated the reform (and not water shortage per se), Shenzhen's mayor created the WAB after the municipality's rapid growth during the 1980s and early 1990s. Industrial and urban building expansion created a serious shortage of potable water in the city, threatening to slow down Shenzhen's economic activity. A series of subsequent floods exacerbated the problems and were, in part, connected to the hasty construction of canals and wastewater treatment plants that were built without coordination with other parts of the water system.

Responding to these events, the local government passed an emergency water regulation and created the municipality's WAB. The bureau immediately took charge of all construction of water-related projects, including clean drinking-water plants, wastewater and sewage treatment plants, dikes for flood control, and other infrastructure projects. The bureau also took responsibility for creating and executing all of Shenzhen's water-related activities including those for industrial supply, wastewater cleanup, and agricultural use. Urban jurisdictions in China usually include surrounding agricultural land. Deliveries to agriculture, industry, and urban residents were all under the control of a single entity. By all accounts, shortly after the creation of the bureau, Shenzhen's water supply and flood prevention improved dramatically.

Since the success of the establishment of the WAB in Shenzhen, the MWR has encouraged the plan throughout China (Ministry of Water Resources, 1999). Through mid-1999, 160 counties had established WABs, although the extent of the authority and success that has been realized varies. By 2007, the percentage of counties that established WABs reached 59%. One problem is that although the reform technically merges the different organizations, the divisions still remain, albeit contained under a single roof. For example, officials affiliated with the former WRB are often concerned that the new unit would take too much water from agriculture, while those from the former UCC may view the new system as designed to remove lucrative water revenues from their control. Without the creation of a clear set of allocation rules and guidelines, these types of conflicts will continue to exist and may prevent the development of well-functioning WABs.

Although unifying urban and rural water management is difficult, the benefits of the system can be significant. In Baoding Prefecture, where a WAB was created in 1997, the WAB built a 30 km 1.5 m pipeline from a former WRB reservoir to the UCC's clean water plant. In this case, the reform created a practical solution to the region's water problems. The city received much-needed high-quality water. The irrigation district, which had had trouble using all of its water for agriculture due to a decaying delivery system, was happy to have the additional investment and a new cash-paying customer. Farmers, who sometimes had been implored to take water deliveries from the irrigation district, turned their attention to groundwater sources, which in this particular area were relatively abundant.

Broadening the authority of a single regional water authority also has helped address certain environmental problems. Drawdown by upstream IDs and increased industrial waste ended up affecting the ecological balance of Hebei's largest lake, Baiyang Dian. In the early 1990s, the lake was severely polluted, unable to support either large-scale aquaculture or tourism. Counties below the lake were also reluctant to use irrigation water during certain seasons because of high concentrations of toxic chemicals. In response, the provincial WRB took administrative control of the lake and intervened in the water allocation plans of three prefectures that affected or were affected by the lake. A new canal was constructed leading from one of the large reservoirs, which actually had seen its service area shrink over the years. With access to new flows of water, the province greatly improved water quality in the lake, and the fishing and tourism industries rebounded. Provincial officials claim that, although only a small part of the newly raised revenues from the lake were used to pay for the additional water flows, the irrigation district was revitalized by the payments.

Options for Future Reform

While reforms that unify water management authority have helped to allocate water more rationally among users, the formal extension of water rights may provide for even more effective water allocation. A workable system of water rights, however, also requires sound legal institutions to enforce contracts and resolve conflicts. Presently, the transfer of water licenses or water-use rights is technically prohibited in China because all water is state-owned property (although water transfers do happen under certain circumstances, as indicated in the examples above). Reforms that establish more secure water rights, and making these rights tradable, will increase the flexibility and rationality of water allocation in China and may even increase rural incomes and hasten the development of rural areas (Rosegrant and Binswanger, 1994). The efficacy of water markets, however, will depend heavily on establishing a transparent and independent legal system to enforce contracts and resolve disputes.

CONCLUSIONS AND POLICY IMPLICATIONS

Over the last several decades, China has achieved remarkable gains in agricultural and industrial production by successfully diverting its limited water resources; however, in many important agricultural areas of northern China, the exploitation of existing water resources has gone beyond sustainable levels. Policymakers in

China are responding to this situation by establishing policies and institutions to encourage better water management, but have yet to tackle the main problems of inadequate enforcement of water policies and regulations and conflicting interests at all levels of the political chain. It is still unclear whether China can adapt to a world where water is relatively scarce while continuing to both maintain current levels of agricultural production and increase levels of industrial production.

To evaluate how China's new policies and institutions will affect agriculture requires a better understanding of the resultant improvements in surface water storage and conveyance infrastructure and the extent to which these improve the reliability of surface water deliveries, especially at the ends of the water delivery systems. Without a reliable delivery system, and stronger policy enforcement mechanisms to better align individual incentives with those of national policymakers, farmers will continue to rely on raw groundwater and eschew more water-effective irrigation methods or conservation practices. Moreover, understanding how changes in upstream IDs affect downstream users is critical to determining the overall effect of water management investments on the hydrological system and China's economy.

In later chapters, we present more rigorous evaluations of some of the more recent policy changes, the potential they hold for inducing water conservation, and the effects they will have on China's agricultural production.

REFERENCES

Hong, L., Li, Y., Deng, L., Chen, C., Dawe, D., Barker, R., 2001. Analysis of changes in water allocations and crop production in the Zhanghe irrigation system and district, 1966–98. In: Proceedings of International Workshop on Water-saving Irrigation for Paddy Rice, Wuhan, China, March 23–25, 2001.

Huang, Q., Rozelle, S., Msangi, S., Wang, J., Huang, J., 2008. Water Management Reform and the Choice of Contractual Form in China. Environment and Development Economics 13, 171–200.

Lohmar, B., Wang, J., Rozelle, S., Dawe, D., Huang, J., 2003. China's Agricultural Water Policy Reforms: Increasing Investment, Resolving Conflicts, and Revising Incentive. United States Department of Agriculture, Economic Research Service, Agriculture Information Bulletin Number 782, Washington DC.

Ministry of Water Resources, Institute of Water Resources and Hydropower Research, 1999. Study on Real Water Saving, World Bank Financed Research Project. #7107256, Draft Report, Beijing.

Minstry of Water Resources, 1998. Water Pricing Reform Status in China. Ministry of Water Resources, Beijing, China.

Nickum, J., December 1998. Is China living on the water margin? China Quarterly 156.

Nyberg, A., Rozelle, S., 1999. Accelerating China's Rural Transformation. World Bank.

Park, A., Rozelle, S., Wong, C., Ren, C., 1996. Distributional consequences of reforming local public finance in China. The China Quarterly 147, 751–778.

Rosegrant, M., Binswanger, H., 1994. Markets in tradeable water rights: potential for efficiency gains in developing country water resource allocation. World Development 22 (11), 1613–1625.

Smil, V., 1993. China's Environmental Crisis: An Inquiry into the Limits of National Development. M.E. Sharpe, Armonk, NY, USA.

Stone, 1993. Basic agricultural technology under reform. In: Kueh, Y., Ash, R. (Eds.), Economic Trends in Chinese Agriculture: The Impact of Post-Mao Reform. Clarendon Press, Oxford, UK, pp. 311–359.

Valencia, M., Dawe, D., Moya, P., Pabale, D., 2001. Water fees for irrigated rice in Asia. International Rice Research Notes 26 (2), 78–79.

Wang, J., 2000. Innovation of Property Right, Technical Efficiency and Groundwater Irrigation Management. Chinese Academy of Agricultural Sciences (Ph.D. thesis).

Wang, J., Huang, J., Rozelle, S., Huang, Q., Blanke, A., 2007a. Agriculture and groundwater development in northern China: trends, institutional responses and policy options. Water Policy 9 (S1), 61–74.

Wang, J., Huang, J., Xu, Z., Rozelle, S., Hussain, I., Biltonen, E., 2007b. Irrigation management reforms in the yellow river basin: implications for water saving and poverty. Irrigation and Drainage Journal 56, 247–259.

Wang, J., Huang, J., 2002. Groundwater management and tubewell technical efficiency. Journal of Water Sciences Advances 13 (2), 259–264.

Wang, J., Huang, J., Rozelle, S., 2000a. Property right innovation and groundwater irrigation management. Journal of Economic Research 4, 66–74 (in Chinese).

Wang, J., Huang, J., Rozelle, S., 2000b. Property Right Innovation and Technical Efficiency – Case Study of Groundwater Irrigation in Hebei. Center for Chinese Agricultural Policy. Working Paper #wp-00–e22.

Wang, J., Xu, Z., Huang, J., Rozelle, S., 2005. Incentives in water management reform: assessing the effect on water use, productivity and poverty in the Yellow River Basin. Environment and Development Economics 10, 769–799.

World Bank, 1997. At China's Table. World Bank, Washington, DC.

Xiang, Q., Huang, J., 2000. Property right innovation of groundwater irrigation system and cropping patterns change. Management World 5, 163–168 (in Chinese).

CHAPTER

Water Survey Data

4

Reliable water data is the foundation of any sound water management decision. We cannot manage what we do not know much about. However, despite the importance of dependable data, there is a global decline in the efforts on water data collection during the past decades. For example, during the fifth World Water Forum that took place in September 2009, András Szöllösi-Nagy, Director of the Division of Water and Secretary of UNESCO's International Hydrological Program, noted that "today, we have 30% less data about Africa's hydrology than we did 20 years ago." At the same forum, data was referred as "the poor cousin" in the water sector.

Despite the importance of China's water problems, not many economists have worked on this area. A major barrier to research is the lack of water data in China. Agriculture is still the largest water user in China's economy. However, the agricultural sector is characterized by millions of small farms that are less than 1 ha in size. This makes data collection extremely difficult. No government agencies in China collect farm level water use data. As a result, most studies and debates about China's water problems are based on anecdotes and case studies. Without empirical evidence that comes from nationally representative data, it is impossible to have an accurate understanding of issues at hand.

Our research interests in China's water problems are not impeded by the lack of data. Over the years, we have conducted a series of surveys that we have designed to study China's water economy, in general, and address questions about the effectiveness of irrigation practices and agricultural water management, in particular. In this chapter, we describe the sample sites and the sampling strategies of the surveys. Details on the data collected are described in the data section of each chapter.

In all the surveys we have conducted, in order to reduce the differences in the quality of the data among our villages, we mobilized an intensive data collection effort using well-trained and highly motivated enumerators. Our field survey was finished in a short time. We usually sent a separate survey team to each province and finished data collection in the field within a month. The enumerators we recruited were Master's and PhD students from universities (eg, Peking University, Remin University of China, China Agricultural University, Nanjin Agricultural University, and Inner Mongolia Agricultural University). Enumerators went through both a training course taught by us and field practices in rural villages before they finally participated in the survey.

Managing Water on China's Farms. http://dx.doi.org/10.1016/B978-0-12-805164-1.00004-X

CHINA WATER INSTITUTIONS AND MANAGEMENT (CWIM) SURVEY

Micro-level data sets that focus on the social and economic side of water resources and behavior of water users are very limited. Longitudinal data are even less available. The China Water Institutions and Management (CWIM) panel survey is probably the most unique of its kind. The survey was conducted in three rounds. The first round was carried out from the middle of December 2001 to the middle of January 2002 and collected data and information on 1990, 1995, and 2001. Data in 1990 and 1995 are recall data. The second round collected data and information on 2004. The third round was carried out in April 2008 and collected data on 2007.

That is, in 2001, 2004, and 2007, we tracked 80 randomly selected villages in northern China, one of the most water-short areas worldwide.

The CWIM survey covers three provinces: Hebei, Henan, and Ningxia provinces. The sample areas cover two of the nine major river basins in China (Fig. 4.1).

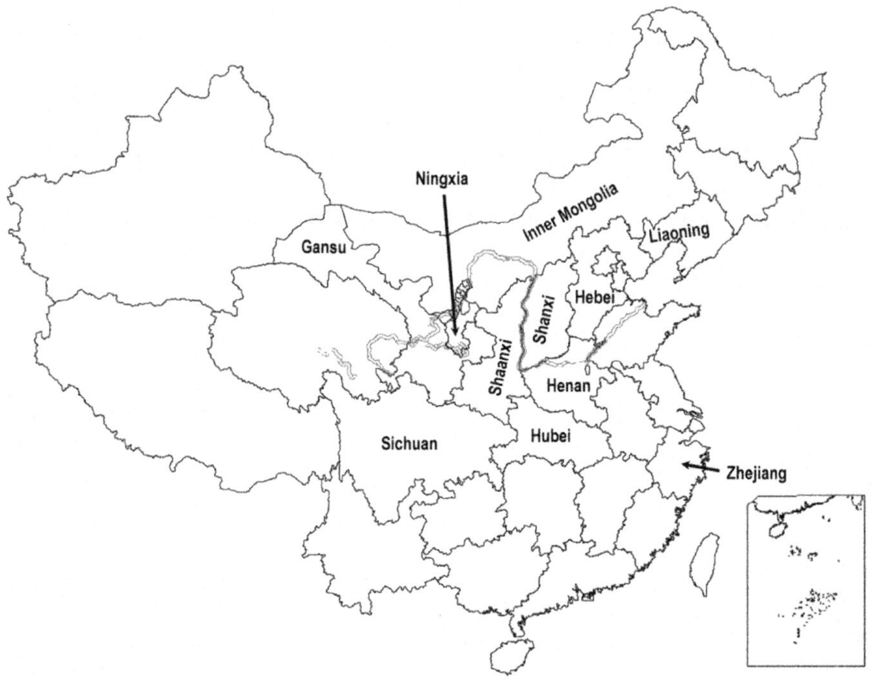

FIGURE 4.1

Sample provinces in our surveys.

CWIM: Hebei, Henan, and Ningxia

NCWRS: Inner Mongolia, Hebei, Henan, Liaoning, Shaanxi, and Shanxi

CNRS: Hebei, Liaoning, Shaanxi, Zhejiang, Hubei, and Sichuan

Bank Survey: Gansu, Hubei, and Hunan

Hebei takes up most of the Hai River Basin (HRB). Ningxia and Henan provinces are located in the upper reaches and lower reaches of the Yellow River Basin (YRB), respectively. Hebei has the highest rates of groundwater irrigation in the country, with 78% of irrigated land relying on groundwater. In contrast, most of Ningxia is irrigated using surface water from the YRB. Henan is a mixed case. Some parts are irrigated using groundwater, with other parts using surface water.

The villages were randomly chosen according to their geographic locations, which reflect the differences on water endowment. For example, in Hebei province, one county (Xian County) is located along the coastal belt, the most water scarce area of China; the other county (Tang County) is located along the inland belt, an area with relatively abundant water resources since it is next to the mountains in the western part of Hebei province, and the third county (Ci County) is located in the region between the coast and mountains. In Henan and Ningxia Provinces, we chose one irrigation district (ID) in the upstream (having better water access condition) and one in the downstream (having relatively poor water access condition). The two IDs in Ningxia Province are Weining Irrigation District and Qingtongxia Irrigation District. The IDs in Henan Province are People's Victory Irrigation District and Liuyuankou Irrigation District. After the IDs were selected, we randomly chose sample villages from a census of villages in the upper, middle, and lower reaches of the canals within the IDs. For example, in Ningxia, within the larger ID (Qingtongxia), which is comprised of nine counties, we chose one county from the upper, middle, and lower sections of the ID. In the smaller ID (Weining), we included both counties that make up the majority of the district's command area. From within each county we randomly selected two townships, from which we randomly chose two villages to include in our sample.

In each village, we designed three kinds of separate survey questionnaires, one for village leaders, one for water managers (tubewell and canal managers), and one for farmers. Since most villages only have one pattern of surface water management (collective, contracting management, or water user association), generally one canal manager was interviewed. For groundwater management issues, we randomly selected three tubewell managers within the village. Questionnaires for village leaders and water managers cover many issues, such as physical and socioeconomic conditions of villages, characteristics of canal and tubewells, supply reliability of surface and groundwater resources, irrigation water use by crop, and water management and policies. Finally, we randomly chose four households within each village to conduct detailed household surveys. After getting the basic information about each household's plot (including information about the number of irrigations per plot, each plot's soil type, the distance to the farmer's home, and whether or not the plot was hit by a disaster), the enumerators chose two plots from each household for more careful investigation on their crop production output, input, and irrigation behaviors. In order to reduce the differences in the quality of the data among our villages, we mobilized an intensive data collection effort, using well-trained and highly motivated enumerators.

NORTH CHINA WATER RESOURCE SURVEY (NCWRS)

The NCWRS survey was conducted in December 2004 and January 2005, collecting data and information for 1995 and 2004. The NCWRS survey interviewed village leaders from 400 villages in Inner Mongolia, Hebei, Henan, Liaoning, Shaanxi, and Shanxi provinces (Fig. 4.1). Our sample villages also represent all or part of four major river basins: the HRB, the lower and middle reaches of the YRB, the northern bank of the Huai River Basin, and the Songliao River Basin in the northeast.

We used a stratified random sampling strategy in order to generate a sample that was representative of northern China. We first sorted counties in each of our regionally representative sample provinces into one of four water scarcity categories, then randomly selected two townships within each county and four villages within each township. In Hebei province, where county level groundwater overdraft statistics are available, the scarcity categories were defined according to a Ministry of Water Resources publication that categorized provinces by scarcity, almost certainly based on the degree of annual overdraft. In the remaining provinces, the scarcity indices were defined according to the percentage of irrigated area as follows: very scarce (between 21% and 40%), somewhat scarce (between 41% and 60%), normal (more than 61%), and mountain/desert (less than 20%). Of the latter strata, we sampled only one county; in each of the others, we sampled two or three counties.

An extended version of the village leader instrument in CWIM survey was used, incorporating the village portion of the well operator and canal manager instruments. The scope of the survey was quite broad, with the questionnaire including more than 10 sections. Among the sections, there were those that focused on the nature of rural China's water resources, the most common types of wells, and pumping technology. There also were blocks that were focused on understanding China's water problems, the response taken by the government (eg, water policies and regulations) and the way that institutions emerged when water was scarce (eg, tubewell privatization). The survey collected data on many variables for 2 years, 2004 and 1995. By weighting our descriptive and multivariate analyses with a set of population-based weights (large villages in large townships in large counties were weighted more heavily in the analysis), we are able to generate point estimates for all of northern China.

CHINA NATIONAL RURAL SURVEY (CNRS)

The CNRS includes a randomly selected, nationally representative sample of 60 villages in six provinces (Hebei, Liaoning, Shaanxi, Zhejiang, Hubei, and Sichuan) of rural China (Fig. 4.1). To reflect accurately varying income distributions, in each province we randomly selected one county from each income quintile, as measured by the gross value of industrial output (GVIO). The survey team then randomly selected two villages from each county and used village rosters to randomly select 20 households, including both those with their residency permits (*hukou*) in the village and those without. The survey included a total of 1199 households.

In choosing our sample, our objective was to choose a sample that contained a range of observations on households from poor to rich. Although we could have used rural per capita income to stratify the sample, there are several problems with doing so. Rural per capita income measures based on the information from the annual Household Income and Expenditure Survey of China's National Bureau of Statistics are only calculated at the provincial level (since only a fraction of counties, townships, and villages are included in the sample). An alternative source of per capita income data, the government's annual census of villages that is reported up through the government hierarchy, does create estimates of rural per capita income for most villages. However, there are serious reporting problems with these measures (Park and Wang, 2001). According to our experience, rich villages sometimes tend to underreport income per capita and poorer villages sometimes tend to overreport. If this is so, the distribution of per capita incomes based on these data would be artificially compressed and provide a less powerful stratification scheme. In response to these shortcomings, we use per capita GVIO since we believe that industrial output in a township or village is more observable and, hence, likely to be more accurately measured. Moreover, GVIO is highly correlated with rural incomes (Rozelle, 1996). Finally, in the townships (rarely) and villages (sometimes) in which GVIO is not available, we have found it is relatively easy for officials to rank townships in terms of their level of industrialization.

In this book, we use the CNRS data to examine the relationship between irrigation, agricultural production and rural income in chapter Irrigation, Agricultural Production, and Rural Income. The survey collected a wide range of information on each household's production activities, and included a special block that focused on collecting by-plot information. For each plot, households recounted crops that were grown during the sample year, crop yield, and the plot's irrigation status (was it irrigated by surface water, by groundwater, conjunctively by surface water and groundwater, or rainfed). The survey collected data on rural household income that can be disaggregated into cropping, off-farm, and other income sources such as earnings from livestock, rent earnings, asset sales, and pensions. The CNRS survey also collected a number of other plot-specific attributes such as soil quality and household characteristics, which has allowed us to control for a large set of factors in our analysis.

BANK SURVEY

This survey was conducted in 2006 at three World Bank Water User Association (WUA) sites spread across three provinces: Gansu, Hubei, and Hunan (Fig. 4.1). Each of the sites was situated in one ID in each province—except in Gansu in which we worked in three IDs.

After the sites were identified we randomly selected the study villages. To do so, in each province we first randomly selected 10 villages from the list of the Bank WUA candidate villages. Henceforth, these are called the Bank WUAs or Bank villages. After these were chosen we then asked ID and water bureau personnel to identify a sublist of all villages from the list of non-Bank candidate villages which were in the same physical proximity as the Bank WUAs. We also had them identify a sublist of collectively managed villages that were in the same physical proximity as the Bank WUAs. Our definition of "same physical proximity" was 10 km or the closest set of 10 villages (if there were not any within the circle with a 10 km radius). From these two sublists (in each of the three sets of IDs) we chose five villages—which made a total of 15 non-Bank WUAs (in 15 non-Bank villages) and 15 collectively managed villages. In total we chose 60 water management institutions in 60 villages, which were made up of 30 Bank WUAs, 15 non-Bank WUAs, and 15 collectively managed villages. It should be noted that we use villages as the unit of survey, even though WUAs are sometimes made up of multiple villages. Since we randomly selected villages within the WUA, we obtained a representative view of the WUA. The Bank Survey used this design to make the survey consistent with the CWIM and NCWRS surveys.

In each village we implemented a survey aimed at understanding the organization of water management institutions in each village and the impact that it may have had on agricultural production and incomes. There were two groups of respondents. The first group was comprised of village leaders and the head of the WUA executive committee, all of whom are the most familiar with water management issues at the village level. The second group was comprised of farmers whose agricultural production is directly influenced by water management in the village. For the farmers' survey, we adopted the pattern of focus group discussion, with five farmers in each group.

REFERENCES

Park, A., Wang, S., 2001. China's poverty statistics. China Economic Review 12, 384–398.

Rozelle, S., 1996. Stagnation without equity: patterns of growth and inequality in China's rural economy. The China Journal 35, 63–92.

SECTION

II

Groundwater Management

CHAPTER

5 Evolution, Determinants, and Impacts of Tubewell Ownership/Management

Faced with increasing demands and limited surface water supplies, farming communities in the North China Plain began to turn to groundwater in the late 1960s. Since then, under the directive of the central government, tubewells have developed quickly. By 1997, agricultural producers were extracting groundwater from 3.5 million tubewells and irrigating nearly 15 million ha, mainly in the Hai River Basin and lower reaches of the Yellow River (Ministry of Water Resources, 2000).

Unfortunately, the rise of the groundwater economy is not without cost. Reliance on groundwater extraction has led to falling water tables. For example, in Feixiang County, a county located in the upstream part of the Fuyang River Basin, the shallow groundwater table fell by 0.6 m per year in the 1980s and 1.3 m per year in the 1990s (Fig. 5.1, Panel B). Even greater rates of decline of the shallow ground water table occurred in the middle and downstream parts of the basin. Excessive water withdrawals and falling water tables have caused land subsidence in some rural areas, cones of depression under some cities, and deteriorating water quality near the coast (Hebei Hydrological Bureau and Water Environmental Monitoring Center, 1999).

Despite the growing water crisis, agricultural production in the North China Plain has not declined. Agricultural yields of wheat and maize rose by nearly 15% to 30% during the 1990s (China National Bureau of Statistics, 2002). Output grew by 20% to 35%. Facing rising demand for new cash crops from domestic and export markets, farmers also have begun to shift from staples into cash crops, even though such crops often require more intensive use of water and more precise timing.

In examining the apparent contradiction between the looming water shortage and booming agricultural production, researchers have observed the rapid rise of private tubewell ownership in North China. Visitors to China's countryside are bombarded by innumerable advertisements for private tubewell drilling services. Some scholars regard the privatization of tubewells as a measure that has improved groundwater management efficiency and increased agricultural productivity (Cai, 1985; Dong and Zhang, 1994; Chen et al., 1997), while others blame private well owners for the fall in North China's groundwater table (Chen and Liu, 2000; Wang, 2004a,b). Although it is certainly possible that the shift of tubewell ownership and management responsibilities to private individuals has helped ease some of the key constraints that agricultural producers have faced in recent years, it also is

Managing Water on China's Farms. http://dx.doi.org/10.1016/B978-0-12-805164-1.00005-1

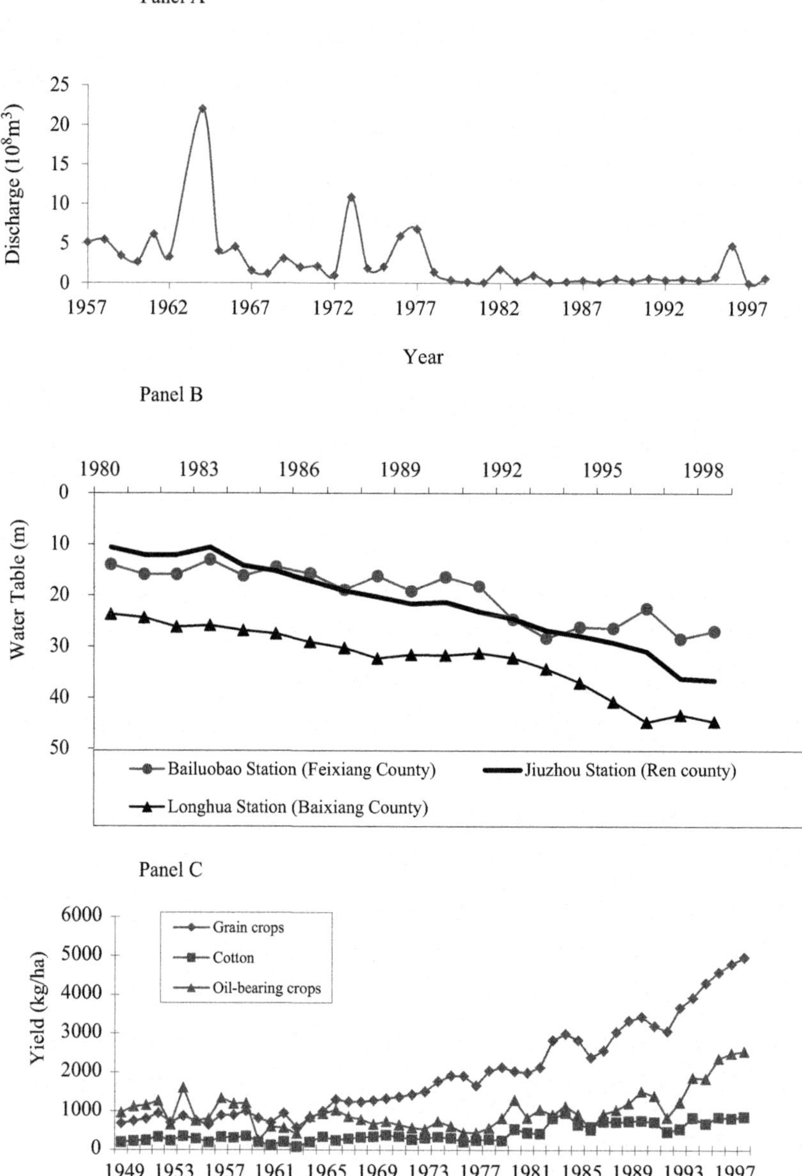

FIGURE 5.1

River discharge, groundwater table, and crop yields in Hebei Province. Panel A: river discharges at Aixinzhuang Hydrometric Station, Hebei province. Panel B: shallow groundwater table in three hydrometric stations, Hebei province. Panel C: crop yields in Handan, Hebei province.

possible that the privatization of pumping services in an open-access environment (like that of groundwater in the North China Plain) could lead to an inefficiently rapid depletion of the region's water resources. Surprisingly, very little research has systematically studied the impact of the tubewell privatization trend on regional productivity.

In this chapter, we describe some of the factors contributing to tubewell privatization, and then explore how the shift in tubewell ownership in the North China Plain has affected the region's agricultural sector. We first present descriptive analysis of survey data from rural China, using the results to frame our hypotheses regarding the impact of tubewell privatization on China's agricultural sector. Next, we conduct regression analysis to determine the precise effects of privatization on several determinants of the rural economy, including cropping decisions, rural incomes, and the water table.

THE EVOLUTION OF TUBEWELL OWNERSHIP

We have identified two types of tubewell ownership: collective and private. If the village's leadership council owns the tubewell, we define it as collective, otherwise the tubewell is defined as private. There are two types of private tubewells. If a tubewell belongs to a single individual or family, we call it an individual tubewell. Other private tubewells are owned by groups of individuals. Since in many of the groups the individual members are assigned shares that indicate the investment stake that each member has in the tubewell, the groups are often called shareholding groups and their tubewells are called shareholding tubewells.

Tubewell ownership in our study areas has shifted sharply from collective to private. The policy constraints that originally limited private activities have gradually been relaxed. In fact, in recent years policy makers have often encouraged individual farmers to become engaged in the groundwater economy. According to the China Water Management Survey (CWMS) data, collective ownership declined from 51% to 19% from 1995 through 2004. Over the same period, private ownership increased from 49% to 81%. The North China Water Resource Survey (NCWRS) data covers a wider, more diverse region, but exhibits a similar pattern. According to NCWRS, 46% of tubewells were collectively owned in 1995. By 2004, only 31% were collectively owned. Moreover, except for Hebei province, individual ownership is the main form of private ownership. Shareholding appears to be an institutional arrangement prominent only in Hebei province.

The shift in tubewell ownership has come mostly from the expansion in the number of new private wells (Table 5.1). Due to the fall of the groundwater table and lack of maintenance on pumps and engines, a number of collective tubewells became inoperable over the past two decades. In fact, the absolute number of collective tubewells fell between 1990 and 2004 (column 1, rows 5–8), while the number of private wells increased rapidly (column 2, rows 5–8). Again, NCWRS data show a similar pattern. Although in the NCWRS sample the number of collective wells increased

Table 5.1 Changes of Tubewell Number and Shifts in Tubewell Ownership in Hebei and Henan Province, 1990–2004

Year	Collective vs. Private		
	Total	Collective	Private
Share of Tubewells (%)			
1990	100	51	49
1995	100	40	60
2001	100	22	78
2004	100	19	81
Number of Tubewells			
1990	1464	742	722
1995	1854	740	1114
2001	3656	836	2820
2004	3357	614	2743

Data source: Authors' surveys in 48 randomly selected villages in Hebei and Henan Province.

slightly (by 26%) from 1995 to 2004, this was eclipsed by the drastic increase (233%) in the number of private wells.

While most of the rise in private tubewells was *denovo* (a local term indicating that the investment and management of the well drilling and installation was completely done by private farmers), the data show that an increasing share of private wells came from the privatization of preexisting collective wells (Table 5.1). In the 2004 NCWRS survey, enumerators asked respondents to provide two pieces of information about each tubewell: the initial investor and the current owner. In fact, the question asked specifically about alienation rights, that is, if the tubewell was sold, who would make the decision and who would receive the proceeds of the sale. According to the findings of the survey in 1995, of the 2672 tubewells that had been originally invested in by the collective, 358 of them (or 13%) had been sold to shareholding groups (5.5%) or private individuals (7.5%). In 2004, the rate of privatization had risen. Of the 3464 wells that had originally been invested in by the collective, 821 (or 23%) had been privatized (19% to individuals). Despite the rise in privatization, however, privatized wells still made up a small fraction of all private wells (12% or 821 of 7140). From this, it is clear that *denovo* private wells have accounted for the majority share of the rise of wells since the mid-1990s.

DETERMINANTS OF PRIVATIZATION

Scholars in recent years have analyzed the determinants of institutional innovation both theoretically and empirically. For example, White (1995) finds that government policies, the degree of democratization, and financial market liberalization play important roles in institutional change. Otsuka (1995) shows that in much of the empirical literature, environmental and population factors, government policies, and other socioeconomic variables are the main determinants of institutional change. Tang (1991) and Uphoff (1986) have identified three kinds of factors that influence water management organization in particular: physical and technical characteristics of the resource, characteristics of the group of users, and attributes of institutional arrangements.

While little empirical work has focused specifically on groundwater, the international literature has identified several factors that affect tubewell ownership. Based on a case study of tubewell ownership innovation in Pakistan, Meinzen-Dick (1996) concluded that the emergence of private tubewells is mainly due to changes in groundwater and surface water utilization, farm scale, and population intensity. Shah (1993) showed how the emergence of institutions that encourage water sales in villages has levered the rise in private tubewell ownership. Barker and Molle (2002) point to the increased availability of reliable pumping technology and hint that this has been a factor in increasing the rise of private tubewell ownership in South and Southeast Asia. Despite the importance of groundwater in China's agriculture and its rapid evolution over time, almost no work has attempted to analyze the factors that have affected tubewell ownership in China.

Drawing on our data from Hebei and Henan, we find that several factors are associated with the shift of tubewell ownership from collective to private. Most strikingly, factor endowments, especially water, are correlated with ownership changes (Table 5.2). In villages in which water is scarce, tubewell ownership has privatized quickly (column 1 vs. column 2, rows 1–4). Although the patterns

Table 5.2 Relationship Between Tubewell Ownership and Resource Endowment and Policy Measures in Hebei and Henan Province, 1990–2004

		Water Scarcity	**Policy Intervention**		**Village Fiscal Health**
Year	**Share of Private Tubewells (%)**	**Groundwater Table (Meter)**	**Villages Receiving Investment Subsidies for Water Projects (%)**	**Villages Receiving Bank Loans for Water Projects (%)**	**Per Capita Village Real Fiscal Income (Dollars)**
1990	49	9	2	0	3.3
1995	60	11	9	2	3.4
2001	78	14	23	14	3.4
2004	81	20	12	7	1.8

Data source: Authors' surveys in 48 randomly selected villages in Hebei and Henan Province.

in the descriptive statistics do not prove causality, they are consistent with the idea that private tubewells may have emerged in response to North China's growing water scarcity.

The survey also finds that villages' fiscal health also affects the privatization of tubewell ownership. Since the rural reforms of the early 1980s, the fiscal revenue position of most villages has declined, especially relative to that of the individual producer. With declining village fiscal income, the capacity for village leadership councils to invest in tubewells has diminished. NCWRS data demonstrate that, between 1995 and 2004, there was a negative relationship between the share of private tubewells and village fiscal income (Table 5.2). When per capita village fiscal income declined (from \$3.26 per capita to \$1.81 per capita), the share of private tubewells increased.

Given this relationship between village finances and privatization, it is likely that government programs to encourage investment by individual farmers and village leaders also may have influenced the pattern of tubewell ownership. Officials have implemented two main policies—fiscal subsidies and loans—that affect tubewell ownership decisions. While the subsidy programs mainly support the investment efforts of individual farmers, banks typically target the special loans to village leaders. Because of the targeting rules of the two policies, we expect that in areas that have had relatively large fiscal subsidy programs, there should have been more of a shift toward private ownership. Likewise, in those areas with an active bank loan program, the access to special investment funds of the village leadership council may be keeping the collective active in maintaining or expanding their tubewells. Curiously, the fall in tubewell ownership between 2001 and 2004 mirrors a reduction in government support (fiscal and financial), supporting this hypothesis (Table 5.2, columns 1 vs. 5 and 6).

FRAMEWORK FOR UNDERSTANDING THE EFFECT OF PRIVATIZATION ON THE RURAL ECONOMY

Tracking the precise impact of the rise of private tubewells on irrigated area, production and income is difficult for several reasons. It is difficult to separate out the effect of rising groundwater use from privatization because there are two ways that the rise of the private sector can play a role. First, there is a new well effect. In other words, there is a privatization effect that will increase the number of tubewells. Had there not been the private sector, there would have been less groundwater use and less impact on irrigated areas. In addition, the production and income impacts of this effect is potentially quite large—essentially the productivity and income difference between irrigated and nonirrigated farming (see Huang et al., 2006, for evidence for the yield and revenue effects associated with increasing irrigated area). Secondly, there is also an efficiency effect that potentially affects production and income. Conditional on having a tubewell, this effect is measured as the difference between production and income that arises when a tubewell is operated by a private individual rather than the collective.

The international literature also suggests another effect, the crowding out effect. Although privatization is often seen to lead to efficiency gains, there could be negative distributional effects. When the collective is in charge, it is possible that the criteria (and terms) by which water is distributed are different than when private individuals are making decisions based on a profit maximizing basis. Hence, it is possible that although efficiency rises, total yields and/or income could fall. Because we lack information on this effect, it is not discussed here. The work of Wang et al. (2007), however, suggests that there has not been a large negative distributional effect of privatization.

The effect on the groundwater level in farming areas is also difficult to isolate because there are different channels—sometimes offsetting—through which increased privatization can affect overdrafting. When trying to identify the effects on the water table it is first necessary to understand that true water use is going to rise in one of two main ways in North China—firstly, by the increased evapotranspiration (ET) associated with the expansion of sown area or the change in the mix of the crops being cultivated (henceforth, the ET effect) or secondly, by increased flow of unused irrigated water into sinks or into rivers systems that flow into the sea (henceforth, the sink effect). Following the discussion in the previous paragraph, both the new well effect (especially) and efficiency effect (more indirectly) can affect sown area and lead to an increase in the ET effect.

There is one more way that the groundwater level can fall without the expansion of sown area. If the long-term weather patterns changed, there could also be an acceleration in the lowering of the water table without an expansion of sown area. We are assuming that the duration of our study period is too short to have been influenced by such a fundamental set of effects. It is possible that by switching irrigation sources from surface water to groundwater, the reduction of water available to the farming area would lead to a fall (or an acceleration in the fall) of groundwater levels in the farming area. Assume for example, a farming district was being irrigated by a surface water system that originally drew its water from a reservoir. If instead the water was diverted to a city in some other region (that pumped its effluent into a river that could not be reused or went to the sea), and if the farmers in the farming district began to rely on groundwater, there could be a (new) fall in groundwater levels. A similar effect would occur if the surface water canal system or reservoir became unusable and the water originally sent to the farming district stayed in the mountains (and either evaporated or went into a sink) or continued flowing in the river and flowed into the sea. Henceforth, the study calls this the switching effect (from surface water to groundwater).

PRIVATIZATION AND THE WATER TABLE

According to our strategy of analysis, there are only a small number of ways that the expansion of the private groundwater economy may affect the water table. Following the same set of assumptions made in section "Framework for Understanding the Effect of Privatization on the Rural Economy", one of the main effects of the emergence of private tubewells is due to the new well effect. In any area that the rise of groundwater use (from the increased area sown and increased intensity of production in newly irrigated areas, both of which will increase ET) is not being offset by an equal volume of recharge, groundwater levels will fall. In fact, in 78% of both the newly irrigated villages and in the previously irrigated villages that experienced newly expanded irrigated area, the groundwater level is falling. Hence, it appears as if at least one of the reasons for China's falling groundwater levels is that new tubewells are increasing the sown area.

The water table may also have been affected by the switching effect. In fact, in villages in which the rise in tubewell-supplied area exceeds the rise in total irrigated area (implying that there was decrease in surface water irrigation), 65% of villages experienced a fall in groundwater level. Of these villages, 33% are falling faster than the average rate. Unfortunately, our data does not explain whether the fall in surface water irrigated area was caused by deterioration of the surface water system or from the diversion of surface water to other sources.

Significantly, according to the current analysis, with the exception of the new well and switching effects (discussed above), there are not many other ways that the rise of private tubewells affect the water table. For example, the current data show that when the share of private tubewells increased from 49% to 81% between 1990 and 2004, the share of sown area under wheat cultivation decreased by 3% (Table 5.3). At the same time, the share of area under maize and cotton increased. Most importantly, the area devoted to other crops (mostly horticulture crops) rose by 50% (from 5% in 1990 to 10% in 2004). Although descriptive statistics from the CWMS data also indicate that changes in tubewell ownership may have led to shifting cropping patterns, it is unclear how much of the change in cropping patterns is due to the rise of private tubewells. There certainly are many other factors that affect cropping patterns. In fact, when using multiple regression analysis, part, but not all, of the shifts in cropping patterns over time are associated with the rise of private tubewells. Wang et al. (2005a) suggest that it may be the greater efficiency of

Table 5.3 Relationship Between Tubewell Ownership and Cropping Patterns and Yields in Hebei and Henan Province, 1990–2004

Year	Share of Private Tubewells	Share of Sown Area (%)				Crop Yield (kg/ha)		Real Farmer Income[a] (Dollars)
		Wheat	Maize	Cotton	Other Cash Crops	Wheat	Maize	
1990	49	44	27	6	5	4155	4650	98.55
1995	60	45	26	6	6	4515	5010	146.43
2001	78	45	26	9	6	4890	5625	221.40
2004	81	41	30	10	10	5295	5490	273.40

[a] *1990 price.*

Data source: Authors' surveys in 48 randomly selected villages in Hebei and Henan Province.

private tubewell owners that allows for the cultivation of more season-oriented crops (such as horticulture crops).

However, even if all of the observed shifts in cropping patterns were due to the rise of private tubewells, it is unclear if there would be much effect on the water table. According to technical manuals based on extensive experimentation in many different parts of China (Chen et al., 1995), there is little difference in annual water use (mostly ET) for the major crops in the region. The differences between wheat, maize, cotton, and horticulture crops are, on average, less than 10% to 20%. Hence, the overall effect of any cropping pattern shift must be small.

Likewise, the efficiency effect of the rise of private water use on the water table also appears to be small. In multiple regression analysis (reported in Wang et al., 2005b, 2009), one of a series of equations examines the determinants associated with the lowering of the water table. In the current analysis with the CWMS data (1990–2001), there is a small positive effect of the rise of private tubewells on the water table. In the analysis using the NCWRS, the effect is negative. However, in each case the magnitude of the effect is so small that it would account for at most only a small fraction of the new well and switching effect.

PRIVATIZATION AND AGRICULTURAL PRODUCTIVITY

Although it is possible that the groundwater economy would have expanded as fast as it actually did had the private sector not been involved, it is hard to imagine it doing so. In the NCWRS sample villages from throughout northern China, in total the collective had only invested in and operated around 4000 tubewells between the time groundwater started being exploited and 2004. In contrast, the private sector invested in more than 4000 tubewells in the 9 years between 1995 and 2004. In short, while it is impossible to know for certain, it seems likely that the rise in private tubewells has increased irrigated area (or at least kept it from deteriorating or growing less fast).

Assuming that the private sector was a driving force behind the rising number of tubewells, decomposing the rise in the cultivated area that new irrigation covers demonstrates that there is both a new well effect and a switching effect. The decomposition analysis divides the total rise in the area irrigated by tubewell into three components: (1) the rise in newly irrigated area in newly irrigated villages (that is villages that previously grew crops that were not irrigated); (2) the net expansion in irrigated area in villages that previously had irrigated area; and (3) the rise in irrigated area which replaced the irrigated area that originally was supplied by surface irrigation (also in villages that previously had irrigated area). According to the data, of the rise of area irrigated by tubewells (about 8000 ha; from 20,725 ha in 1995 to 28,542 ha in 2004), about 4000 ha (approximately half) are in newly irrigated villages. This rise in irrigated area occurred in 46 newly irrigated villages (on average about 87 newly irrigated ha per village). There was also 667 ha of newly irrigated area from surface water.

The rest of the increased tubewell-supplied irrigated area occurred in the 219 villages that previously had irrigated area. In these villages, although the total expansion of area irrigated by tubewells was 3000 ha, the net increase in irrigated area was only 2000 ha. Data also suggest that there were at least two projects that increased surface water-supplied irrigated area by 333 ha. Hence, this means that there were 1667 ha of newly expanded irrigated area from groundwater. Although it is almost certain that not all of the 5667 ha (4000 ha in newly irrigated villages and 1667 ha in previously irrigated area) can truly be counted as a new well effect (the amount of irrigated area that would not have been irrigated had there not been private tubewells), at least a part (and probably a large part) of the 5667 ha is due to the new well effect. The rest of the expanded tubewell-supplied irrigated area, by deduction, must be due to the switching effect. In the NCWRS sample villages (2333 ha) the amount of area that was originally irrigated by surface area before being supplied by groundwater is equal to the difference between the rise in area supplied by tubewells in previously irrigated areas (4000 ha) and the amount of the net increase in newly expanded tubewell-supplied irrigated area (1667 ha).

Descriptive statistics from CWMS data also indicate that changes in tubewell ownership have led to shifting cropping patterns (Table 5.3). Although there certainly are many other factors that affect cropping patterns, data from CWMS show that when the share of private tubewells increased from 49% to 81% between 1990 and 2004, the share of sown area under wheat cultivation decreased by 3% (columns 1 and 2, rows 1–4). At the same time the area devoted to maize and cotton increased (columns 3 and 4). Most importantly, the area devoted to other crops (mostly horticulture crops) rose by 40%, from 6% to 10% (column 5).

The relationship between well privatization and crop yields is less clear. It is true that the descriptive data illustrate that yields increase over time as private tubewell ownership increase (Table 5.3, columns 6 and 7). There are, of course, many reasons (such as new technology) why yields may have risen. The correlation between tubewell ownership is less strong when examining the rates of increase of yields; yields rise less rapidly than that of private tubewell ownership (rows 1–4). The lack of correlation also can be found when analyzing the relationship between yields and tubewell ownership across villages (rows 5–8).

Descriptive statistics also indicate that as the percentage of privately owned tubewells increases, farmers increase their earnings. When the share of private tubewells increased from 49% to 81% between 1990 and 2004, the real income of farmers also increased from $98.55 to $273.40. In summary, then, our descriptive analysis suggests that the rise of the private groundwater economy has had positive effects on both crop yields and income in rural China between 1990 and 2004. If even some of the rise in groundwater use is due to a new well effect, then it can be said that the evolution of the groundwater economy has helped increase crop yields and incomes. Nevertheless, a number of other factors also may be contributing to the observed income increases, and so before drawing any conclusions, multivariate analysis is needed.

ESTIMATION RESULTS OF ECONOMETRIC MODELS

We have specified one set of econometric models to analyze: (1) the determinants of tubewell ownership; (2) possible effect of privatization of tubewell ownership on groundwater table; (3) impact of tubewell ownership on cropping patterns and income; and (4) the impact of tubewell ownership on crop yields. The descriptions of these models are included in Methodological Appendix to Chapter 5.

DETERMINANTS OF TUBEWELL OWNERSHIP

Our results show that increasing water scarcity affects the evolution of tubewell ownership. The coefficient on the groundwater table level variable is positive and significant. All other things held constant, when the groundwater table falls and water becomes scarce, tubewell ownership shifts from collective to private. Our findings can be interpreted as support for the induced innovation hypothesis, a hypothesis that has been found to be true in many studies outside of water management. Changes of natural resource endowments will induce institutional innovation.

Although the robustness of the coefficients on the water scarcity variable in the tubewell ownership equation suggests that endogeneity may not be a major statistical problem, examining the water scarcity equation demonstrates the statistical validity of our instrumentation strategy (Table 5.4, column 2). The depth of the water table in 1990 has high explanatory power in determining the depth of the water table in later years. In addition, the Chi-square test used in the second part of the exclusion restriction test shows that the instrumental variables have no independent explanatory function in the tubewell ownership equation. The test statistics is 0.001. In other words, the Hausman-Wu exclusion restriction test demonstrates that our instrument is valid.

In addition to pressures provided by scarce resource endowments, the government's policies also have influenced tubewell ownership, although different programs have had different impacts (Table 5.4). The coefficient of the fiscal subsidy variable is positive and significant, suggesting that fiscal subsidies for tubewell investment have promoted the ownership of tubewells by private individuals. In contrast, the coefficient of bank loan is negative (column 1, rows 5 and 6), indicating that targeted loans from banks have encouraged the expansion of the collective ownership of tubewells. Both of these shifts are as expected.

OWNERSHIP IMPACTS ON THE LEVEL OF THE GROUNDWATER TABLE

The coefficient on the tubewell ownership variable in the water scarcity equation also is of interest in its own right (Table 5.4, columns 2, row 1). Our results suggest that the groundwater table is lower in villages with more private tubewell ownership. In other words, the finding indicates that the shift to private ownership does lead to a more rapid fall in the groundwater table than villages that have collective tubewell ownership.

Table 5.4 Regression Analysis Using Two Stage Least Squares of the Determinants of Tubewell Ownership and Its Impact on Cropping Pattern and Yield

	(1) Dependent Variable (Stage 1): Share of Private Tubewells	(2) Dependent Variable (Stage 2): Log of Groundwater Table	Dependent Variable (Stage 2): Share of Sown Area				Dependent Variable (Stage 2): Crop Yield per Hectare		(9) Dependent Variable (Stage 2): per Capita Income
			(3) Wheat	(4) Maize	(5) Cotton	(6) Other Cash Crops	(7) Wheat	(8) Maize	
Tubewell Ownership									
Share of private tubewells (predicted from column 1)		0.024 (7.19)***	−3.008 (2.23)**	2.780 (1.83)*	−3.315 (1.96)**	1.937 (1.09)	182.077 (1.05)	−7.136 (0.03)	81.232 (0.50)
Share of private tubewells[a] year dummy in 2004		0.001 (0.79)	−0.057 (3.05)***	0.023 (1.07)	0.101 (4.27)***	0.059 (2.39)***	−2.365 (0.61)**	6.924 (1.29)	6.807 (2.98)***
Water Scarcity									
Log of groundwater table (predicted from column 2)	14.093 (3.48)***		118.850 (2.00)**	−108.415 (1.82)*	123.674 (1.86)*	−70.312 (1.01)	−6794.028 (0.99)	1550.184 (0.17)	−2336.079 (0.36)
Log of groundwater table in 1990		0.517 (4.39)***							
Policy Interventions									
Dummy of fiscal subsidies for tubewell investment	13.117 (1.72)*								
Dummy of bank loans for tubewell investment	−4.493 (0.48)								

Other Control Variables									
Per capita village fiscal income	−0.269 (1.82)*		0.034 (0.83)	−0.057 (1.24)	0.080 (1.56)	0.063 (1.17)	11.481 (2.00)**	10.595 (1.28)	7.060 (1.43)
Per capita land area	−281.979 (1.61)		31.217 (0.73)	75.267 (1.56)	24.440 (0.46)	160.210 (2.85)***	4928.073 (0.87)***	27,762.01 (2.60)***	11,752.12 (2.26)**
Share of surface water irrigation	0.089 (0.81)	−0.001 (0.37)	0.080 (1.31)	−0.225 (3.27)**	0.112 (1.47)	0.015 (0.19)	−9.399 (1.20)	1.731 (0.17)	−6.653 (0.90)
Water quality (1 = good 0 = bad)		−0.092 (0.70)	4.302 (1.35)	−7.429 (2.07)	12.468 (3.12)***	−5.642 (1.35)	−297.785 (0.73)	75.869 (0.14)	−182.402 (0.47)
Share of labor force with higher than primary schooling	0.147 (0.77)	−0.012 (3.51)***	1.398 (2.27)**	−1.272 (1.83)*	1.467 (1.90)**	−0.725 (0.89)	−83.458 (1.04)	23.718 (0.22)	−12.220 (0.16)
Share of nonagricultural labor force	0.054 (3.23)***		−0.000 (0.01)	−0.013 (2.23)**	0.037 (5.80)***	−0.007 (1.06)	1.346 (2.09)**	−0.379 (0.40)	0.092 (0.15)
Distance to road	0.269 (0.22)	−0.027 (1.20)	3.484 (2.13)**	−3.429 (1.87)*	3.504 (1.71)*	−2.065 (0.96)	−225.672 (1.08)	161.478 (0.58)	−104.482 (0.53)
Number of firms		−0.002 (0.25)							
Constant	23.610 (0.85)	0.743 (2.36)***	−66.141 (1.20)	143.770 (2.32)**	−138.699 (2.00)**	47.889 (0.66)	10,546.81 (1.48)	−825.111 (0.09)	692.437 0.10
Observations	158	158	158	158	158	158	158	138	158
Adjusted R-squared	0.67	0.78	0.65	0.87	0.73	0.49	0.70	0.61	0.68

Absolute value of z statistics in parentheses. * Significant at 10%; ** significant at 5%; *** significant at 1%.*

[a] *Village dummies were included, but are not reported to save space.*

OWNERSHIP IMPACTS ON AGRICULTURAL PRODUCTION

Our results also show that the evolution of tubewell ownership has led to systematic adjustments in the cropping patterns of our sample farmers in the North China Plain (Table 5.4, columns 3–6). The coefficients on the tubewell ownership variable (share of private tubewells or intercrossing variables multiplied by ownership and year dummy) in maize, cotton, and other cash crop equations are positive and significant (columns 4, 5, and 6, rows 1 and 2). In contrast, the coefficient is negative and significant in the wheat equation (column 3). In other words, when the share of private tubewells in a village rises, farmers in our sample shift sown area from wheat and other grain crops to maize, cotton, and other cash crops (mainly horticulture crops). Given the greater demand by horticultural producers for timely water deliveries, our results from the sown area equations might be interpreted as meaning that the shift to private tubewell ownership has facilitated the expansion of high-valued crops that have special water needs. In addition, faced with increasing water scarcity, farmers have begun to shift out of water-intensive crops (such as wheat, which requires up to five irrigations per year) to less water-intensive but relatively high-valued crops (such as maize and cotton, which use one to three irrigations per year). Our results may indicate that with change of tubewell ownership from collective to private, farmers have attached more importance to increasing the value of water use.

In contrast, our results show that there is no significant relationship between tubewell ownership and crop yields (Table 5.4, columns 7 and 8). The coefficient of the tubewell ownership variable is not significant in either the wheat or maize yield equation. It indicates that despite increasing water scarcity, due to changes in tubewell ownership, agricultural productivity will not be adversely influenced.

OWNERSHIP IMPACTS ON RURAL INCOME

Finally, the analysis demonstrates that farmers earn more money with the shift to private tubewell ownership (Table 5.4, column 9). The coefficient of tubewell ownership is positive and significant. While we do not know the exact reason for the rise in income, the result is consistent with the idea that the shift in cropping patterns, from lower-valued wheat to higher-valued cash crops, is leading to an increase in income.

CONCLUSIONS AND POLICY IMPLICATIONS

In this chapter, we have sought to understand the evolution of tubewell ownership in the North China Plain and its effect on production and groundwater levels. The results show that since the early 1980s collective ownership of tubewells has largely been replaced by private ownership. At present, private tubewell ownership has become the dominant form of ownership in many regions. Most private tubewells are still owned jointly by several individuals as shareholding tubewells.

Changes of natural resource endowments have been shown to lead to changes in the commonly observed forms of institutions, consistent with the induced innovation hypothesis (as also commonly found in other developing economies). With falls in the groundwater table and reduced deliveries of surface water resources, water has become scarce in the North China Plain.

Fiscal and financial policies have also played important roles in the evolution of tubewell ownership. Since fiscal subsidy programs have been designed to directly extend funding to individual farmers for tubewell investment, these fiscal measures have promoted the emergence of private tubewells. In contrast, targeted bank loan policies that mainly have provided bank loans to village leadership councils for tubewell investment have slowed down tubewell privatization.

Our findings also demonstrate that the privatization of tubewells has promoted the adjustment of cropping patterns while having no adverse impact on crop yield. Such results are consistent with the hypothesis that when tubewell ownership shifts from collective to private (shown in this chapter) and water is more efficiently managed (shown in Wang et al., 2002), producers are able to cultivate relatively high-valued crops, which in some cases demand greater attention of tubewell owners. Specifically, our results show that after privatization, farmers have expanded sown area of water sensitive and high-value crops, such as maize, cotton, and non-cotton cash crops (mainly horticulture crops). It is perhaps because of the rising demand for horticulture crops that some private individuals have become interested in investing in tubewells. When combined with the rising efficiency of groundwater services that are associated with private tubewells, we may have discovered at least part of the reason why agricultural production has increased when facing increasing water scarcity. Institutions have evolved that have conserved scarce resources while allowing producers to continue to produce.

Our research indicates that, consistent with the concerns of some observers, the privatization of tubewells has accelerated the fall in the groundwater table. Because of the positive effect privatization has had on income, however, we still believe that policy makers should continue to support the privatization of tubewells in the North China Plains. Nevertheless, measures should be taken to address the falling groundwater table. Because a return to collective tubewell ownership would entail giving up the efficiency gains achieved over the last decades, other policies—such as pricing and regulatory measures—may need to be used to combat the deterioration of China's groundwater. In fact, given the increased pressure to move into higher valued crops, despite increasing resource scarcity, it seems that the shift to private tubewells will continue. Given their greater efficiency, encouragement of this trend may be warranted, but only when combined with measures to slow the deterioration of the groundwater resource.

REFERENCES

Barker, R., Molle, F., 2002. Perspectives on Asian irrigation. In: Paper Presented at the Conference on Asian Irrigation in Transition-responding to the Challenges Ahead, 22–23 April 2002 Workshop. Asian Institute of Technology, Bangkok, Thailand.

Cai, X., 1985. Contracting tubewell by farm household. Journal of China Water Conservancy 1, 25 (in Chinese).

China National Bureau of Statistics, 2002. China Statistical Yearbook. China Statistic Publishing House, Beijing.

Chen, J., Liu, Y., 2000. Development of North West China. http://www.cass.net.cn/zhuanti/y_xibei/ (accessed 15.01.05.) (in Chinese).

Chen, X., Chen, D., Tang, J., Xia, H., Yue, M., 1997. Property right reform of rural water conservancy in Henan. Journal of China Water Conservancy 2, 10–11 (in Chinese).

Chen, Y., Guo, G., Wang, G., Kang, S., Luo, H., Zhang, D., 1995. Water Demand and Irrigation of Major Crops in China. China Water Resources and Hydropower Publishing House, Beijing.

Dong, J., Zhang, Y., 1994. One good measure: developing shareholding tubewell in Guangzong county. Journal of China Water Conservancy 1, 15–16 (in Chinese).

Hebei Hydrological Bureau and Water Environmental Monitor Center, 1999. Hebei Water Resources Assessment. Research report prepared by Hebei Hydrological Bureau and Water Environmental Monitor Center, Shijiazhuang, China.

Huang, Q., Rozelle, S., Lohmar, B., Huang, J., Wang, J., 2006. Irrigation, Agricultural Performance and Poverty Reduction in China. Food Policy 31, 32–52.

Meinzen-Dick, R., 1996. Groundwater Markets in Pakistan: Participation and Productivity. Research Reports 105. International Food Policy Research Institute, Washington, DC.

Ministry of Water Resources, 2000. China Water Resources Bulletin. Available from URL: http://www.chinawater.net.cn/cwsnet/gazette-new.asp (accessed 10.01.02.).

Otsuka, K., 1995. Land Tenure and Forest Resource Management in Sumatra: Analysis Issuers and Research Policy for Extensive Survey. (MP 11), IFPRI, Washington DC.

Shah, T., 1993. Groundwater Markets and Irrigation Development: Political Economy and Practical Policy. Oxford University Press, Bombay, India.

Tang, S., 1991. Institutional arrangements and the management of common-pool resources. Public Administration Review 51 (1), 42–51.

Uphoff, N., 1986. Local Institutional Development: An Analytical Sourcebook with Cases. Kumarian Press, West Hartford, CT.

Wang, S., 2004a. Recent Trends in Poverty in PRC. ADB Institute, Sharing Development Knowledge about Asia and the Pacific. http://www.adbi.org/discussion-paper/2004/01/04/83.poverty.targeting/recent.trends.in.poverty.in.prc/ (accessed 28.01.05.).

Wang, X., 2004b. Water Scarcity in Hebei. http://www.yzdsb.com.cn/20040315/ca338813.htm (accessed 03.01.05.) (in Chinese).

Wang, J., Huang, J., Huang, Q., Rozelle, S., Floyd-Walker, H., 2009. The evolution of China's groundwater governance: productivity, equity and changes in the level of China's aquifers. Quarterly Journal of Engineering Geology and Hydrogeology 42, 267–280.

Wang, J., Huang, J., Rozelle, S., Huang, Q., Blanke, A., 2007. Agriculture and groundwater development in Northern China: trends, institutional responses, and policy options. Water Policy 9 (S1), 61–74.

Wang, J., Huang, J., Blanke, A., Huang, Q., Rozelle, S., 2005a. The development, challenges and management of groundwater in rural China. In: Giordano, M., Shah, T. (Eds.), Groundwater in Developing World Agriculture: Past, Present and Options for a Sustainable Future. International Water Management Institute (forthcoming).

Wang, J., Huang, J., Rozelle, S., 2005b. Evolution of tubewell ownership and production in the North China Plain. Australian Journal of Agricultural and Resource Economics 49 (2), 177–195.

Wang, J., Huang, J., Rozelle, S., 2002. Groundwater management and tubewell technical efficiency. Journal of Water Sciences Advances 13 (2), 259–263 (in Chinese).

White, A., 1995. Conceptual Framework: Performance and Evolution of Property Rights and Collective Action. (MP 11), IFPRI, Washington, DC.

CHAPTER

Development of Groundwater Markets in China

6

From the 1980s to the early part of the 21st century, groundwater has begun to play an increasingly important role in irrigation in China, especially in northern China where per capita water availability is less than one-twentieth of the world average (Liu and He, 1996). Although surface water dominated China's irrigation development in the 1950s and 1960s, since the end of 1960s groundwater gradually has become the primary source of irrigation water. According to official statistics, between 1965 and 2003, the number of tubewells increased from 0.2 million to 4.7 million (Ministry of Water Resources and Nanjing Water Institute, 2004; Ministry of Water Resources, 2003). Nearly all of the tubewells (95%) are in northern China. Today, these tubewells provide about 68% of the total irrigation water in northern China (Wang et al., 2007).

The rise of groundwater in China not only fueled an expansion of sown area and rising production (Huang et al., 2005), as the reliance on groundwater has increased, China's groundwater economy has become characterized by a growing water crisis (Wang et al., 2007). Using data from a field survey conducted in six provinces in northern China, Wang et al. (2006) found that from 1995 to 2004, the water table is falling sharply in more than half of the regions in northern China. As a result, there has been a concomitant rise in the cost of sinking a tubewell. In parts of the North China Plain the shallow water table has been dropping at a rate of more than 1 m per year (Ministry of Water Resources, 2002). The deep water table has fallen faster, declining at a rate of more than 2 m per year in some areas (Wang et al., 2005). The cost of sinking and operating a tubewell has at least doubled in many parts of northern China (Huang et al., 2005).

During the 1990s with the falling of groundwater table, the ownership of tubewells also has begun to evolve. Before the rural reforms in the 1980s, most tubewells were owned and operated by the collective. For a variety of reasons, including the decline in the strength of the collective and the increased freedom of individuals to invest in their own farms, soon after the economic reforms began in the early 1980s the ownership of China's tubewell began to shift sharply from collective to private (Wang et al., 2005; Shah et al., 2004). The number of private tubewells increased from almost nothing in the 1970s to nearly 40% by 1990. The shift to private tubewell ownership continued during the 1990s and beyond. For example, in 1995 collective ownership accounted for 58% of tubewells in the average groundwater using village in northern China (Wang et al., 2007). By 2004, private tubewells rose to 70%.

Managing Water on China's Farms. http://dx.doi.org/10.1016/B978-0-12-805164-1.00006-3

While the rise of private tubewells has been shown to lead to more efficient use of water, higher levels of irrigated area and more complex cropping systems (Wang et al., 2005, 2006), it has also made access to irrigation water an increasingly important issue. During the Socialist era (1950s through the 1970s) when local leaders were in charge of allocating groundwater in almost all villages, the equitable distribution of groundwater was not an issue. However, as tubewells have been installed and begun to be operated by private individuals, and as tubewells have begun to be sunk to deeper levels (making the real price of water rise), concern has arisen that not all farmers may have equal access to groundwater (Meinzen-Dick, 1996). It is possible that the farmers that have access to capital are the ones that are more likely to sink and manage tubewells. It is also possible that because of this, these better endowed farmers have much better access to water than those without tubewells. If so, it is possible that part of the gains from increased efficiency that accrues from the rise of private tubewell ownership is being offset by rising inequities in the distribution of water and the associated gains.

The rise of private tubewells, however, does not have to lead to inequities if groundwater markets emerge and function well. While little has been written on groundwater markets in China, outside of China groundwater markets have long existed and recently have attracted the attention of researchers. For example, markets in groundwater have been found in many parts of South Asia. In 1975 a World Bank study in Pakistan reported nearly 30% of tubewell owners sold part of their pumpage to other farmers (Shah, 2000). In the early 1990s Pant (1991) found that 86% of the households in eastern Uttar Pradesh purchased water for irrigation; in central and western Uttar Pradesh 65% of farm households purchased water. In the 1990s, studies in Pakistan by Strosser and Meinzen-Dick (1994) and Meinzen-Dick (1996) have found groundwater markets pervasive. While the analysis of many issues are complicated and the findings of many studies controversial (meaning more study is needed on the management of groundwater markets), the South Asian experience has shown that groundwater markets in many places have provided opportunities for the farmers without tubewells to get access to water (Shah, 1993; Strosser and Meinzen-Dick, 1994; Mukherji, 2004; Sharma and Sharma, 2004).

Despite the observations by field workers regarding the similarities between the rise of groundwater markets in China and those of South Asia, almost no empirical studies have been done on the development of China's groundwater markets. In fact, searching of the literature has found that there is almost no reference in any work on groundwater markets in either Chinese or English. Despite the absence of research, policy makers and scholars have begun to raise a series of questions. How prevalent are groundwater markets in northern China? More specifically, what proportion of the tubewells participates in selling water? How much of their water do they sell? What are the characteristics of groundwater markets in northern China? Who are the buyers? Finally, why do we observe water markets in some villages but not in others? In other words, what are the determinants of groundwater markets in northern China and what are their effects on the equity of access to water?

The overall goal of this chapter is to answer these questions in order to develop a better understanding of the development of groundwater markets in northern China. To do so, we focus on getting the data right and providing a profile of groundwater markets and their determinants in northern China. To meet this goal, four specific objectives are addressed. First, we describe the evolution of groundwater markets in northern China. Second, we explore their characteristics. In doing so, China's groundwater markets are compared to those that have emerged in South Asia. Third, we measure the determinants of groundwater markets in northern China and try to understand why they have emerged in some villages but not in others. Finally, we seek to understand what types of farmers are selling water and what types of farmers are buying water in an effort to understand the equity implications of emerging groundwater markets.

Due to limitations on the nature of the data and the length of this chapter, we do not examine the impacts of the rise of groundwater markets. While it is extremely important to measure the effects of groundwater markets on cropping patterns, crop productivity, and the water table, it is beyond the scope of this chapter. In addition, in the past there has been other work done on groundwater markets and a paper by Mukherji (2004) provides an excellent review of the literature that does not need replication here. According to Mukherji, there are a number of papers that have looked at the groundwater markets from an institutional point of view (Palmer-Jones, 1994; Kajisa, 1999; Dubash, 2000, 2002). In this paper, however, we take a neoclassical economist's approach, as have a number of other economists (eg, Shah, 1985, 1989, 1991, 1993).

GROUNDWATER MARKETS WITH CHINESE CHARACTERISTICS

Groundwater markets are defined as localized, community-level arrangements through which owners of tubewells sell pump irrigation services to other farmers of the village and neighboring villages (ie, they sell water to other farmers from their wells for use on crops). This chapter is only going to examine "private" water markets; or, the nature of groundwater markets that are being driven by individuals and groups of individuals that sink tubewells. In adopting such a definition, we are assuming that when village leaders (the collective) provide water to villagers, it is being done under nonmarket conditions and is not a groundwater market transaction.

This section measures the degree of the development of groundwater markets in terms of both breadth and depth as well as describes their characteristics. According to Shah's (2000) definition, "*Breadth* can be conceptually defined as the proportion of the farming community that is participating in water trade, as buyers or sellers or both; *depth* can be conceptually defined as the quantitative significance of water transactions in the economies of individual farm households of a community." In the rest of the chapter the breadth of groundwater markets is measured by two indicators. One indicator is the share of villages that have any degree of groundwater market activity. The second indicator is the share of tubewells from which the tubewell owner is selling water to water-buying households. Depth is measured by the "share of the volume of water" sold to water-buying households that is pumped from tubewells that are selling water.

When using breadth indicators, groundwater markets have developed quickly in northern China. According to the North China Water Resource Survey (NCWRS), in 1995 groundwater markets had emerged in only 9% of the sample villages (Table 6.1, column 1, row 1). However, by 2004 there were groundwater markets in 44% of the villages (column 2). During the same period the share of tubewells from which owners sold water also increased. In 1995 water was sold from only 5% of tubewells; by 2004, however, this number increased to 18% (row 2). In addition, when using indicators of the depth of groundwater markets, the CWIM survey shows that by 2004, groundwater market activities were dominating the tubewell pumping activities of those farmers-cum-tubewell owners that were selling water (row 3). We also note that the share of water sold reduced from 80% in 1995 to 77% in 2004 (Table 6.1, row 3). This is either due to the increase in the number of wells selling water or a reduction in demand of water with the increase in number of tubewells in the region. According to our data, either of these interpretations is plausible. Between 1995 and 2004, both the number of wells selling water and the total number of wells has increased. In the 68 sample villages from the NCWRS, the number of wells selling water increased from 75 in 1995 to 342 in 2004; at the same time, the total number of wells also increased from 1472 to 1967.

Although there has been a lot more attention given to the study of groundwater markets in South Asia (because groundwater markets have traditionally been quite widespread), the data show that China is catching up quickly. For example, a number of studies suggest that groundwater markets have become quite pervasive in Pakistan (Strosser and Meinzen-Dick, 1994; Meinzen-Dick, 1996). These studies indicate that 30% to 60% of tubewell owners in Pakistan sell water. In India and other

Table 6.1 Development of Breadth and Depth of Groundwater Markets in China, 1995, 2004

	1995	2004
Breadth		
Share of villages having groundwater markets (%)	9	44
Share of tubewells selling water (%)	5	18
Depth		
Share of water sold (%) [conditional on tubewell owner selling water]	80	77

Data source: Data in row 1 and row 2 are from authors' survey of 68 randomly selected villages in 4 provinces (Hebei, Henan, Shanxi, and Shaanxi) of the North China Water Resource Survey (NCWRS); Data in row 3 are from authors' survey of 50 randomly selected tubewells in 2 provinces (Hebei and Henan) of China Water Institutions and Management (CWIM) survey. We do not use data from all of the sample villages of the two surveys since the information in the table is conditioned on villages that use groundwater to irrigate and that have private tubewells.

neighboring countries, Shah (2000) shows that when tubewell owners sell water, they sell from 40% to 70% of the volume of water they pump. Hence, while these numbers may not be exactly comparable, assuming that the estimates are correct (ie, water is being sold from 18% of tubewells in northern China; and when water-selling households sell water, they are selling 77% of the water from their tubewells), the development of groundwater markets in northern China is approaching levels that are being observed elsewhere in the world.

CHARACTERISTICS OF GROUNDWATER MARKETS IN NORTHERN CHINA

Although groundwater markets in northern China have evolved, there are at least three characteristics which appear to be shared by groundwater markets in northern China and South Asia. First, almost all groundwater markets in both places are informal. According to Shah (1993), a water market is informal when transactions between water-selling and water-buying households are done without legal sanction. In other words, farmers buy and sell water without a contract and their oral commitments cannot be adjudicated in a court of law. According to the data, there were zero written contracts covering water sales among agents in northern China; the same is true according to the literature from India.

Second, groundwater markets in both northern China and South Asia are almost always localized. According to Shah (1993), the localized nature of water markets is almost universal. In the survey data in China, water transactions also are mostly limited to water-selling and water-buying households that live and work in the same village. In fact, only 6% of water-selling tubewell owners (and a smaller share of the volume of water that they pump) sell water to farmers in other villages.

Third, groundwater markets in both northern China and South Asia are largely unregulated. In Shah (1993), unregulated means the government exercises no direct influence on the functioning of the market. Shah (1993) finds little evidence of any intervention by any level of government in India. Based on the NCWRS survey data, there were only formal regulations on the books about any aspect of groundwater markets in less than 25% of villages (eg, a price ceiling on the amount that a water-selling household can charge). Although somewhat higher than the case of India, during our field work and during interviews with tubewell owners, enumerators almost have never encountered a case in which the tubewell owner was constrained by a government regulation; village leaders and tubewell operators almost never were aware that there was any attempt by upper level officials to influence the functioning of water selling and buying. In the villages that claimed there was a formal regulated ceiling on the price of water, typically, the regulation was said to have been set by the price bureau. When such an announcement was made, in many cases the regulation was mainly targeted at all water users in the county (including, often especially, at water uses in industry and municipal water districts). Because of this, in some villages, even though leaders said that there was a regulation "on the books," it did not always mean that the rule was "implemented" in practice in the village.

While there are a number of similarities, it appears as if the different environments within which groundwater markets have evolved in northern China and South Asia have produced several differences in the nature of groundwater markets. First, although field work in many villages has found evidence of impersonal markets, groundwater markets in parts of South Asia appear not to be fully impersonalized. For example, Shah (1993) states that transactions between water-buying and water-selling households are impersonal in many cases. In India this means that water-selling households in many villages do not distinguish among various buyers in terms of price at which they sell water and the quality of service provided. However, other studies find that groundwater markets can be personalized. For example, studies in Pakistan (Jacoby et al., 2004), Bihar, India (Wood, 1995), and other regions in South Asia (Ballabh et al., 2002; Pant, 2003) report that sellers charge some buyers one price and other buyers another price. In other words, after observing what "kind of a person" a potential buyer is, the price of water is set.

In contrast, in the case of northern China, groundwater markets are almost fully impersonal. Based on our survey of 30 village leaders, we found that within villages, only 7% of water selling tubewell owners charge different prices for different types of buyers. In addition, in our survey of the tubewell owners, not one reported that they charged different prices for different types of buyers.

Second, the patterns of payment within groundwater markets in China and South Asia sometimes are different. In South Asia, water-buying households often provide labor or offer a share of their crop's harvest in exchange for water (Shah, 2000). In northern China, however, water sold in groundwater markets is almost always paid for on a cash basis. Such differences may be primarily related to the different land tenure arrangements that dominate both countries. For example, in China the ownership of cultivated land belongs to village collectives. Since the early 1980s, with the implementation of household production responsibility in rural China, cultivated land was allocated relatively evenly to each farmer within a village. As such, it implies that each farmer has land use rights, though they have no land ownership. However, in South Asia, land ownership is private and many farmers have no land to be used for their production. These landless farmers have to rent land from land owners or provide their labor to land owners. Therefore, in the same way that they pay their rent with a share of the labor (ie, through a sharecropping contract), water-buying households in South Asia also often provide labor or a share of their crop's harvest in exchange for water. In contrast, water-buying households in China do not exchange their labor for water in such a way and in all cases in our sample (100%) pay cash for water.

Another important difference between China and South Asia is the way in which electricity is priced. This, too, may have a major impact on the way groundwater markets work. For instance, in many Indian states, electricity is priced on a flat rate basis and this does not always reflect the scarcity value of water. In China, however, electricity meters are in place, electricity prices are mostly at market rates, and the cost of pumping (and consequently price of water) reflects mostly the scarcity value of water (because it reflects the depth from which the water is pumped).

Also, in India rural electrification is poor and, hence, many farmers depend on diesel-driven pumps, and this may create a different configuration of groundwater markets than would appear when there are electric pumps.

There may be one final difference between China and India. Whereas there are many institutional aspects of groundwater markets that differ between villages that are relatively poor and villages that are relatively rich in India, the same is not true in China. In fact, according to our data, most institutional features do not differ in a statistically significant way between villages that are relatively well off in China and those that are relatively poor. We show this by dividing the sample into two groups based on the village's per capita income. Our data show that income differs dramatically between rich and poor villages; average income in the richer villages was 250% higher than that in the poorer villages (Table 6.2, row 1). Despite the large

Table 6.2 Characteristics of Groundwater Markets in Poor and Rich Villages in China, 2004

	Poor Villages[a]	Rich Villages[a]
Village characteristics		
Per capita income (dollars)	121.89	309.67
Groundwater markets		
Level of Development (breadth/depth)		
Share of private tubewells selling water (%)	44	37
Share of water sold (%)	74	91
Policy interventions		
Share of villages having subsidy (%)	5	4
Share of villages receiving bank loans (%)	7	7
Share of villages requiring well-drilling permits (%)	66	70
Nature of Groundwater Markets		
Informal (share of villages not having contract for selling water, %)	100	100
Localized (share of water-selling tubewells selling water to farmers in the villages, %)	87	100
Unregulated (share of villages not having price ceilings, %)	80	74
Impersonal (share of water-selling tubewells owners that charge same prices for all types of buyers, %)	95	100
Patterns of payment (share of water-selling tubewells that pay in cash for selling water, %)	100	100

[a] *Poor villages are those villages in which per capita income is less than $196.50 per capita; rich villages are those villages in which per capita income in more than $196.50 per capita.*

Data source: Authors' survey of 68 randomly selected villages in 4 provinces (Hebei, Henan, Shanxi, and Shaanxi) of the NCWRS and 13 randomly selected tubewells selling water in Hebei province of CWIM. We do not use data from all of the sample villages of the two surveys since the information in the table is conditioned on villages that use groundwater to irrigate and that have private tubewells.

differences in income, there is almost no difference (statistically) in the development of groundwater markets, the nature of policy interventions, or the characteristics of groundwater markets between these two types of villages. For example, in the rich villages 37% of tubewells sell water. This share is almost same as that in the poor villages (in which 44% of tubewells sell water; row 2). Therefore, there is evidence that the development of groundwater markets in both rich and poor villages is more or less the same. In addition, the share of villages having bank loans or the share of villages that are affected by other policy interventions is also the same in both poor and rich villages (rows 4–6). Finally, in both rich and poor villages, groundwater markets are unregulated, informal, localized, and impersonal; payment for water (or the way in which transactions are settled) is the same (rows 7–11).

GROUNDWATER MARKETS, TUBEWELL OWNERSHIP, AND RESOURCE SCARCITY

In the small number of papers internationally that have sought to understand the determinants of groundwater markets, a number of factors arise consistently and can be used as a basis for generating hypotheses that we can test using the survey data from northern China. For example, Shah (1993) descriptively shows that the availability of water resources, the scale of irrigation technology, and the extent of land fragmentation are correlated with the rise of groundwater markets. Strosser and Meinzen-Dick (1994) set up a theoretical framework that posits (among other factors) the depth of the groundwater table and the population density of a community are important factors affecting groundwater markets. Shah (1993) and many others, such as Mukherji (2004), also indicate the importance of policy intervention in promoting or constraining the development of groundwater markets. Therefore, we would expect that if there were regulations, they would constrain the expansion of the number of tubewells. In the case of other policy intervention variables (eg, the existence of bank loan and grant programs targeting those that invest in pumps and tubewells), when there are government grant and loan programs one would expect more tubewells and, as such, greater groundwater market activity. In other words, the observations of the researchers working on South Asia's groundwater markets suggest that groundwater markets are arising at least in part in response to the nature of the technology needed for sinking a well, the degree of resource scarcity—both for land and water—and policy. If these observations and conjectures are picking up more general relationships, then based on the relationships, a set of testable hypotheses can be generated: When the cost of sinking a well rises (either due to the falling groundwater table or the relative competitiveness of larger tubewells/pump sets) or when the attractiveness of sinking a well at a given cost declines (due to the fact that a farmer may have an increasingly small parcel of land that is not able to utilize the entire command area of a tubewell investment, or due to a policy intervention by the government), groundwater markets can be expected to emerge.

According to descriptive statistics based on the survey data from northern China, empirical evidence for the hypotheses is present. For example, the data indicate that the development of groundwater markets maybe correlated with water resource scarcity (Table 6.3). When the water table falls in the NCWRS sample villages over time (from 28 to 38 m; column 3), the share of tubewells from which water is being sold is higher (column 1). When dividing the villages in the sample by the share of tubewells from which water is sold into four groups, there is a positive correlation between the amount of groundwater market activity and level of the groundwater table (columns 1 and 3, rows 3–6). Likewise, when dividing the tubewells in the sample by the share of water sold in three groups, there is a positive relationship between the amount of groundwater market activity and level of the groundwater table (Table 6.4, columns 1 and 4). One explanation of these trends is that when the groundwater table is lower, the cost of sinking a tubewell is higher, which could keep some farmers from investing in their own tubewells even though they have a high demand for irrigation services. Alternatively (although mainly in a relative sense), it could be that the lower the groundwater table, the larger the size of the optimal tubewell/pump set. In villages with larger tubewells/pump sets, other factors (including land size) held constant, there is less of a need for all farmers to have their own tubewells. We want to emphasize that in this part of this section of the chapter we are only examining correlations with our descriptive data. We are

Table 6.3 Relationship Between Development (Breadth) of Groundwater Markets and Tubewell Ownership/Resource Endowment in China, 1995, 2004

		Tubewell Ownership	Water Scarcity	Land Scarcity
	Share of Tubewells Selling Water (%)	**Share of Private Tubewells (%)**	**Groundwater Table (m)**	**Per Capita Arable Land (ha)**
Grouped by Year[a]				
1995	5	50	28	0.12
2004	18	81	38	0.10
Grouped by Share of Tubewells Selling Water[b]				
0	0	68	28	0.11
0–30	12	46	45	0.12
30–90	57	70	48	0.11
90–100	100	100	48	0.09

[a] *The number of observations used for each row in rows 1 and 2 is 68.*

[b] *The number of observations used for each row in rows 3–6 is n = 100 (row 3); n = 10 (row 4); n = 8 (row 5); and n = 18 (row 6). Data are averages for two sample years.*

Data source: Authors' survey in 68 randomly selected villages in 4 provinces (Hebei, Henan, Shanxi, and Shaanxi) of NCWRS. We do not use data from all of the sample villages of the NCWRS survey since the information in the table is conditioned on villages that use groundwater to irrigate and that have private tubewells.

Table 6.4 Relationship Between Development (Depth) of Groundwater Markets and Tubewell Ownership/Resource Endowment in China, 2001

		Tubewell Ownership		Water Scarcity	Land Scarcity
	Share of Water Sold (%)	Share of Individual Tubewells (%)	Share of Shareholding Tubewells (%)	Groundwater Table in 1995 (m)	Per Capita Arable Land (ha)
Grouped by Share of Water Sold[a]					
0	0	19	81	13.6	0.120
0–90	48	44	56	11.1	0.091
90–100	97	100	0	17.6	0.089

[a] *The number of observations used for each row in rows 1–3 is n = 32 (row 1); n = 9 (row 2); and n = 9 (row 3).*

Data source: Authors' survey of 50 randomly selected tubewells in 2 provinces (Hebei and Henan) of CWIM. We do not use data from all of the sample villages of the CWIM survey since the information in the table is conditioned on villages that use groundwater to irrigate and that have private tubewells.

not suggesting causality. The most that we can say with descriptive statistics is that we are showing data trends that are consistent with the hypotheses. We include a multivariate analysis in the next section.

Likewise, the data provide similar support for the hypotheses when looking at the relationship between groundwater activity and land scarcity (Table 6.3, columns 1 and 4). Between 1995 and 2004 the average size of land per capita for the sample villages fell from 0.12 to 0.10 ha (rows 1 and 2). Coupled with the observed rise in the share of tubewells that are selling water, the descriptive data are consistent with the idea that when farm size gets smaller, households have less of a need to invest in their own tubewells. This could be one reason behind the rise in groundwater markets. The same trends appear when grouping observations either by the share of tubewells from which water is sold (rows 3–6) or by share of water sold (Table 6.4, columns 1 and 5).

Trips to the field and discussions with local officials and water users, as well as the survey data, also raise the prospect of one factor unique to China that may be behind the emergence of groundwater markets in the sample. According to the data, the private ownership of tubewells is correlated strongly with the development of groundwater markets. Groundwater market activity is higher as the share of private wells has risen over time (from 50% to 81%; Table 6.3, columns 1 and 2, rows 1 and 2). Likewise, when observations are grouped according to the water sales activity of the tubewells, the share of private wells also rises sharply (from 68% to 100%; rows 3–6). Specifically, when the share of individual tubewells increases, the share of water sold is higher (Table 6.4, columns 1 and 2). Interviews with village leaders and tubewell owners and managers revealed that the main force driving this correlation appears to be the incentives that private tubewell owners face, which encourage them to produce earnings from their tubewell investments. Hence, in studies of China, an additional testable hypothesis arises from descriptive statistics and observations in the field: Groundwater market activities rise as the share of private wells in a village increases.

ESTIMATION RESULTS OF ECONOMETRIC MODELS

This section seeks to test more rigorously the determinants of groundwater markets. The chapter is interested in identifying the factors that explain why some villages have groundwater markets and others do not; this analysis is needed because the findings will help us better understand the forces that are creating the swell in groundwater market activity. This is important since it may help us predict as China's villages confront the rising economic and environmental pressures of the nation's development process (eg, forces such as the steady shift toward private ownership of productive assets and increasing resource scarcity), whether or not institutions will emerge that will allow farm households to gain access to water, one of the most critical resources that they need for production. To analyze the determinants of groundwater markets in terms of breadth and depth of groundwater markets, we have specified two econometric models. The descriptions of these two models can be found in Methodological Appendix to Chapter 6.

DETERMINANTS OF THE BREADTH OF GROUNDWATER MARKETS

In estimating equations (1) with the survey data, the econometric estimation performs well (Table 6.5, column 1). Most coefficients of the control variables have the expected signs and a number of the coefficients are statistically significant. For example, the coefficient of well-drilling permit regulation variable is negative and statistically significant (column 1, row 8). Even if we drop these policy intervention variables (column 2) or if we drop the conservation technology variable (column 3), econometric estimation still performs well and there are few differences among the estimation results.

More importantly, when examining the variables of interest, research results show that the change of tubewell ownership from collective to noncollective induces the development of groundwater markets. The coefficient on the share of noncollective tubewells variable is positive and significant (Table 6.5, row 1, columns 1–4). All other things held constant, when the share of the noncollective tubewells in a village increases, the share of tubewells selling water increases. If the share of noncollective tubewells increases by 10%, the share of tubewells selling water will increase by nearly 3% (column 4). Although we cannot infer causality, this result shows the correlation between privatization and the rise of groundwater markets. One explanation of this is that in villages with more privately owned tubewells, there is more of an incentive to sell water. Another explanation of this relationship is that in villages with a higher percent of private tubewells, it may

Table 6.5 Regression Analysis of the Determinants of Development (Breadth and Depth) of Groundwater Markets in China (Using Tobit Estimator)

	Dependent Variable: Share of Tubewells Selling Water				Share of Water Sold
	(1)	(2)	(3)	(4)	
Tubewell ownership					
Share of private tubewells	0.183 (3.86)***	0.286 (7.70)***	0.180 (3.83)***	0.286 (7.40)***	
Dummy of individual tubewell					0.389 (4.33)***
Water and land scarcity					
Log of groundwater table	0.003 (3.82)***	0.006 (5.06)***	0.003 (3.81)***	0.006 (4.96)***	
Log of groundwater table in 1995					0.008 (2.01)**
Log of per capita cultivated land	−0.900 (2.39)**	−1.036 (3.21)***	−0.909 (2.40)**	−1.036 (3.10)***	−4.745 (3.50)***
Policy interventions					
Dummy of fiscal subsidies for tubewell investment	0.051 (0.46)		0.041 (0.38)		−0.121 −1.58
Dummy of bank loans for tubewell investment	0.065 (0.59)		0.066 (0.60)		0.484 (3.02)***
Dummy of well-drilling permission regulation	0.116 (3.09)***		0.117 (3.08)***		0.045 −0.46
Other control variables					
Dummy of adopting water delivery pipes	−0.025 (0.64)	0.008 (0.23)			−0.093 −0.94
Per capita net income of farmers	−0.000 (0.18)	−0.000 (0.88)	−0.000 (0.24)	−0.000 (0.85)	0.000 (1.94)*
Constant	−4.257 (3.68)***	−3.853 (4.74)***	−4.204 (3.66)***	−3.918 (4.76)***	−2.943 (3.34)***
Observations	136	136	136	136	50
Chi-square	35.19	96.41	35.30	94.29	46.37

Coefficients are marginal effects; absolute value of z statistics in parentheses. significant at 10%; ** significant at 5%; *** significant at 1%.*

Data source: Data in the model "Share of tubewells selling water" come from authors' survey in 68 randomly selected villages in 4 provinces (Hebei, Henan, Shanxi and Shaanxi) in 2 years (1995 and 2004) of CNWRS. Data in the model "Share of water sold" come from authors' survey in 50 randomly selected tubewells in 2 provinces (Hebei and Henan) of CWIM. We do not use data from all of the sample villages of the two surveys since the information in the table is conditioned on villages that use groundwater to irrigate and that have private tubewells.

be that there is less service provided by the operators of the collective tubewells (for any number of reasons). Therefore, in these villages, for farmers to gain access to water, it would be necessary for farmers to access water from sales from private tubewells.

Resource scarcity is also associated with the emergence of groundwater markets. Although it could have been that deeper water tables mean higher water prices and less demand, in fact, the coefficient on the depth to groundwater table is positive and significant (Table 6.5, row 3, columns 1–4). When the groundwater table drops by 1 m, the share of tubewells selling water will increase by about 1% (column 4). Hence, the alternative interpretation is consistent with the findings: in areas in which the groundwater table is deep, farmers demand for water from groundwater markets is higher (relative to providing water from one's own well). In simplest terms, if these results are indicative of underlying causal relations, the findings are evidence of the hypothesis that in villages with scarce water resources, groundwater markets develop more quickly.

Research results also show that land pressure has increased the breadth of groundwater markets. The coefficient on the per capita arable land variable is negative and statistically significant (Table 6.5, row 5, columns 1–4). If per capita cultivated land in the villages falls by 0.1 ha, the share of tubewells selling water will increase by 10% (column 4). In other words, the results imply that with the decrease of per capita land resources, the share of tubewells selling water has increased. In order to understand the influence of household size on the breadth of groundwater markets, the cultivated land per capita also has been replaced by the cultivated land per household (farm size) (Appendix Table, column 1, row 3). The results of the new analysis show that the coefficient of this variable also is significant in the regression analysis (as was the original coefficient) of determinants of breadth of groundwater markets (ie, the equation that used "share of tubewells selling water" as the dependent variable; Appendix Table, column 1, row 3). These results imply that when the average holding of land in a village is small, there is more of a tendency for a village's tubewell owners to sell water. What are the implications of these results? They do not mean necessarily that only small households are buying water. In China, one of the effects of the nature of the nation's initial decollectivization movement was that there was not much difference in the size of the land that was allocated to farmers within the village. Therefore, what we are seeing are inter-village differences. As a result, this means that it is in villages that have mostly small households that have more sales as opposed to villages with mostly large households. This feature of China makes it different from other countries, especially those in South Asia.

DETERMINANTS OF THE DEPTH OF GROUNDWATER MARKETS

The econometric estimation also performs well when estimating the depth of groundwater markets (Table 6.5, column 5). The Chi-square is 46, higher than that above explaining the breadth of groundwater markets (column 1, row 13).

Similarly, most of the coefficients of the control variables have the expected signs and a number of the coefficients are statistically significant. For example, the coefficient on the variable of farmer per capita net income is significant (row 10). This means that farmers in villages with higher per capita income sell more.

In addition, similar to the regression results on the determinants of the development of the breadth of groundwater markets, the development of the depth of groundwater markets is significantly associated with tubewell ownership and water and land scarcity. For example, the coefficient on the dummy variable for individual tubewell ownership is positive and significant (Table 6.5, row 2). This means that compared with shareholding tubewells, individual tubewells sell a higher share of their water. In addition, the coefficient on the groundwater table is also positive and significant (Table 6.5, row 4). Hence, in areas in which the groundwater table is deep, farmers demand for water from groundwater markets is higher. In simplest terms, if these results are also indicative of underlying causal relations, the findings are evidence of the hypothesis that in villages with scarce water resources, groundwater markets develop more quickly.

Research results also show that land pressure has intensified the depth of groundwater markets. The coefficient on the per capita arable land variables is significant (Table 6.5, row 5). Therefore, it appears that if agricultural land is scarcer (making it less desirable for an individual farmer to sink his/her own tubewell), the average tubewell operator sells a greater share of water from his/her tubewell.

DO GROUNDWATER MARKETS HELP THE POOR?

Based on the analysis in the previous section, the research results show that groundwater markets in northern China have developed in terms of both their breadth and depth. Household data further indicate the importance of groundwater markets for irrigation in northern China. Results show that more than 70% of the sample households depend on groundwater for irrigation (Table 6.6, rows 1 and 2). However,

Table 6.6 Participation in Groundwater Markets by Farm Households in China, 2004

	2004
Total households	173
Number of household using groundwater	128
Share of households having tubewells themselves to irrigate (%)	34
Share of households getting groundwater irrigation through groundwater markets (%)	20

Data source: Authors' survey data from 46 randomly selected villages in 2 provinces (Hebei and Henan) of CWIM. We do not use data from all of the sample villages of the CWIM survey since the information in the table is conditioned on villages that use groundwater to irrigate and that have private tubewells.

among all of the households using groundwater, only 34% of them own tubewells. In contrast, 20% of households must depend on groundwater markets to gain access to water for irrigation (rows 3 and 4). Although a fairly large number of households still access groundwater from collective tubewells, as the strength of the collective diminishes, it is clear that the role of groundwater markets will become increasingly important in the coming years.

As groundwater markets become increasingly important, it is important to understand whether they are helping or hurting the poor and/or increasing or reducing inequality in rural China. If groundwater markets emerge and function well, the rise of private tubewells does not have to lead to inequities (although it may—the final answer is an empirical one). Elsewhere in the world, research has shown that groundwater markets can be equity enhancing. For example, in Pakistan, Meinzen-Dick (1996) find that groundwater markets improve equity of groundwater use by making water available to small landowners or tenants and younger households—those farmers who are least likely to own tubewells.

Our field surveys in northern China provide similar evidence as that in Pakistan and other countries. According to the survey data, groundwater markets increase opportunities for poorer farmers to access groundwater, thus reducing potential income gaps. Specifically, households in the sample that buy water from groundwater markets are poorer than water-selling households. For example, the per capita income from cropping of water-buying households is only 61% of that of water-selling households; per capita total income of water-buying households is also lower than that of water-selling households (Table 6.7, rows 1 and 2). Such results may imply that, although poor farmers do not have enough money to sink tubewells, they can buy water from markets. If this is the case, groundwater markets almost certainly help the poor.

The data also indicate that groundwater markets benefit older farmers with small land holdings and less education. Households that buy water have smaller holdings of cultivate land than water-selling households. For example, the per capita land area

Table 6.7 Characteristics of Water-Selling and Water-Buying Households in China, 2004

	Water-Selling Households	Water-Buying Households
Per capita cropping income (dollars)	194.56	119.47
Per capita total income (dollars)	349.58	318.50
Per capita cultivated land area (ha)	0.15	0.13
Education level of household head (year)	6.3	5.5
Age of household head (year)	47.6	50.0

The sample sizes of water-selling households and water-buying households are 36 and 25.
Data source: Authors' survey in 46 randomly selected villages in 2 provinces (Hebei and Henan) of CWIM. We do not use data from all of the sample villages of the CWIM survey since the information in the table is conditioned on villages that use groundwater to irrigate and have private tubewells.

of water-buying households is 0.13 ha while the land area of water-selling households is slightly larger, 0.15 ha (Table 6.7, row 3). Such a finding implies that households with small holdings of land, who are unable or choose not to sink a tubewell (and cannot utilize the entire command area of a tubewell investment), are able to buy water through groundwater markets. In addition, research finds that less educated and older farmers depend more on groundwater markets to gain access to groundwater (rows 4 and 5). We also conducted statistical tests (*t*-tests) to understand whether the differences between water-selling and water-buying households are significant statistically. Results show that differences between water-buying and water-selling household are not significant. The *t*-statistics for each variable are: per capita land area is 1.18; education level of the household head is 0.52; and the age of the household head is 1.17. However, these results do not prove causality. If these characteristics influence the decision of buying or selling water, we need to conduct further multivariate analysis.

Whether or not groundwater markets benefit the poor may also be related to the structure of the markets (ie, whether they are monopolistic or competitive). The poor should benefit more when markets are competitive than when water buyers face a single seller. However, analysts do not always agree on how to measure market structure. In other research outside of China, researchers indicate that due to physical and topographical conditions, groundwater markets may be fragmented and could be monopolistic (Dhawan, 1988; Pant, 1991; Shah, 1993; Bagachi, 1995; Campbell, 1995; Kahnert and Levine, 1994; Jacoby et al., 2004). In other work, economists attempt to measure the degree of competition. For example, to do so, following the work of Lerner (1970) and Shah (1993) hypothesized that the ratio of water price to total variable cost can be used as a fairly good indicator of the level for monopoly profits (see Lerner, 1970 for the proof of this proposition). In applying this to the case of India, Shah and Ballabh (1997) found that the ratio was high, about 2.5–3.0 in Mazaffarpur. Based on this, Shah and Ballabh (1997) concluded that groundwater markets were not very competitive. Other researchers do not agree with the approach of Lerner (1970) and Shah (1993). For example, both Fujita and Hossain (1995) and Palmer-Jones (2001) found fairly high ratios (water price to total variable cost) in their studies (averages 2.6). However, unlike Shah and Ballabh (1997), they interpreted the results to mean that, far from being monopolistic, water markets were competitive; the high ratio merely reflects the entrepreneur's risk premium. In other words, based on their observations, water markets were competitive, and the authors explain the high ratio with an alternative explanation.

In order to examine the structure of groundwater markets in China, we follow the lead of other researchers from South Asia. First, following the approach of Shah (1993), we calculate the ratio of water price to total variable cost. According to our data, the ratio ranges from 1.2 to 3.3, with an average of 2.2. More than 70% of tubewells have the ratios which were lower than 2.5. This average level is lower than the findings of Shah and Ballabh (1997) in India (range is from 2.5 to 3.0) and Fujita and Hayami (1995)/Palmer-Jones (2001). Hence, if the low ratio of water price to total variable cost does, in fact, measure competition, from this angle there is evidence that groundwater markets in China are relatively competitive.

In addition, we also examine the data on water prices and compare within village averages of price variations with between village variations. Groundwater markets are localized and most transactions occur among farmers in the same villages. Therefore, if markets are competitive, we expect that most price variation should occur between villages, not from within villages. In fact, we find that the variation in the price of water is mainly due to regional differences among villages, not from within villages. For example, in one village the price of water from one tubewell is more than 3.4 times that from one tubewell in another village. However, within any of our villages, the highest difference among the price observations is only 50%. In 75% of villages, water price differences among tubewells selling water within villages are much smaller. According to our survey, in the villages which have collective tubewells and private tubewells selling water, the price of water from collective tubewells is almost the same as private groundwater markets. On average, the difference between the price of water being sold from collective tubewells and that being sold from private tubewells is less than 15%. In addition, we have not found a significant level of difference between the price of water being sold by private tubewell owners and that being sold by the owners of shareholding tubewells. The shares of total expenditures that are accounted for by water are also relatively homogeneous within villages. Whether a household purchased water from a tubewell operator or whether he/she supplied water from his/her own well, our results show that within the same village there are only narrow differences. These results are also consistent with the findings of other researchers that believe water markets in their study areas are competitive. For example, after controlling for the influence of other factors, Kajisa and Sakurai (2003) found that the variation of water price mainly comes from regional differences, which lead them to the conclusion that groundwater markets are not monopolistic.

Besides analyzing prices directly, we also examine other types of data to provide evidence that supports the finding of the nonmonopolistic nature of groundwater markets. First, we look at profits from selling water. When using our data, we are able to estimate both the fixed and variable costs that are associated with pumping and selling water. Accordingly, our results demonstrate that (even when we do not consider the value of family labor that is used to pump and sell water), profits are generally small.

Second, we look at the number of well operators that sell water and water delivery conditions. Shah (2000) suggests that when wells are sunk in a fairly dense manner, and when there are lined conveyance structures in a village, there is less of a probability that a single seller will have monopoly power and that the price of water will be relatively more competitive. Using this approach with our survey data, we find that in almost all villages, there are many tubewell operators selling water, not just one. On average, in each village, there are 18 tubewells (and 13 private ones) and more than 70% of the private tubewells sell water. Furthermore, the adoption rate of surface pipe (or hoses) in groundwater irrigation regions of northern China (ie, the use of efficiency-enhancing conveyance technologies) is common. Our survey found that more than 70% of tubewell owners use surface pipes to deliver water. The adoption of surface pipes greatly increased the ability of farmers to choose tubewells from which they want to buy water. Therefore, based on these analyses, it seems that groundwater markets in northern China are not monopolistic.

CONCLUSIONS AND POLICY IMPLICATIONS

This chapter has sought to understand the development of groundwater markets in northern China and examine the factors that determine the development of groundwater markets. Using our two data sets, research results provide strong evidence that groundwater markets in northern China have developed in terms of both their breadth (the share of villages in which there is groundwater market activity) and depth (the share of water which the average water-selling tubewell owner sells to others on a market basis). Interestingly, although fewer people have worked on groundwater markets in China compared to countries such as India and Pakistan, which have better documented groundwater markets, groundwater markets in northern China have emerged and are almost equal in pervasiveness.

Our findings also demonstrate that while there are many similarities between groundwater markets in China and other countries, there are also differences. Groundwater markets in northern China have many characteristics similar to those in South Asia; they are informal, localized, and mostly unregulated. At the same time, however, China's markets appear to be less personalized, and transactions in China are done more on a cash basis.

While the multivariate analysis is carried out mostly to understand descriptively the correlates of groundwater markets, there are a number of robust findings that support the hypotheses of interest. The form of ownership appears to be strongly correlated with the emergence of groundwater markets. Groundwater markets also appear in more villages, and tubewell owners sell a higher share of the water from their wells, when the groundwater table is deep (ie, water is scarce) and land is scarce. All of these findings suggest that when the factors that affect supply and demand for groundwater are present, there is a tendency for markets to emerge.

While much of the results are suggestive that groundwater markets are largely self-organizing and unregulated, there does appear to be a role for the state. The findings show that when the government makes it easier for individuals and shareholding groups to get access to capital, and when well owners are not subject to local regulations, there is a greater level of groundwater market activity. Since our research results also show that groundwater markets are not regressive and may, in some cases, be progressive, it is possible that government-sponsored investment and banking programs that allow individuals access to grants and loans to sink tubewells will further promote groundwater markets.

Finally, the research results indicate that groundwater markets in northern China help the poor. Households that buy water from groundwater markets are poorer than water-selling households. Such a finding implies that groundwater markets have provided greater access to groundwater to poor farmers and possibly help reduce income inequalities in rural China.

While it is beyond the scope of the chapter to measure the impact of the emergence of groundwater markets on the groundwater table, it is possible that groundwater markets, to the extent that they encourage the greater use of groundwater, could accelerate the fall of the water table. If this is so, the question remains whether groundwater markets should be encouraged. What are the options? If groundwater markets were suppressed, given our results, it could actually hurt the poor. So what should happen? To avoid hurting the environment (if it is being hurt), instead of directly trying to suppress groundwater markets, alternative policies that control the drawdown of the water table (eg, water pricing policies) should be promoted. In addition, efforts to allow groundwater markets to emerge could help spread the benefits that come with greater access to irrigation.

The analysis in this chapter also has implications that go beyond the water literature. First, the emergence of markets does not have to hurt the poor. In fact, it is possible that they are pro-poor. According to our analysis, in the case of China's groundwater markets, the poor have benefited. Poor households are involved in both the supply and demand sides of the market. This is somewhat different from the emerging groundwater markets in other parts of the world.

Why might this be the case? One thought is that markets work well and are competitive and expand the opportunities for access to resources for the poor (and rich) when they are composed of agents that all have access to a minimum amount of resources—both land and capital—and the market environment is relatively unregulated. In the case of China, all households in each village have land and the government has instituted programs that offer loans and grants to those that want to sink a well. In addition, given the initial investment in water by the government (in the pre- and early reform era), the incomes of most farmers were already high enough to allow some farmers to gain access to enough capital for investment (and so had access to sufficient liquidity) that they were able to afford to buy water when it was provided in a competitive market environment. Therefore, when groundwater markets emerge in such an environment, buyers and sellers can both benefit, and overall access to water can raise production and the welfare of all participants. Such a case, however, may not occur in places in which resources—such as land and capital—are less equitably distributed.

REFERENCES

Ballabh, V., Choudhary, K., Pandey, S., Mishra, S., 2002. Groundwater development and agricultural production: a comparative study of Eastern Uttar Pradesh, Bihar and West Bengal. In: Paper Submitted to IWMI-Tata Water Policy Programme, Anand, Gujarat.

Bagachi, K.S., 1995. Irrigation in India: History and Potentials of Social Management. Upalabdhi Trust for Development Initiatives, New Delhi, India.

Campbell, D., 1995. Design and operation of smallholder irrigation in South Asia. In: World Bank Technical Paper No. 256. The World Bank, Washington, DC.

Dhawan, B.D., 1988. Irrigation in India's Agricultural Development: Productivity, Stability, Equity. Sage Publications, New Delhi, India.

Dubash, N.K., 2000. Ecologically and socially embedded exchange: Gujarat model of water markets. Economic and Political Weekly 35 (16), 1376–1385.

Dubash, N.K., 2002. Tubewell Capitalism, Groundwater Development and Agrarian Change in Gujarat. Oxford University Press, New Delhi, India.

Fujita, K., Hossain, F., 1995. Role of the groundwater market in agricultural development and income distribution: a case study in a Northwest Bangladesh village. The Developing Economies 33 (4), 442–463.

Huang, Q., Rozelle, S., Wang, J., Huang, J., 2005. Irrigation, agricultural performance and poverty reduction in China. Food Policy 31 (1), 30–52.

Jacoby, H., Murgai, R., Rehman, S., 2004. Monopoly power and distribution in fragmented markets: the case of groundwater. Review of Economic Studies 71, 783–808.

Kahnert, F., Levine, G., 1994. Groundwater irrigation and the rural poor: options for development in the Gangetic Basin. In: A World Bank Symposium. The World Bank, Washington, DC.

Kajisa, K., 1999. Contract Theory and Its Application to Groundwater Markets in India (Ph.D. thesis). Michigan State University.

Kajisa, K., Sakurai, T., 2003. Determinants of groundwater price under bilateral bargaining with multiple modes of contracts: a case of Madhya Pradesh, India. Japanese Journal of Rural Economics 5, 1–11.

Lerner, A.P., 1970. Principles of Welfare Economics. Augustus M. Kelley, New York.

Liu, C., He, X., 1996. Water Issues in the 21th Century. Science Publishing House, Beijing, China.

Meinzen-Dick, R., 1996. Groundwater Markets in Pakistan: Participation and Productivity. Research Reports 105. International Food Policy Research Institute, Washington, DC.

Ministry of Water Resources and Nanjing Water Institute, 2004. Groundwater Exploitation and Utilization in the Early 21st Century. China Water Resources and Hydropower Publishing House, Beijing.

Ministry of Water Resources, 2002. China Water Resources Bulletin. Ministry of Water Resources, Beijing.

Ministry of Water Resources, 2003. China Water Resources Bulletin. Ministry of Water Resources, Beijing.

Mukherji, A., 2004. Groundwater markets in Ganga-Meghna-Brahmaputra Basin: theory and evidence. Economic and Political Weekly 31, 3514–3520.

Palmer-Jones, R., 1994. Groundwater markets in South Asia: a discussion of theory and evidence. In: Moench, M. (Ed.), Selling Water: Conceptual and Policy Debates over Groundwater Markets in India. VIKSAT-Pacific Institute-Natural Heritage Institute, Ahmedabad, Pakistan.

Palmer-Jones, R., 2001. Irrigation Service Markets in Bangladesh: Private Provision of Local Public Goods and Community Regulation. http://www.sasnet.lu.se/palmer_jones.pdf.

Pant, N., 1991. Groundwater issues in Eastern India. In: Meinzen-Dick, R., Svendsen, M. (Eds.), Future Directions for Indian Irrigation: Research and Policy Issues. International Food Policy Research Institute, Washington, DC.

Pant, N., 2003. Key trends in groundwater irrigation in the eastern and western regions of Uttar Pradesh. In: Paper Submitted to IWMI-Tata Water Policy Programme, Anand, Gujarat.

Shah, T., 2000. Groundwater markets and agricultural development: a South Asian overview. In: GWP, Pakistan Water Partnership, Proceedings of Regional Groundwater Management Seminar, October 9–11, 2000, Islambad, pp. 255–278.

Shah, T., 1993. Groundwater Markets and Irrigation Development: Political Economy and Practical Policy. Oxford University Press, Bombay, India.

Shah, T., 1991. Water markets and irrigation development in India. Indian Journal of Agricultural Economics 46 (3), 335–348.

Shah, T., January, 1989. Groundwater Grids in the Villages of Gujarat: Evolution, Structure, Working and Impacts. Wamana, pp. 14–29.

Shah, T., 1985. Transforming groundwater markets into powerful instrument of small farmer development. In: ODI Irrigation Management Network Paper No. 11d. Overseas Development Institute, London, UK.

Shah, T., Ballabh, V., 1997. Water markets in North Bihar: six village studies in Muzaffarpur District. Economic and Political Weekly 32 (52), A183–A190.

Shah, T., Giordano, M., Wang, J., July, 2004. Irrigation institutions in a dynamic economy: what is China doing different than India? Economic and Political Weekly 31, 3452–3461.

Sharma, P., Sharma, R., 2004. Groundwater markets across climatic zones: a comparative study of arid and semi-arid zones of Rajasthan. India Journal of Agricultural Economics 59 (1), 138–150.

Strosser, P., Meinzen-Dick, R., 1994. Groundwater markets in Pakistan: an analysis of selected issues. In: Moench, M. (Ed.), Selling Water: Conceptual and Policy Debates over Groundwater Markets in India. VIKSAT-Pacific Institute-Natural Heritage Institute, Ahmedabad, Pakistan.

Wang, J., Huang, J., Blanke, A., Huang, Q., Rozelle, S., 2007. The development, challenges and management of groundwater in rural China. In: Giordano, M., Villholth, K.G. (Eds.), The Agricultural Groundwater Revolution: Opportunities and Threats to Development. Cromwell Press, Trowbridge, UK, pp. 37–62.

Wang, J., Huang, J., Huang, Q., Rozelle, S., 2006. Privatization of tubewells in North China: determinants and impacts on irrigated area, productivity and the water table. Hydrogeology Journal 14, 275–285.

Wang, J., Huang, J., Rozelle, S., 2005. Evolution of tubewell ownership and production in the North China Plain. Australian Journal of Agricultural and Resource Economics 49 (2), 177–196.

Wood, G.D., 1995. Private provision after public neglect: opting out with pumpsets in North Bihar. In: Presented at the International Conference on Political Economy of Water in South Asia: Rural and Urban Action and Interaction, Held at Madras Institute of Development Studies, January 5–8.

CHAPTER

7

Impacts of Groundwater Markets on Agricultural Production in China

From the early 1980s to the end of the 2000s on the North China Plain, farmers have changed the way they access groundwater irrigation services. Before the 1980s, farmers only accessed irrigation services from collective tubewells, which were owned and operated by the leaders of each village (Wang et al., 2005). With the onset of reforms during the early 1980s, private individuals were encouraged to take responsibility for providing irrigation services, and with the declining groundwater tables in many areas, tubewell ownership has steadily shifted from collective to private. By 2004 the share of private tubewells reached 70%.

With the rise of private tubewells, there has been an emergence of irrigation service markets in many regions (Zhang et al., 2008). While almost no farmers purchased water from other farmers in the 1990s, by 2004 there were active groundwater markets in 44% of villages on the North China Plain. This development has enabled farmers in many villages to obtain water from other farmers through irrigation service markets.

The emergence of irrigation services enabled by the development of groundwater sources, especially in South Asia, has attracted the attention of researchers. Pant (1991) found that 86% of the households in eastern Uttar Pradesh purchased water for irrigation services. In central and western Uttar Pradesh, 65% of farm households purchased irrigation services. Shah et al. (2006) showed that the share of sample villages reporting activity in local groundwater irrigation service markets varies from 9% to 100%, and that more than 50% of villages in India have groundwater irrigation service markets. Strosser and Meinzen-Dick (1994) and Meinzen-Dick (1996) found that markets for irrigation services from groundwater are widespread in Pakistan.

Despite the interest in groundwater markets among researchers, there has been a noticeable absence of work measuring the impacts of this new phenomenon on agricultural production and incomes in rural communities. Many important research questions remain unanswered. For example, farmers who participate in markets for irrigation services might use less water per hectare if they must pay a higher price for water than those who have their own tubewells. Because they use less water, farmers who utilize groundwater markets might not irrigate sufficiently, and their yields might be lower than those obtained by farmers who own tubewells or obtain water from a collective well. Using multivariate analysis, Meinzen-Dick (1996)

Managing Water on China's Farms. http://dx.doi.org/10.1016/B978-0-12-805164-1.00007-5

found that groundwater irrigation service markets adversely affect the income of some users and negatively affect income distribution.

There might also be positive effects from the emergence of irrigation service markets. If there is no collective well, and if farmers cannot afford their own wells, irrigation service markets can provide farmers with access to groundwater, thus enabling them to achieve higher yields. Shah and Ballabh (1997) found that water buyers obtained higher yields than water sellers in all six villages in their sample in North Bihar in India.

However, much early work on irrigation service markets is focused only on South Asia. More importantly, many empirical analyses have not controlled the unobserved (or unmeasurable) factors that can affect both the increased availability of irrigation service markets and the outcomes, such as higher yields. Researchers studying the impacts of groundwater markets must isolate the net effects of the markets on the outcomes of farmers. The results observed in earlier studies might be attributed to the challenge of isolating causes and effects, particularly when considering differences in regions and study areas.

With increasing water scarcity and declining groundwater tables, understanding the impacts of the ways that farmers access irrigation services from groundwater has become an important policy issue in China. Therefore, the goal of this chapter is to measure the impacts of the changes in the ways that farmers access irrigation services on water use, yields, and income. We begin by assuming that farmers who irrigate obtain higher yields and earn higher incomes, which is consistent with results in Huang et al. (2006). We then ask: does it matter "how" people obtain access to groundwater?

ACCESS TO GROUNDWATER AND WATER USE, YIELDS, AND INCOME

Each of the three ways to access groundwater are important for farmers on the North China Plain. Of all households, 47% gained access to irrigation from collective tubewells in 2004 (Fig. 7.1). The number of households buying water through markets (23%), however, was less than the number of households gaining irrigation from their own tubewells (30%—Fig. 7.1).

IMPACT ON CROP WATER USE AND YIELDS

Groundwater markets on the North China Plain could reduce crop water use. For example, if farmers buy groundwater through markets to irrigate wheat, water use per hectare is 3241 m^3, 9% lower than those farmers who use water from their own tubewells (3571 m^3) (Fig. 7.2, panel A). In addition, the level of water use through markets was 11% lower than relying on collective tubewells.

These results are consistent with the findings when we only compare access to irrigation services within our sample villages. Within one village, when comparing

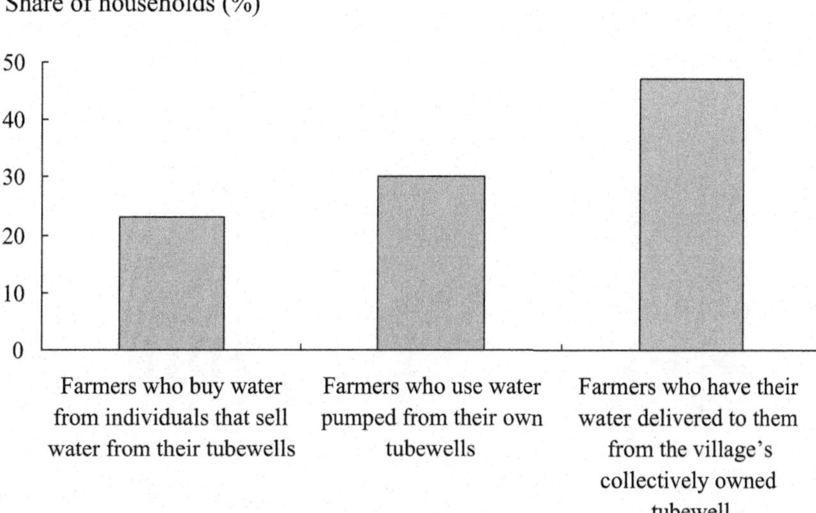

FIGURE 7.1

Alternative ways of gaining access to groundwater for sample farmers on the North China Plain, 2004.

Data source: Authors' survey.

the two types of farmers, those farmers relying on their own tubewells use 12% more water than farmers buying water from markets. In addition, those farmers using collective tubewells had 35% higher water usage than farmers who purchase water from groundwater markets.

Why do farmers who buy water use less water? One reason may be that farmers who purchase water pay more for their water. If so, they would have an incentive to reduce water use. Compared with farmers depending on their own tubewells or collective tubewells, farmers irrigating crops through irrigation service markets for groundwater have higher outlays for their water (Fig. 7.3).

Similar to the analysis above, the most accurate way to compare water prices among farmers who gain access to groundwater in different ways is to compare the water prices within a single village. Water buyers pay more than other farmers who do not depend on groundwater irrigation service markets. Because of this, it is reasonable to expect that farmers who purchased their water from groundwater markets will use water differently and produce different yields (and possibly earn different levels of income) than other farmers (Wang et al., 2007).

Because of this, crop yields fall slightly with the decrease in water use of farmers buying water from irrigation service markets for groundwater. For example, if farmers irrigated wheat with water purchased from groundwater markets, per hectare wheat yields are lower—though not significantly—than those farmers irrigating from their own tubewells (1%) or those depending on collective tubewells (8%)

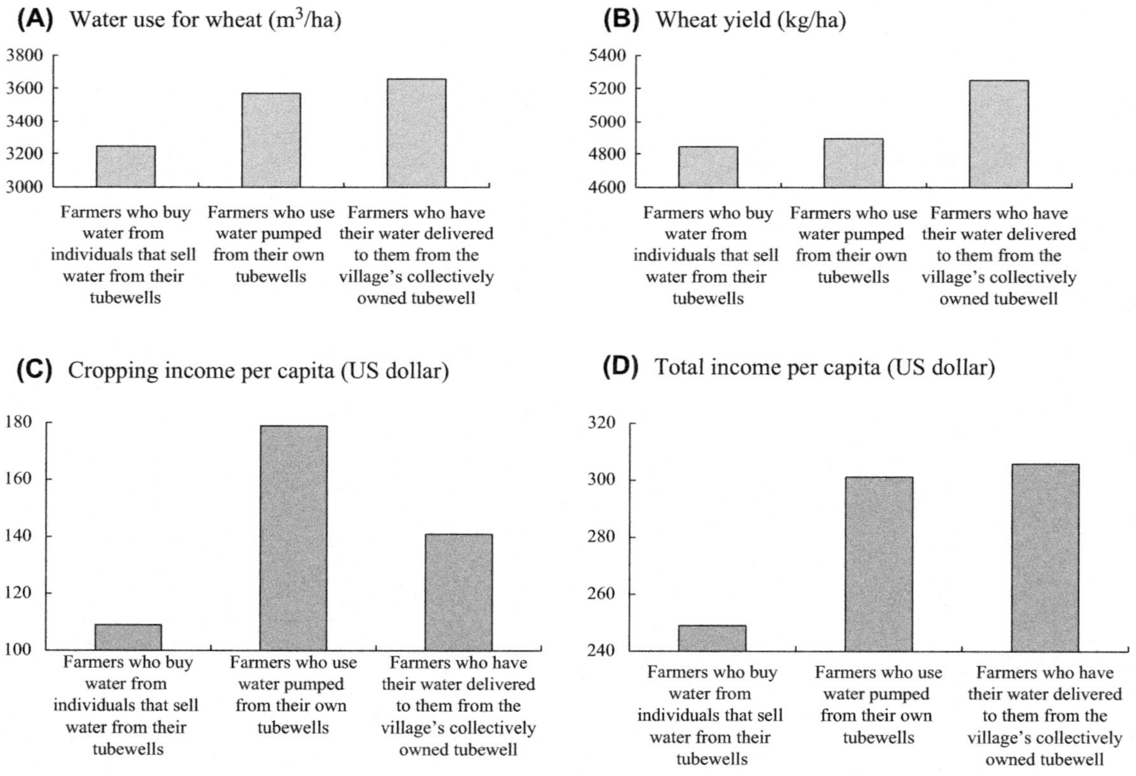

(A) Water use for wheat (m^3/ha)

3800
3600
3400
3200
3000

Farmers who buy water from individuals that sell water from their tubewells
Farmers who use water pumped from their own tubewells
Farmers who have their water delivered to them from the village's collectively owned tubewell

(B) Wheat yield (kg/ha)

5400
5200
5000
4800
4600

Farmers who buy water from individuals that sell water from their tubewells
Farmers who use water pumped from their own tubewells
Farmers who have their water delivered to them from the village's collectively owned tubewell

(C) Cropping income per capita (US dollar)

180
160
140
120
100

Farmers who buy water from individuals that sell water from their tubewells
Farmers who use water pumped from their own tubewells
Farmers who have their water delivered to them from the village's collectively owned tubewell

(D) Total income per capita (US dollar)

320
300
280
260
240

Farmers who buy water from individuals that sell water from their tubewells
Farmers who use water pumped from their own tubewells
Farmers who have their water delivered to them from the village's collectively owned tubewell

FIGURE 7.2

Relationship between alternative ways to gain access to groundwater and water use, yields, and income of sample farm households on the North China Plain. Panel A: relationship between irrigation service markets and crop water use. Panel B: relationship between irrigation service markets and crop yield. Panel C: relationship between irrigation service markets and cropping income. Panel D: relationship between irrigation service markets and total income.

Data source: Authors' survey.

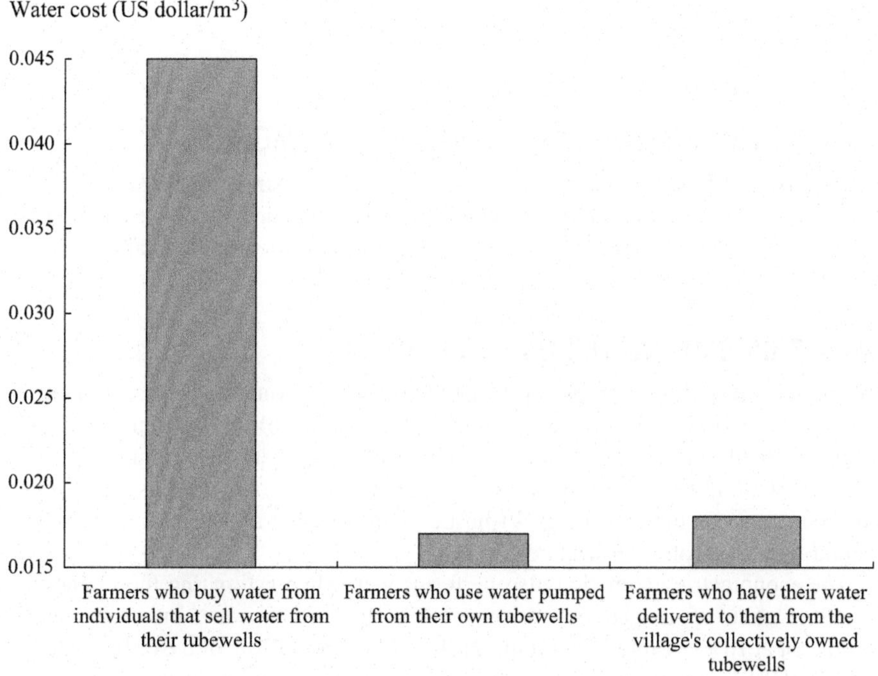

FIGURE 7.3

The cost of water (per cubic meter) that gain access to groundwater through different ways on the North China Plain.

Data source: Authors' survey.

(Fig. 7.2, panel B). Within the same village, the same results show that wheat yields of water buyers are lower, but not significantly so.

IMPACT ON FARMER INCOME

Irrigation service markets for groundwater possibly have a negative effect on the income of farmers who buy water from groundwater markets. For example, per capita cropping income for water buyers is $109.07, 61% of that of tubewell owners ($179.20) and 77% than that of farmers getting irrigation from collective tubewells ($141.23) (Fig. 7.2, panel C and panel D). Within the same village, the impact of differences between cropping incomes of those relying on groundwater markets with other farmers shows consistent results.

Although interesting, our descriptive analysis has shortcomings. The effects of other variables on farmer income are not controlled for. Therefore, it is too early to conclude that irrigation service markets for groundwater have any significant effects on water use, yields, or income. A multivariate analysis is necessary to better understand the relationship between the ways in which farmers access groundwater and farmer yields and income.

ESTIMATION RESULTS OF ECONOMETRIC MODEL

In order to identify the impact of the various ways of accessing groundwater on crop water use, crop yields, and farmer income, we have specified a set of econometric models. The descriptions of these models are included in Methodological Appendix to Chapter 7.

IMPACT ON CROP WATER USE AND YIELDS

In estimating the effect of accessing groundwater on crop water use, the econometric estimation performs well (Table 7.1, column 1). Several of the coefficients of the control variables have the expected signs and are statistically significant. For example, we find that after holding constant other factors, households using high shares of water-saving technologies (plastic piping—either above ground or underground) use less water per hectare.

The econometric estimation also performs well when estimating the impact of the emergence of irrigation service markets for groundwater on crop yields significant at 10%—Table 7.1, column 2). The R-square statistic of the ordinary least squares (OLS) version of the equation is 0.61. In addition, as in the estimation, several coefficients of the control variables have the expected signs and are statistically significant at 1%. For example, the coefficient on the production shock variable is negative and significant (Table 7.1, column 2, row 21).

Importantly, our results show that water use decreases for farmers who buy water from groundwater markets compared with farmers who have their own tubewells or use collective wells. The coefficient of the variable measuring the emergence of groundwater markets is negative and significant (Table 7.1, column 1, row 1). Hence, farmers who buy groundwater from private tubewell owners use less water for wheat than tubewell owners. Interestingly, the coefficient of the variable for collective tubewells, although negative, is not significant (column 1, row 2). Such results are consistent with our descriptive statistics.

So what is causing this? One explanation is that farmers who buy water through groundwater markets have greater incentives to reduce crop water use because they pay more for water (as seen in the discussion above). Tubewell owners are willing to use more water, because the cost per unit is smaller.

We did not include "Water Price" in the Water Use model (Table 7.1, column 1) due to concern regarding multicollinearity. Specifically, we were concerned that "Water Price" and "Irrigation service markets for groundwater" were highly correlated, as private water sellers systematically sold water at higher price levels.

Table 7.1 Regression Analysis of the Impact of the Emergence of Irrigation Service Markets for Groundwater on Crop Water Use, Crop Yield, and Farmer Income

	Log of Water Use per Hectare for Wheat	Log of Wheat Yield per Hectare	Cropping Income per Capita	Total Income per Capita
Irrigation Service Markets for Groundwater				
Buying water from private tubewell (1 = yes; 0 = no)	−0.340 (1.65)[a]		84.249 (0.05)	−718.512 (0.34)
Using water from collective tubewell (1 = yes; 0 = no)	−0.424 (0.97)		2305.948 (1.51)	861.595 (0.44)
Production Inputs				
Log of water use per hectare		0.022 (0.44)		
Log of labor use per hectare		−0.066 (1.37)		
Log of fertilizer use per hectare		0.134 (2.49)[b]		
Log of value of other inputs per hectare		0.105 (2.40)[b]		
Production Environment				
Share of village irrigated area serviced by groundwater	−0.315 (1.18)	0.148 (1.22)	437.095 (0.74)	169.110 (0.23)
Village water scarcity indicator variable	0.155 (1.82)[a]	0.014 (0.30)	−102.536 (0.34)	−215.973 (0.56)
Household Characteristics				
Age of household head	0.051 (0.83)	−0.002 (0.11)	22.576 (0.31)	54.391 (0.60)
Age of household head, squared	−0.001 (0.95)	0.000 (0.25)	−0.053 (0.07)	−0.384 (0.37)
Education of household head	−0.014 (0.67)	0.003 (0.31)	−59.787 (1.19)	42.633 (0.67)
Area of plot	−1.088 (1.91)[a]	−0.371 (1.66)[a]		
Number of plots per household	−0.003 (0.17)			

Continued

Table 7.1 Regression Analysis of the Impact of the Emergence of Irrigation Service Markets for Groundwater on Crop Water Use, Crop Yield, and Farmer Income—cont'd

	Log of Water Use per Hectare for Wheat	Log of Wheat Yield per Hectare	Cropping Income per Capita	Total Income per Capita
Population of household	0.063 (1.74)[a]			
Arable area per capita of household			9412.560 (3.69)[c]	6123.917 (1.89)[a]
Plot Characteristics				
Loam soil	−0.004 (0.03)	0.040 (0.70)		
Clay soil	0.069 (0.61)	0.115 (2.13)[b]		
Distance to home	−0.163 (1.26)	−0.097 (1.91)[a]		
Water Saving Technology				
Share of surface or underground channel	−0.275 (2.26)[b]			
Flood irrigation (1 = yes; 2 = no)	−0.108 (0.98)			
Production Shocks				
Yield reduction due to production shocks		−0.015 (10.44)[c]		
County dummy	—	—		
Constant	7.932 (6.12)[c]	6.860 (9.20)[c]	−2017.856 (1.19)	−644.721 (0.30)
Observations	120	140	200	200
Chi^2	55.80	176.20	42.21	24.75
R^2	0.37	0.61	0.10	0.09

Note: Absolute value of z statistics in parentheses.
[a] *Significant at 10%.*
[b] *Significant at 5%.*
[c] *Significant at 1%.*

In contrast, although water use per hectare falls for farmers who buy water from groundwater markets, yields do not fall significantly. The coefficient on the water use variable, while positive, is not significant (Table 7.1, column 2, row 3). However, when farmers buy water from irrigation service markets, they reduce water use per hectare (Table 7.1, column 1, row 1). These two results suggest that, after holding other factors constant, even though farmers who buy their water from groundwater markets use less water, wheat yields are not negatively affected. Observations during our field work suggest that farmers purchasing water may waste less water.

IMPACT ON INCOME

The emergence of irrigation service markets for groundwater on the North China Plain does not have a negative effect on income. In the cropping income and total income equations, the coefficients on the groundwater market variable are not statistically significant (Table 7.1, columns 3 and 4, row 1). Hence, when holding other factors constant, compared with tubewell owners (and farmers who buy water from collectively managed wells), the income of farmers who buy water from groundwater markets will not be lower.

These results can be extended. In another chapter, we found that irrigation service markets for groundwater in China have provided better (and new) access to groundwater for poorer farmers. In our sample, households purchasing water from irrigation service markets for groundwater are poorer than households owning their own tubewells and selling water (Zhang et al., 2008). We conclude that irrigation service markets for groundwater on the North China Plain have made positive contributions to improving the welfare of the poor in rural areas.

CONCLUSIONS AND POLICY IMPLICATIONS

Many farmers on the North China Plain purchase water from private owners of tubewells. Many of these farmers pay more per cubic meter for their water than farmers who have their own tubewell or those with access to water from collectively owned wells. This situation generates concern that farmers who gain access to water through emerging groundwater markets might use less water and, as a consequence, produce lower yields and earn less income.

Our results suggest that farmers who buy water from local groundwater markets use less water than farmers who have their own tubewells or use collective tubewells. However, yields do not diminish. In addition, there is no measurable negative effect on income. Our findings imply that as water in China becomes scarcer, necessitating increased water efficiency, the emergence of markets for groundwater may be an effective way to provide irrigation services.

With the results of Zhang et al. (2008), our results show that leaders should consider supporting privatization and encouraging the development of groundwater markets. Such developments might reduce water demands without reducing either production or incomes. Generally, when farmers pay more for water, they exert effort to save water while maintaining current yields.

We consider this research to be a starting point for additional work on the subject of water savings in agriculture. Further studies are needed to better understand the linkages between farm-level objectives, water prices, irrigation methods, and hydrology. The emergence of private water sales from tubewells presents an additional, interesting dimension to an already challenging research agenda.

REFERENCES

Huang, Q., Rozelle, S., Lohmar, B., Huang, J., Wang, J., 2006. Irrigation, agriculture performance and poverty reduction in China. Food Policy 31 (1), 30–52.

Meinzen-Dick, R., 1996. Groundwater markets in Pakistan: participation and productivity. In: Research Reports 105. International Food Policy Research Institute, Washington, DC.

Pant, N., 1991. Ground water issues in eastern India. In: Meinzen-Dick, R., Svendsen, M. (Eds.), Future Directions for Indian Irrigation: Research and Policy Issues. International Food Policy Research Institute, Washington, DC.

Shah, T., Ballabh, V., 1997. Water markets in North Bihar: six village studies in Muzaffarpur District. Economic and Political Weekly 32 (52), A183–A190.

Shah, T., Singh, O.P., Mukherji, A., 2006. Some aspects of South Asia's groundwater irrigation economy: analyses from a survey in India, Pakistan, Nepal Terai and Bangladesh. Hydrogeology Journal 14, 286–309.

Strosser, P., Meinzen-Dick, R., 1994. Groundwater markets in Pakistan: an analysis of selected issues. In: Moench, M. (Ed.), Selling Water: Conceptual and Policy Debates over Groundwater Markets in India. VIKSAT-Pacific Institute-Natural Heritage Institute, Ahmedabad, Pakistan.

Wang, J., Huang, J., Rozelle, S., 2005. Evolution of tubewell ownership and production in the North China Plain. Australian Journal of Agricultural and Resource Economics 49 (2), 177–196.

Wang, J., Huang, J., Rozelle, S., Huang, Q., Blanke, A., 2007. Agriculture and groundwater development in northern China: trends, institutional responses, and policy options. Water Policy 9 (S1), 61–74.

Zhang, L., Wang, J., Huang, J., Rozelle, S., 2008. Development of groundwater markets in China: a glimpse into progress to date. World Development 36 (4), 706–726.

SECTION

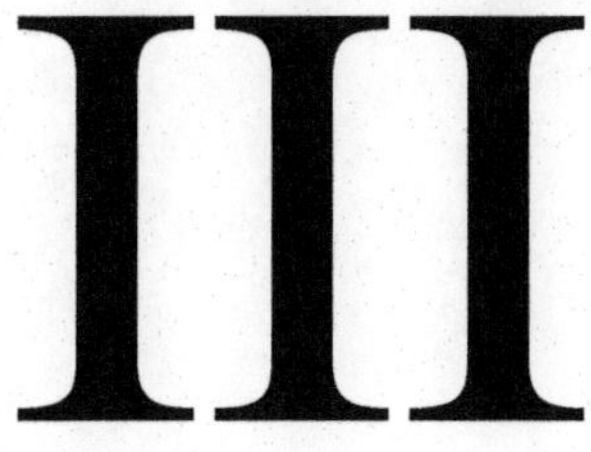

Surface Water Management

CHAPTER

8 Water User Associations and Contracts: Evolution and Determinants

Despite the seriousness of China's water problem, the government has responded slowly in addressing growing water shortages (Wang et al., 2007). In contrast to the government's slow response, however, water users have been more active. Just as groundwater users have stepped in to address some of the management issues left by local governments crippled by fiscal deficits, so are surface water users taking the initiative in creating new institutions to manage surface irrigation water. Several institutions, created by farmers and village leaders, have emerged in the wake of rising scarcity (Chen, 2002; Fang, 2000).

Internationally, since the 1980s, many developing countries have transferred surface water irrigation management responsibilities from the government to farmer organizations or other private entities (Vermillion, 1997). Water User Associations (WUAs), have been shown to be effective in raising the efficiency of irrigation, increasing incomes, and helping poor farmers (World Bank, 1993). However, beyond a handful of internationally funded sites, little is known about WUAs and other forms of water management institutions, especially inside China.

The limited number of existing studies on surface water management reforms in China mainly focus on narrow sample populations. Zhang (2001) studies only World Bank project sites. Wang et al. (2005b) examine only four irrigation districts in two provinces. There is little, if any, nationwide empirical research analyzing whether or not institutional reforms help alleviate China's water crisis, or how the reforms have affected the welfare of farmers.

Our goal in this chapter is to increase our understanding of newly emerging water institutions (WUAs and/or contracting) and the identifying factors that lead to their creation in one community but not in another. To meet these goals, we pursue three objectives. First, we document the evolution of existing and new water management institutional forms over time and across provinces throughout northern China. Second, we describe the characteristics of WUA governance and compare them to traditional collectively managed irrigation systems and contracting. Finally, we analyze the determinants of the emergence of these institutions throughout northern China to understand the role of water scarcity, the size of a village's irrigation system, government policy, and other village characteristics in water management reform. Analysis in this chapter uses data from the China Water Institutions and Management (CWIM) survey, and the North China Water Resource Survey (NCWRS).

Managing Water on China's Farms. http://dx.doi.org/10.1016/B978-0-12-805164-1.00008-7

TRENDS IN NORTHERN CHINA'S WATER MANAGEMENT REFORM, 1995 TO 2004

Based on our field surveys, surface water is managed in three general ways. If the village leaders (ie, the village council) directly take responsibility for water allocation, canal operation and maintenance (O&M), and fee collection, the village's surface irrigation system is said to be run by *collective management*, the system that essentially had allocated water in most of China's villages during the People's Republic period. In this chapter, we refer to the collective management system as the traditional system. In contrast, a *WUA* is in principle a farmer-based, participatory organization that manages the village's surface irrigation water. In WUAs, a board, supposedly elected by villagers, manages the village's surface water and facilitates farmer participation. *Contracting* is a system in which the village leaders contract the village's canal out to an individual, who manages the canal in return for a payment that may or may not be related to the size of water savings the individual can achieve. We consider WUAs and contracting to be reform-oriented management systems.

CHANGES OVER TIME

According to our data, WUAs and contracting systems have gradually emerged in northern China between 1995 and 2004. However, tracking these changes is complicated by the changing nature of China's water resources (Table 8.1). In 1995, of the 481 sample villages, 235 had surface water irrigation (column 1, row 9). During the survey, the enumerators learned that of the 235 villages, 30 of them had stopped using surface water by 2004 (row 8). During the same period, 17 villages used surface water in 2004 but not in 1995 (row 10). In total, 205 villages (235 minus 30) used surface water in both 1995 and 2004.

When examining the villages that used surface water in both 1995 and 2004, our data reveal a clear tendency in the ways that villages are reforming their water management structure (Table 8.1, row 1). Of the 181 villages that were being managed under collective management in 1995, only 143 were still managed in this way in 2004 (columns 1 and 2). This means 38 villages (21%) implemented some form of water management reform.

The reform efforts from 1995 to 2004 were split almost exactly between shifts to WUAs and contracting (Table 8.1, row 1, column 3 and 4). Villagers in 14 villages chose to create WUAs (column 4). Villagers in 18 villages shifted into contracting. There were also six villages that reformed only part of their village's surface water system or chose a mix of WUAs and contracting (column 5 to 8).

While the trends in northern China's villages are clearly reform-oriented, it is interesting to note that in villages that had already reformed by 1995, there is some evidence that villagers are continuing to experiment with different institutional forms and are not afraid of going back to collective management. For example, of the eight villages that had created WUAs to manage their surface water systems

Table 8.1 Transition Matrix of Changes in the Forms of Surface Water Management in Northern China Between 1995 and 2004

	Forms of Surface Water Management in 1995	(1) Number of Sample Villages	Forms of Surface Water Management in 2004						
			(2) Collective Management	(3) Water User Association (WUA)	(4) Contracting	(5) WUA and Collective Management	(6) WUA and Contracting	(7) Contracting and Collective Management	(8) WUA, Contracting and Collective Management
1	Collective management	181	143	14	18	2	2	1	1
2	WUA	8	2	5		1			
3	Contracting	11	1		9			1	
4	WUAs and collective management	1		1					
5	WUAs and contracting	1		1					
6	Contracting and collective management	2	1		1				
7	WUAs, contracting, and collective management	1		1					
8	Villages that shut down SW irrigation between 1995 and 2004	30							
9	Total villages with SW irrigation in 1995	235	–	–	–	–	–	–	–
10	Villages with new SW irrigation created between 1995 and 2004	17	14	1	1	1			

Source of data: 2004 NCWRS and 2001–04 CWIM panel.

in 1995, three of them had either discontinued or partially discontinued the experiment by 2004 (Table 8.1, row 2, column 1, 2, and 5). Two of the 11 villages that chose contracting systems in 1995 decided to either fully or partially go back to collective management by 2004 (row 3, column 1, 2, and 7). These shifts into and out of WUAs and contracting may indicate that water management reform is not universally successful. This is of concern to national leaders worried about whether surface water management reform is suitable for China's villages.

The emergence of water management reform is not closely tied with the creation of new irrigation systems (Table 8.1, row 10). In the 17 villages that used surface water in 2004 but not in 1995, 14 (82%) chose to be managed under the collective management system. Only three chose to implement WUA or contracting. This percentage of reformed management villages (among newly irrigated villages—18%) is lower than the overall average.

So how should one interpret the trend of water management reforms between 1995 and 2004 in northern China? Our data show that the changes were significant (Fig. 8.1). The share of collective management declined from 90% in 1995 to 73% in 2004. WUAs and contracting have developed at about the same pace. By 2004, 10%

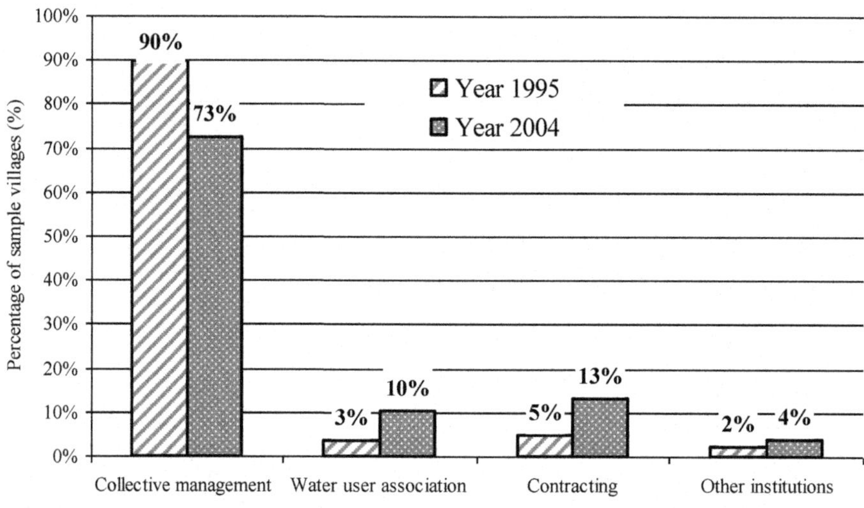

FIGURE 8.1

Changes in water management institutions from 1995 to 2004. Note: Other institutions includes the four types of mixed institutions: (1) water user association combined with collective management; (2) water user association combined with contracting; (3) contracting combined with collective management; (4) water user association, contracting, and collective management.

Source of data: 2004 NCWRS and 2001–04 CWIM panel.

of villages managed their surface water through WUAs and 13% through contracting. The mixed systems also rose from 2% to 4% between 1995 and 2004. While collective management still is the dominant form of management, 27% of villages in northern China had been affected by water management reform by 2004.

DIFFERENCES ACROSS SPACE

Although the overall trend since 2005 was the shift from collective management to WUAs or contracting, water management reform varies significantly across the seven sample provinces, showing that the nature of reforms is not universal (Table 8.2). For example, the use of collective management between 1995 and 2004 has fallen in five sample provinces: Inner Mongolia, Ningxia, Liaoning, Shanxi, and Henan (row 2 to 4, 6, and 8, column 1). In Shaanxi and Hebei (row 5 and 7, column 1), however, the use of collective management actually increased. Moreover, even among the five reforming provinces, there were striking differences. In Inner Mongolia the share of collective management fell sharply from 89% in 1995 to 44% in 2004. In Ningxia the share of collective management also fell sharply from 78% in 1995 to 31% in 2004. Villages in the other three provinces (Liaoning, Shanxi, and Henan) reformed significantly less. The share of villages under collective management fell by only 5 to 10 percentage points.

Beyond the differences across the villages regarding their decision to reform or not, the direction of reform also varied among provinces (Table 8.2). Most poignantly, when examining the nature of reforms in Inner Mongolia, one of the two actively reforming provinces, in almost all the sample villages that had reformed by 2004, villagers decided to manage their surface water systems through WUAs instead of contracting (15 of 19, row 2, columns 2 to 4). In contrast, villages in the other actively reforming province, Ningxia, mostly chose contracting. There were also differences in other provinces. Villagers in all of the reforming villages in Shanxi and Hebei chose contracting, while five of the six reforming villages in Shaanxi chose WUAs.

Although some of the differences in water management may be due to differences in local villages, the dramatic differences among provinces suggest that government policy may also play an important role. For example, in 2000, Ningxia provincial water officials issued several documents that encouraged local governments to proceed with water management reform (Wang, 2002). Regional water officials exerted considerable effort to promote water management reform in several experimental areas. In contrast, in Hebei (a nonreforming province) when we approached provincial officials about WUAs, no one knew anything about them. The sharp shift away from collective management in provinces whose leaders were supportive of reforms shows that these policy measures were effective in pushing reform forward. For example, contracting has risen rapidly after provincial officials relaxed the previous restrictions against contracting out the village's irrigation system.

Table 8.2 Number of Villages That Are Under Different Water Management Forms in Northern China in 1995 and 2004

		(1) Collective Management		(2) WUA		(3) Contracting		(4) WUA and Collective Management		(5) WUA and Contracting		(6) Contracting and Collective Management		(7) WUA, Contracting, and Collective Management	
		2004	1995	2004	1995	2004	1995	2004	1995	2004	1995	2004	1995	2004	1995
1	North China	161 (72.5)	211 (89.8)	23 (10.4)	8 (3.4)	29 (13.1)	11 (4.7)	4 (1.8)	1 (0.4)	2 (0.9)	1 (0.4)	2 (1.1)	2 (0.9)	1 (0.5)	1 (0.4)
2	Inner Mongolia	15 (44.1)	33 (89.2)	13 (38.2)	3 (8.1)	4 (11.8)	1 (2.7)	2 (5.9)							
3	Ningxia	10 (31.3)	25 (78.1)	4 (12.5)	1 (3.1)	14 (43.8)	3 (9.4)	1 (3.1)		2 (6.3)	1 (3.1)		1 (3.1)	1 (3.1)	1 (3.1)
4	Liaoning	37 (90.2)	44 (95.7)	2 (4.9)	1 (2.2)	2 (4.9)	1 (2.2)								
5	Shaanxi	39 (86.7)	31 (83.8)	4 (8.9)	3 (8.1)		1 (2.7)	1 (2.2)	1 (2.7)			1 (2.2)	1 (2.7)		
6	Shanxi	20 (80.0)	27 (96.4)			5 (20.0)	1 (3.6)								
7	Hebei	20 (83.3)	19 (82.6)			4 (16.7)	4 (17.4)								
8	Henan	20 (95.2)	32 (100)									1 (4.8)			

Figures in parentheses are the percentages of sample villages that have the water management institution identified by the column head. Table 8.2 counts all sample villages that used surface water in 2004 and all sample villages that used surface water in 1995. Because of this, the number of villages is different from those in Table 8.1. The total number of villages in 1995 and in 2004 may not add up to be the same either because some villages stopped surface water irrigation while others created new surface water irrigation between 1995 and 2004.

Source of data: 2004 NCWRS and 2004 CWIM.

DIFFERENCES IN GOVERNANCE AMONG WATER MANAGEMENT INSTITUTIONS

The shift in China's water management institutions demonstrates that rural villages are at least in part following policy directives that are developed and issued from upper-level governments. The differences among regions, however, show that the nature of reform is not universal. The regional differences are likely characteristic of reform in China, where local governments often are given considerable room in choosing the exact form and timing of changes (Jin et al., 2005). In this section, we examine the differences among the governance approaches in villages that have chosen different forms of water management: collective management, WUAs, and contracting.

OPERATION AND MAINTENANCE RESPONSIBILITIES

According to our data, when villages report their water systems are being run under a certain institutional form, there are major shifts of responsibilities (Table 8.3). When a village claims its surface water system is being managed under the collective management system, nearly all water management activities are carried out by the village leadership (rows 1 to 5). In all villages, canal maintenance, coordination of water delivery, and water fee collection are fully the responsibility of the village council. In a small portion of villages (22%), farmers themselves must operate the sluice gates according to a schedule set by village leaders. Some villages (33%) also depended on township or irrigation district (ID) officials if there was a dispute that required resolution.

When a WUA is formed, most responsibilities are transferred to the WUA board (Table 8.3, rows 6 to 10), which is fully responsible for some of the main tasks including the operation of sluice gates, water fee collection, and conflict resolution. In half of the villages with WUAs, the WUA was responsible for canal maintenance, but in the other half, the village leadership maintained responsibility. According to our observations, this is done in some WUAs because in some areas of rural China a village is divided into several small groups or *xiaozu* (and the plots of households in each small group are located together); while canal maintenance in some villages is carried out by small group members, the WUA is a village-wide organization and so the village leadership is still needed to coordinate among the small groups. WUAs also share responsibilities with village leaders and farmers for coordination of water delivery.

The organization of water management activities is more complicated when contracting is used (Table 8.4, rows 11 to 15). In contracting villages, the village council helps the contractor conduct all of the different activities, in part because contractors may lack the ability to conduct some activities (eg, canal maintenance). In addition, some contractors are unable to act as a disinterested party when conducting other activities, such as dispute resolution.

Table 8.3 Division of Responsibilities (Water Management Activities) Under Different Water Management Institutional Forms in Northern China Sample Villages, 2004

Water Management Institution		Water Management Activity	% of Sample Villages in Which a Water Management Activity Is Carried Out by:				
			Village Council	Water User Association	Contractor	Contractor and Village Council	Others[a]
Collective management ($n = 161$)	1	Canal maintenance	100				
	2	Operation of sluice gates	78				22
	3	Coordination of water delivery	100				
	4	Water fee collection	100				
	5	Conflict resolution	67				33
Water user association ($n = 23$)	6	Canal maintenance	50	50			
	7	Operation of sluice gates		100			
	8	Coordination of water delivery	25	50			25
	9	Water fee collection		100			
	10	Conflict resolution		100			
Contracting ($n = 29$)	11	Canal maintenance	35		25	35	5
	12	Operation of sluice gates			90	5	5
	13	Coordination of water delivery	10		75	5	10
	14	Water fee collection	10		80	10	
	15	Conflict resolution			50	35	15

[a] *Others include farmers, irrigation district, and township government.*
Source of data: 2004 NCWRS and CWIM.

Table 8.4 Incentives Faced by Canal Managers in Northern China Sample Villages, 1995 and 2004

	Percentage of Sample Villages in Which Canal Managers Have Been Provided With Incentives (%)	
	2004	**1995**
Collective management	0	0
Water user association	32	14
Contracting	73	27

Source of data: 2004 NCWRS.

INCENTIVES

Although there are many similarities between international experiences and those of China, even in its early phase, policy documents emphasize certain things that make China's water management reform unique. Above all, water officials have promoted the idea of using incentives to make water management reform more effective. The use of incentives is not new in the context of China's overall economic reform effort. Reformers frequently have relied on incentives to induce agents to exert more effort, allocate resources more efficiently, and enter into new economic activities (Naughton, 1995). The household responsibility system was based on new profit incentives for farmers (Lin, 1992). Fiscal reforms gave local leaders strong incentives to begin township and village enterprises (Walder, 1995), while grain reforms gave grain bureau personnel incentives to commercialize commodity trading (Rozelle et al., 2000).

With the past success of using incentives in various reforms, water officials hope similar policy efforts would improve water management in China. In many of the new reform efforts, water managers are supposed to be provided with monetary rewards if they achieve water saving objectives. In our analysis, village leaders are the managers of collectively managed irrigation systems. The chair of the WUA board is the manager in villages with WUAs, while the contractor is the manager in contracting villages. If a canal manager is provided with earning incentives, he or she can claim all or part of the profits from the operation of a canal. Usually a canal manager is either paid a portion of the water fees collected or a portion of the residual profit from canal operation.

The differences in incentives faced by managers vary across institutional forms and over time (Table 8.4). In none of the collectively managed villages were village leaders paid a bonus or given any portion of the residual revenues in either 1995 or 2004. In contrast, managers in WUA villages faced better incentives. In 14% of the villages with WUAs, WUA managers were provided with incentives. The share rose to 32% in 2004. Despite the rising trend, however, WUA managers in more than two-thirds of villages still faced no incentives of any kind.

In contrast, the incentives faced by contractors rose rapidly and reached a high proportion of villages by 2004. Although only 27% of contracting villages offered contractor-cum-managers financial incentives in 1995, the proportion reached 73% in 2004. Clearly, the provision of incentives distinguishes contracting significantly from collective management or WUAs.

PRACTICE AND PRINCIPLE: PARTICIPATION

So far, we have only reported the results from the NCWRS. Although this is a comprehensive survey, it relies only on the opinion of village leaders. In such a survey, it is impossible to check the validity of some answers due to the lack of an alternative source of information. In the CWIM survey, we interviewed multiple stakeholders including village leaders, canal managers, and farmers. Separate sets of survey instruments were used and the answer of each respondent was held in confidence from all others. Hence, we have more accurate information to assess the practices of reformed irrigation management systems and whether they vary from principle. In particular, we can assess the level of participation in water management by different stakeholders. In our survey, we inquired about three major dimensions of participation: how much farmers participated in the process of the establishment of the reform process; whether or not farmers were involved in the selection of the managers; and whether or not farmers were invited to attend regular meetings. These three aspects cover nearly all the major decision-making in water management institutions.

According to our data, participation is not part of either collective management or contracting. Traditionally, many government services in China are conducted from the top down, with little consultation or participation of farmers (Zhang et al., 2003). Although collectively managed services, such as those provided by collectively run water organizations, in principle are supposed to be determined by the entire collective, in reality, village leaders have managed their villages largely based on the authority they have derived from higher-level officials. In our sample villages, we find that farmers participated little (and mostly not at all) in collectively run water management organizations. Similarly, contracting only involved transferring control and income rights to an individual. Village leaders usually decided whether to contract out canals or not. Farmers often played no role in the transition. Only in a few villages did farmers participate in electing the contractor.

Even when the reforms that led to the creation of WUAs explicitly attempted to encourage farmer participation, practice often varied from principle. In the CWIM survey areas, farmers had little voice in deciding the establishment of WUAs or appointing the WUA board members. For example, our data show that, on average, only about 12.5% of WUAs involved farmers in the decision on their establishment. In fact, most farmers (70%) did not know they were members of a WUA. If the motivation of promoting WUAs is to increase participation by farmers, the canal system in these villages is only being nominally managed by WUAs.

Farmers also were seldom encouraged to participate in other aspects of water management. Based on our sample, *none* of the WUA board members were actually elected by farmers. Only 25% of WUAs allowed farmers to participate in selecting managers. As a result, in most cases (70% of the WUAs), the WUA board was comprised of the village leadership. In a few cases (30% of the WUAs), village leaders appointed a chair or manager to conduct the daily duties of the WUAs. In many of these WUAs, however, the managers had close ties to the village leadership. For example, the manager was frequently a former village leader or a relative of the current leader. Moreover, although 80% of WUAs conducted regular meetings, only in 25% of the WUAs were farmers invited to participate.

Compared with collective management and contracting, however, our data show that management under WUAs was more transparent. Nearly 40% of WUAs shared three types of information about irrigation management with farmers, including the total amount of water fees collected, the volume of water actually delivered by the irrigation district to the village, and the actual area irrigated. About 50% of the WUAs shared two of the three types of information with farmers.

WHY ARE WUAs AND CONTRACTING USED IN SOME AREAS BUT NOT OTHERS?

The data described above suggest that although many villages in northern China are reforming their water management institutions, reform is happening in some places but not in others. In this section, we seek to identify the determinants of water management reform—first descriptively and then using multivariate analysis.

DESCRIPTIVE ANALYSIS

Our data suggest that the nature of a village's water resources may play an important role in reform. Contrary to common perception, however, it is not villages with the most severe water problems that are reforming. In our survey, we asked about the number of years between 1993 and 1995 (a time before we measured whether or not there was reform) in which there was not enough water in canals to satisfy irrigation demands in the village. WUA villages reported on average there were only 0.45 years of water shortages during the three-year period, while there were 0.92 years in collectively managed villages (Table 8.5, rows 1, column 1 and 2). The difference in the degree of water availability was statistically significant. We also asked village leaders directly if they think water resources in their villages were scarce in 1995. Although the differences were not statistically significant, water was also less scarce in WUA villages (row 2). Furthermore, WUAs and contracting seemed to be more likely to appear in villages with access to both surface water and groundwater (row 3).

The finding that reform seems to occur in areas with relatively more available water resources has important implications. It suggests that although water

Table 8.5 Characteristics of Water Resources, Canals, Village Leaders, and Villages Under Different Water Managerial Forms in Northern China Sample Villages

		(1) Collective Management	(2) Water User Association	(3) Contracting	(4) Other Institutions[a]
Number of Sample Villages[b]		**111**	**22**	**24**	**7**
Characteristics of Water Resources in the Village					
1	Water availability (Number of years that there was not enough water in canals between 1993 and 1995)	0.92 (1.24)	0.45** (0.80)	0.73 (1.07)	0.64 (1.18)
2	Village water scarcity indicator variable (1 = water was scarce in the village in 1995, 0 = otherwise)	0.1 (0.30)	0.05 (0.21)	0.04 (0.20)	0 (0.00)
3	Conjunctive use (percentage of land that was conjunctively irrigated by surface water and groundwater in 1995, %)	8.97 (25.13)	12.09 (27.15)	14.38 (33.73)	3.14 (7.47)
Characteristics of Canals in the Village					
4	Canal lining (percentage of the total length of tertiary canals that was lined in 2004, %)	22.1 (51.17)	13.3 (25.45)	4.56* (3.17)	10.8 (18.55)
5	Canal length (total length of tertiary canals in the village in 2004, km)	6.76 (10.68)	14.25*** (13.82)	30.22*** (65.32)	8.6 (8.71)
6	**Policy dummy** (1 = government promoted water user association or contracting and 0 = otherwise)	0.25 (0.44)	0.77*** (0.43)	0.67*** (0.48)	0.86*** (0.38)

7	**Cropping pattern**: Share of sown area in rice in 1995 (%)	21.69 (29.12)	17.36 (29.28)	12.92* (19.30)	16.84 (22.44)
Socioeconomic Characteristics of Villages					
8	Income per capita in 1995 (Yuan, in log form)	7.04 (0.67)	7.39** (0.37)	7.09 (0.38)	7.34 (0.50)
9	Percentage of migrants (%, share of village labor force that out migrated in 1995)	8.75 (11.08)	11.83 (17.17)	10.36 (13.53)	12.29 (14.67)
10	Percentage of self-business households (%, 1995)	7.45 (9.81)	4.1* (4.23)	3.02** (4.17)	4.75 (6.99)
Characteristics of Village Leaders					
11	Age of the party secretary (year)	47.86 (7.38)	49.14 (6.15)	46.54 (6.16)	48.14 (3.08)
12	Level of education of party secretary (years of schooling)	9.21 (2.54)	9.36 (2.28)	8.33* (2.58)	9.43 (3.64)
13	Job dummy (1 = the main job of the party secretary was NOT agriculture in 2004)	0.13 (0.33)	0.14 (0.35)	0** (0.00)	0.14 (0.38)
14	Water management experience of the party secretary (years)	5.19 (6.90)	6.91 (6.85)	4.85 (5.46)	5.29 (5.96)
Village Demography					
15	Level of education of villagers (%, share of 1995 labor force that had education above high school)	1 (2.71)	2.7*** (4.11)	6.91*** (7.25)	10*** (9.24)
16	Number of households in the village in 2004	444 (376)	461 (178)	429 (167)	389 (201)
Location Dummies					
17	Irrigation district location dummy (1 = village located downstream of an irrigation district in 2004 and 0 = otherwise)	0.42 (0.50)	0.5 (0.51)	0.33 (0.48)	0.71* (0.49)

*Standard deviations of variables are reported in parentheses. Asterisks indicate the difference in the mean of a variable between villages under collective management and villages under other institutions is statistically significant. * denotes the difference is significant at 10%; ** significant at 5%; *** significant at 1%.*

[a] *Other institutions includes the four types of mixed institutions: (1) water user association combined with collective management; (2) water user association combined with contracting; (3) contracting combined with collective management; (4) water user association, contracting, and collective management.*

[b] *Villages that used surface water for irrigation in both 1995 and 2004 are included in the analysis. Hebei Province and Henan Province are not included since there are no many variations in water management institutions among villages in these two provinces.*

Source of data: 2004 NCWRS and CWIM 2001–04 panel.

management reform is being encouraged by policy makers as a solution to China's water crisis, WUAs and contracting are not being used in the most water-short places. Although we do not know why, it may be a matter of feasibility of implementation. If water is too scarce, there might be little scope for water savings. Consequently, no contractor would be willing to take the contract, as no water savings would mean a lower income.

Descriptive analyses also suggest that the quality or the complexity of the irrigation infrastructure seem to matter in the choice of management form. When a larger share of the canal in a village is lined with concrete, the village's irrigation system seems more likely to remain collectively managed. Specifically, when a village is collectively managed, on average, 22% of the canal system is lined (Table 8.5, row 4, column 1). In WUA villages, it is only 13% and is even lower in contracting villages (4.5%). On average, WUAs and contractors manage longer canals (14 km and 30 km, respectively) than collectively managed systems (6.7 km, row 5, column 1–3). Hence, leaders appear to be more willing to contract out or turn an irrigation system over to a WUA when the canal system is of poorer quality or more complex. From the contractor's and WUA members' points of view, such an irrigation system might provide more scope for improvement, since there might have been more waste when it was collectively managed.

Consistent with findings in previous analyses on privatization of groundwater (Wang et al., 2005a) and other water reforms, policy appears to play an important role in encouraging water management reform. In 77% of WUA villages and in 67% of contracting villages, provincial officials or irrigation district personnel conducted extension campaigns to encourage water management reforms (Table 8.5, row 6). In contrast, extension efforts for water management reform were conducted in only 25% of villages that remained collectively managed. The differences in the descriptive statistics were significant.

Other factors appear to be associated with the adoption of WUAs. For example, wealthier villages, those with more migrants, and those with fewer self-employed businesses seem more likely to adopt WUAs (row 8–10). WUA villages also tended to have older leaders and better educated villagers. These findings are consistent with the idea that it may be not until a village is relatively better off and its members are more active in the off-farm sector that they begin to look for more effective ways to manage water. Alternatively, it might be that residents in such villages are more open to change.

MULTIVARIATE ANALYSES

Multivariate analysis is used to identify the factors that have contributed to institutional reforms. The dependent variable is a dummy variable that equals one if the village has adopted either a WUA or contracting to manage water. The logit regression is used. The multivariate analysis generates results similar to those from the descriptive analyses. For example, the results suggest that, ceteris paribus, villages that were relatively water abundant were more likely to reform (or adopted noncollective forms of irrigation management). Specifically, during the three-year period, if a village suffered more years without enough water in their canals, the tendency was to *not reform* (Table 8.6, row 1, column 1). When villages had conjunctive water resources in 1995 (which might be interpreted as having relatively more abundant resources, ceteris paribus), reform is also more likely (row 3). Although water management reforms are spreading, our results demonstrate that reforms tend to occur in areas with more available water resources rather than areas in crisis. Leaders and farmers in these villages might not want to change in fear of making a current water crisis even worse, which might happen in the case of experimenting with a new managerial form. In addition, there might be no obvious benefit of change where water resources are so scarce and irrigation systems are relatively small. Clearly, if policy makers want to use water management reform to help solve China's water crises in water short areas, additional policies will be required.

Also consistent with the descriptive analyses, villages appear to reform water management when they have larger and more complex systems. The positive and significant coefficient on canal length means that villages with longer canal systems have a greater propensity to reform (row 5). These findings are consistent with those in Huang et al. (2008) that WUAs and contracting are more likely to be chosen in villages with more complex canal systems. In such systems, managing the canal system is a labor intensive and management intensive task that requires a manager with substantial motivation, which incentives or a participatory body could generate.

In addition, the estimated coefficient on the policy dummy variable is positive, large in magnitude, and statistically significant (row 6, column 1), even though we have included a set of provincial dummies that capture any province level effects. This suggests that much of the policy effect is subprovincial at the county level or the township level. There is enough heterogeneity within each province that when WUAs and contracting are promoted in some villages, but not in others, there is a different rate of response within the province.

Several other factors are also of interest. For example, while the coefficient on the income variable is positive, it is not statistically significant (row 8). Only the level of education of villagers is statistically significant (row 15). Perhaps villages with higher levels of education are more willing to reform because they are more flexible, or because they think WUA or contracting could provide them with more benefits of some sort. Education could also be correlated with several other factors, such as a better pool of candidates for contractors or WUA managers.

A second regression is also run with the dependent variable being a dummy that equals one if the village adopted WUA and zero otherwise (Table 8.6, column 2). The signs, magnitudes, and levels of statistical significance for most of the estimated coefficients are similar to those in column 1 of the same Table. Two estimated coefficients are different. The positive and significant coefficients on the income per capita variable and the age of the party secretary variable suggest that WUAs have emerged in villages that were relatively well off or in villages with leaders

Table 8.6 Logit Regression Explaining Factors That Contributed to Water Management Institutional Reform

		(1) Dependent Variable: Reform Dummy (=0 if Collective Management and = 1 if Otherwise)[a]	(2) Dependent Variable: Water User Association Dummy (=1 if Water User Association Exists in the Village)[b]
Characteristics of Water Resources in the Village			
1	Water availability (number of years that there was not enough water in canals between 1993 and 1995)	−0.555** (2.08)	−0.842* (1.73)
2	Village water scarcity indicator variable (1 = water was scarce in the village in 1995, 0 = otherwise)	0.145 (0.11)	−0.483 (0.41)
3	Conjunctive use (percentage of land that was conjunctively irrigated by surface water and groundwater in 1995, %)	0.028*** (2.91)	0.031* (1.83)
Characteristics of Canals in the Village			
4	Canal lining (percentage of the total length of tertiary canals that was lined in 2004, %)	−0.006 (0.67)	0.000 (0.02)
5	Canal length (total length of tertiary canals in the village in 2004, km)	0.050** (2.19)	0.013 (0.84)
6	**Policy dummy** (1 = government promoted water user association or contracting and 0 = otherwise)	1.794*** (2.69)	3.090*** (2.92)
7	**Cropping pattern**: share of sown area in rice in 1995 (%)	0.009 (0.65)	0.052* (1.91)
Socioeconomic Characteristics of Villages			
8	Income per capita in 1995 (Yuan, in log form)	0.617 (0.95)	3.738*** (2.59)

9	Percentage of migrants (%, share of village labor force that out migrated in 1995)	0.010 (0.47)	0.011 (0.47)
10	Percentage of self-business households (%, 1995)	−0.034 (0.81)	−0.056 (0.76)
Characteristics of Village Leaders			
11	Age of the party secretary (year)	0.050 (1.26)	0.149* (1.83)
12	Level of education of party secretary (years of schooling)	0.016 (0.15)	−0.068 (0.35)
13	Job dummy (1 = the main job of the party secretary was NOT agriculture in 2004)	−0.309 (0.33)	1.446 (1.11)
14	Years of water management experience of the party secretary	0.033 (0.74)	0.003 (0.04)
Village Demography			
15	Level of education of villagers (%, share of 1995 labor force that had education above high school)	0.153** (1.96)	0.058 (0.64)
16	Number of households in the village in 2004	0.000 (0.03)	0.000 (0.18)
Location Dummies			
17	Irrigation district location dummy (1 = village located downstream of an irrigation district in 2004 and 0 = otherwise)	−0.343 (0.61)	−0.013 (0.01)
18	Constant	−9.760** (2.08)	−36.658*** (3.19)
19	Observations[c]	160	137

Villages that used surface water for irrigation in both 1995 and 2004 are included in the regression. Hebei Province and Henan Province are not included since there are no many variations in water management institutions among villages in these two provinces. Shanxi Province is excluded in column two since it does not have any WUA villages.

[a] *Noncollective management includes WUA and contracting.*

[b] *WUA includes villages under WUA or under WUA combined with other institutions.*

[c] *Absolute value of z statistics in parentheses, significant at 10%;** significant at 5%;*** significant at 1%.*

Source of data: 2004 NCWRS and CWIM 2001–04 panel.

that have more experience (row 8 and 11). WUAs are either easier for such villages to organize or WUAs might be in greater demand in such villages.

CONCLUSIONS AND POLICY IMPLICATIONS

We have documented the evolution of existing and new surface water management institutional reforms over time and across provinces. By 2004, more than one-quarter of our sample villages in northern China had replaced collective management with either WUAs or contracting. When the reform took place, there was a shift of responsibility from the collective leadership to either the WUA board or the contractor. The reform-oriented institutions have some features that lead to better water management. For example, more than 70% of the contractors face economic incentives that link their earnings with the quality of irrigation services they provide. Almost half of the WUAs manage irrigation systems in a transparent way, such that board members share management information with users. However, some aspects also call for improvement. Only a small portion of the WUAs are provided any incentives to improve water management. There is little participation by farmers in any of the management activities, even in villages with WUAs, and many WUAs have close ties to the village leadership. Policy makers seeking to deepen reform efforts in China should focus on providing incentives and increasing farmers' participation.

We have also endeavored to explain why some villages have chosen to retain collective management, while others have decided to form WUAs or engage in contracting. Two implications are evident. First, resource availability matters. Regression results show that villages are more likely to reform when water is relatively abundant and when the village canal network is relatively large. Hence, in China's future design of water management reforms, policy makers should be aware that reform may be harder to enact in communities already experiencing a water shortage. Surface water management reform may be best adopted as a preventive measure, before the crisis can escalate. Second, local governments play important roles. When the county level or the township level governments promoted either WUA or contracting, most villages reformed. Thus policy makers should continue to rely on local governments to move the reform in a direction that can address China's water problems faster and better.

In this chapter, we have documented the rapid spread of alternative surface water management programs. It remains to be seen, however, how effective these new management systems are in practice. Management reform is only one piece of the larger puzzle that is China's national water resource network, and it alone is not adequate to address China's water scarcity problems. Institutional reform has not yet occurred in the most water scarce parts of China. Before forcing this new management system on already constrained villages, it is important to conduct a rigorous project evaluation. In the next chapter, we attempt to do just that.

REFERENCES

Chen, L., 2002. Revolutionary Measures: Water Saving Irrigation. The National Water Saving Workshop. Beijing, China (in Chinese).

Fang, S., 2000. Combined with allocating and controlling local water resources to save water (in Chinese). Journal of China Water Resources 439, 38–39.

Huang, Q., Rozelle, S., Msangi, S., Wang, J., Huang, J., 2008. Water management reform and the choice of contractual form in China. Environment and Development Economics 13, 171–200.

Jin, H., Qian, Y., Weingast, B.R., 2005. Regional decentralization and fiscal incentives: federalism, Chinese style. Journal of Public Economics 89, 1719–1742.

Lin, J.Y., 1992. Rural reforms and agricultural growth in China. American Economic Review 82, 34–51.

Naughton, B., 1995. Growing Out of the Plan: Chinese Economic Reform, 1978–1993. Cambridge University Press, New York, USA.

Rozelle, S., Park, A., Huang, J., Jin, H., 2000. Bureaucrat to entrepreneur: the changing role of the state in China's grain economy. Economic Development and Cultural Change 48, 227–252.

Vermillion, D., 1997. Impacts of Irrigation Management Transfer: A Review of the Evidence. International Water Management Institute. Research Report Series No. 11.

Walder, A., 1995. Local governments as industrial firms: an organizational analysis of China's transitional economy. American Journal of Sociology 101, 263–301.

Wang, J., 2002. Field Survey Notes in Ningxia Province. Center for Chinese Agricultural Policy, Institute for Geographical Science and Natural Resource Research, Chinese Academy of Sciences (unpubl.).

Wang, J., Huang, J., Rozelle, S., Huang, Q., Blanke, A., 2007. Agriculture and groundwater development in northern China: trends, institutional responses, and policy options. Water Policy 9, 61–74.

Wang, J., Huang, J., Rozelle, S., 2005a. Evolution of tubewell ownership and production in the North China Plain. Australian Journal of Agricultural and Resource Economics 49, 177–195.

Wang, J., Xu, Z., Huang, J., Rozelle, S., 2005b. Incentives in water management reform: assessing the effect on water use, production, and poverty in the yellow river basin. Environmental and Development Economics 10, 769–799.

World Bank, 1993. Water Resources Management: A World Bank Policy Paper. The World Bank Policy Paper No. 12335.

Zhang, L., Huang, J., Rozelle, S., 2003. China's war on poverty: assessing targeting and the growth impacts of poverty programs. Journal of Chinese Economic and Business Studies 1, 301–317.

Zhang, Y., 2001. Carefully implement the fifth national conference, promoting the development of water resources into a new stage (in Chinese). Journal of China Water Resources 450, 9–10.

CHAPTER

Determinants of Contractual Form

9

In the previous chapter, we showed that the success of water management reform in China is positively correlated with the incentives that are provided to managers. Generally, the village leader (an arm of the local government) provides incentives to private managers (typically one or several villagers) by offering them shares of the profit from the operation of the village's irrigation system. In some villages, this takes the form of leasing the canal network to individuals for a fixed contract fee and allowing the individuals keep all of the earnings net of expenses. In other villages the managers of the water management system are rewarded with part of the water fees that are collected after the irrigation network's expenses have been paid. In villages that provide managers with strong incentives to save water, water use falls sharply while having almost no effect on agricultural production or rural incomes. In contrast, when villages do not provide incentives, managers do not save water.

One question not yet answered, however, is why only a fraction of villages institute managerial incentive structures, given the success of incentive-based systems at reducing water use. To try to explain this puzzling fact, we use a framework that predicts how the contractual form varies with the nature of village's cultivated land, the design of the canal system, and the characteristics of its leaders and the pool of managerial candidates. Because any villager is eligible to take on the role of canal manager, the pool of managerial candidates refers to all the villagers in our sample. Using this framework, we seek to empirically identify the factors that may induce leaders to choose one type of contractual form over another, using both descriptive statistics and multivariate analysis.

The research in this chapter is unique in that we treat the local-level managerial form of China's village irrigation system as endogenous and econometrically measure the factors that determine how water is managed. Unlike most previous studies that are anecdote-based, we use data from our own survey in northern China to describe and econometrically identify the determinants of the managerial form. This is the first empirical village-level study on this subject in China. More importantly, we have incorporated important factors that were not analyzed in previous studies: the characteristics of the key players. In most past studies, discussions of the appropriate institutional forms are usually limited to overcoming the transaction costs of enactment (Easter et al., 1998) or the legal and physical restrictions to their establishment (Radosevich, 1988). Baland and Platteau (1999) are among the few other authors (Dayton-Johnson, 2000a,b; Fujiie et al., 2005) that discuss the impact

Managing Water on China's Farms. http://dx.doi.org/10.1016/B978-0-12-805164-1.00009-9

of local characteristics on the success of proposed institutional arrangements for managing the commons, although none have examined the characteristics of the key players. In our study, we examine the role the leader and the pool of managerial candidates play in the endogenous choice of managerial forms.

The rest of the chapter is organized as follows. We first summarize the existing contractual forms. Next, we briefly explain the model of contractual choices and predictions from the model about the factors that determine how contractual relationships are formed in different types of environments. Using these predictions as a basis for generating testable hypotheses, our empirical work analyzes the relationship between contractual choice and a number of different characteristics of a village: the nature of its cultivated land, the design of its canal system, and the characteristics of the leader and the pool of managerial candidates. In the final section, we conclude.

THREE FORMS OF CONTRACTS IN MANAGING THE CANAL SYSTEM

The managerial contract between the leader and a manager governs not only each party's responsibilities in the operation of canals, but also their rights to claim the residual profits from irrigation services. There are two main types of activities in managing the canal system: *collective activities* and *supervisory activities*. Collective activities include tasks such as maintaining canals, coordinating water deliveries, and resolving water conflicts that involve several households. Collective tasks typically require the leader or manager to coordinate or facilitate negotiations among villagers. It should be noted that collective activities here are distinctly different from the term *collective action* in Ostrom (1990). In our case the leader or manager organizes the actions of villagers in water management activities; in Ostrom (1990), villagers must find ways to cooperate among themselves, often in a self-organizing way. In contrast, supervisory activities are tasks that are relatively effort intensive for the leader or manager. In the operation of a canal system, many supervisory tasks, such as water delivery (eg, the opening and closing of sluice gates), monitoring water allocations, and collecting water fees, are the key elements in the system's efficient and profitable operation. One of the fundamental ways that canal systems vary is that in some communities the leader assumes responsibility for all of the tasks; in other villages the manager assumes responsibility for most of the tasks; and in others, leaders and managers share responsibilities. Not surprisingly, in return for taking on greater (or fewer) responsibilities for implementing the different tasks, leaders and managers are compensated with more (or less) of the earnings of the system.

When sorted according to the ways managerial responsibilities and earnings are divided, irrigation systems in China's villages are managed in three distinct ways. Some villages have chosen to stay with the traditional form of leader-run irrigation management. In these villages, the leader is in charge of organizing both collective activities and supervisory activities. The leader either manages the whole system by himself or hires a foreman for a fixed wage to perform some of the water management tasks under his direct instruction. Importantly, under such a *fixed-wage* contract, the foreman does not have a strong incentive to put out effort since he does not get any share of the profit from the operation of the irrigation system. In fact, because the leader knows this, the foreman is not expected to take much initiative and at most carries out fairly routine activities under the direction of the leader.

In contrast, some leaders have transferred complete control and income rights of the village's canal network to managers on the basis of a *fixed-rent* contract. Under such an arrangement, the individual has effectively become the manager of a private water management company and is provided with full incentives. As long as the manager follows the rules and regulations set by the leader, provides quality irrigation services, and pays the fixed rent (or in terms of China's villages, the contract fee), the leader does not intervene directly in the manager's execution of day-to-day canal management tasks, neither collective activities nor supervisory ones. Finally, in other villages the leader and manager share both responsibilities and earnings from the irrigation tasks, a form of management that is called a *profit-sharing* arrangement. To *maximize the profit* from operating the village's irrigation system (or at least to operate it efficiently), one of the key decisions the leader needs to make is to put the individual that can run the irrigation system most efficiently in charge and to provide that individual with the incentives to do so. In other words, the leader decides how his village's irrigation system is managed by choosing the contract that will return the highest level of profits. Thus the profit-maximization behavior of the leader is consistent with efficiency from a social perspective.

MODELING MANAGERIAL CHOICE IN RURAL CHINA

This section briefly discusses the model in which the leader endogenously chooses the optimal contractual form, which is based on the model of contractual choice in tenancy contracting developed by Eswaran and Kotwal (1985). Technical details of the model and simulation results that lead to the predictions are in the section "Technical Details of the Model of Contractual Choices" in Methodological Appendix to Chapter 9. We make two basic assumptions. First, the leader and the manager have an absolute advantage in being able to perform one or the other of the two activities that are required to run an irrigation system. As the local authority, the leader is efficient at organizing collective activities such as mobilizing the labor of his villagers to clean canals, coordinating irrigation schedules among households, and resolving water conflicts. For example, using his executive authority, the leader can easily (at least in relative terms compared to any individual villager) mobilize households within the village to clean and perform maintenance work on the village's canal system on a seasonal or annual basis. If an individual manager were in charge of maintaining the canals on his own, he would almost invariably have to hire labors from inside and/or outside of the village to help him do so.

In contrast, we assume that the manager is endowed with a superior ability (and more available time) to execute supervisory operations. For example, while there is no reason to believe that leaders could not effectively manage water allocation operations if they had the time, one of our key points is that they often do not. Since the leader is also invariably burdened with many other duties, such as running village enterprises, implementing family planning, maintaining local schools and health facilities, as well as a myriad of other administrative responsibilities assigned by township officials, he almost assuredly is less able to provide the concentrated effort and attention to detail that is needed in certain villages to allocate water and provide other irrigation services. A full-time manager, however, can focus his energy and often can draw on his family members to provide the hours needed to meet the irrigation needs of the farmers in the command area. In addition, managers sometimes have an advantage over leaders in the time-intensive job of collecting water fees. Besides being a time-consuming task (almost always requiring going door-to-door, sometimes many times), households often will use various excuses to avoid or delay payment to the leader or his foreman, frequently taking advantage of the fact that the village is at least partly considered responsible for providing welfare functions to needy villagers. The leader may also be worried that if he forces the villager to pay the water fee, the villager will vote against him during the next election. These tactics are less effective when managers, who themselves are fellow villagers, come to collect payment for services rendered.

Our second key assumption is that the abilities of the parties to organize collective activities and supervisory activities are unmarketable, partly because those services are unobservable and therefore difficult to measure. For example, the leader could hire an individual foreman to supervise water delivery. In theory the leader would like to pay a wage to the foreman on the basis of the amount of effective work that he does. However, it is difficult for the leader to monitor the foreman, and it is impossible to measure the amount of effective work even after the tasks are performed. As a consequence, it is difficult for the leader to provide an incentive for the foremen. For these reasons, it is difficult or impossible to purchase the effort that is needed to execute supervisory tasks on the market. To obtain effective supervisory activities, the leader either has to do them himself or provide self-monitoring incentives to an individual that is able to carry out this task. Likewise, a manager cannot procure the authority that is needed to execute collective activities on the market.

With these assumptions, the unifying mechanism in the model that drives the leader's choice of contractual form to maximize profit from providing irrigation services to villagers is the relative change in the ability of the leader and manager to perform the unmarketable activities that are needed to provide irrigation services. This leads to the first prediction of the model:

> ***Prediction 1****: The dominant contractual form in a given village depends on the relative ability of the villager leader (manager) to perform supervisory (collective) activities. The more experienced the leader (manager) is at supervisory (organizing collective) activities, the more prevalent will be fixed-wage (fixed-rent) contracts.*

Factors that change the relative importance of supervisory activities and collective activities will also influence leaders' contractual choices. Two predictions can be made from the model:

> ***Prediction 2****: The optimal contractual form in a village depends on the nature of its cultivated land, such as the degree of land fragmentation. As land becomes more fragmented, supervisory activities become more valuable, and the leader will be more likely to grant a fixed-rent contract to an individual manager.*

For example, in villages with canal systems that wind intricately through fields of highly fragmented plots that are spread out spatially, the water delivery task is more demanding in that each household may grow different crops on different plots that have different water delivery quantity and timing requirements. Additional efforts are also needed toward the collection of water fees. In such a setting, there is a greater need for the manager to take on supervisory activities. Therefore, a leader will be more likely to offer a fixed-rent contract to a managerial candidate who can more efficiently deliver water and meet the specific water delivery requirements of each household.

> ***Prediction 3****: The optimal contractual form in a village depends on the design of the canal system that determines the degree to which the canal system is maintenance intensive. The less maintenance intensive the canal system, the less important are collective activities, and the more likely it is that the leader will select a fixed-rent contract.*

For example, lined canals typically require less maintenance (or are less *maintenance intensive*). In such a case, collective activities are less important. Since the leader will not play a role in managing the day-to-day operation of canals, there is a relatively greater need for a motivated manager to operate the rest of the irrigation system. Under these circumstances, there should be a greater propensity for the profit-maximizing leader to move away from a fixed-wage contract into a fixed-rent contract.

Finally, the model can show that, holding other factors constant, if the opportunity to find an off-farm job is high for the pool of individuals that potentially could take on the role of manager, the opportunity cost of forgoing other jobs to take on the tasks of managing the village's canals will increase. Then the leader will have to lower the fixed rent or increase the share (required by the managerial candidate) to attract the managerial candidate to take a fixed-rent or profit-sharing contract.

This means it would be less profitable for a leader to offer a fixed-rent or profit-sharing contract. We can summarize this as follows:

> ***Prediction 4:*** *The optimal contractual form in a village depends on the opportunity cost of the leader and the pool of managerial candidates, which is associated with the social-economic environment of the village. The wealthier a village is (or the more opportunities there are to find an off-farm job), the more likely it will be that a fixed-wage contract is chosen.*

VARIABLES

The analysis in this chapter uses the part of the 2000 China Water Institutions and Management Survey (CWIM) survey data that was collected in 32 randomly chosen villages in Ningxia. There are eight villages in our sample that have two canals. So in total we have 40 observations. In order to classify a contract as fixed-wage, we include villages in which there was no one except the village leaders involved in canal management (6% of fixed-wage contracts) and villages in which the foreman (or person—who is not a cadre—responsible for the operation of the canal system) is paid either a truly fixed-wage—88%—or is paid a salary that is less than 10% of the profits of the irrigation system—6%. A contract is categorized as a fixed-rent contract if the manager is the full residual claimant of the earnings of the irrigation system (ie, his share is 100% and the village only receives a lump-sum fee that is paid before beginning of the agricultural season). The remainder of the canal systems is categorized by profit-sharing contracts.

Two variables are used to measure the nature of the cultivated area within which the canal system is operated. The *degree of fragmentation* is measured as the average number of plots of each household in the village. In our survey, a plot is primarily defined as a piece of cultivated land of a farm household that is physically separate from all other cultivated land of the household. In some cases, however, a farmer will cultivate two or more different types of crops on a single physical plot; in such a case a single physical plot will be counted as more than one plot for our analysis. *Water abundance* is measured as an indicator variable that takes the values of 1, 2, 3, or 4. Water abundance is equal to 1 when a village's water resources are considered (by the village leader's subjective evaluation) to be highly scarce and equal to 4 when its water resources are considered to be abundant.

We also created three variables to measure the design of the village's canal infrastructure: *canal lining* (measured as lined if at least 10% of the length of the canal system within the village is lined); the *propensity to silt up* (which is a dummy variable that is equal to 1 if the respondent answered yes to both of the following questions: Does your canal system ever silt up? Is the siltation considered to be serious?); and the *density of sluice gates* (measured on a per kilometer basis as the number of openings in the irrigation system that were designed to move water from the irrigation canal network into the fields of farmers). These three variables are included to measure the relative importance of collective and supervisory activities in each village's irrigation system.

To capture the opportunity cost of the pool of managerial candidates, we construct two variables from the leader questionnaire that asked about the socioeconomic status of the village in detail for years 1990, 1995, and 2001. The opportunity cost of the pool of managerial candidates is captured in part by the *share of off-farm income*. This is measured as the average share of off-farm income in total household income. Alternatively, we measure the opportunity cost as the *proportion of migrants*, which is measured as the percentage of the village's labor force that lives and works outside of the village for at least one month per year. Both variables were measured during 1995 in order to make them exogenous.

Finally, we use standard measures of human capital, education, and age as proxies for the ability of leaders and managers to organize collective activities or take on supervisory activities. The *level of education of the leader* is measured as the number of years of education actually attained by the leader. Better educated leaders may be better suited to run supervisory activities versus those with poorer education. The *age of the leader* captures the level of physical fitness of the leader (which will fall with age) and may be associated with the leader's ability to undertake the strenuous demands of executing supervisory activities during the peak irrigation seasons. The *level of education of the managerial candidates* is measured at the village level as the percentage of the village labor force that has attained at least a high school education. The use of this variable depends on the assumption that managers that come out of such a pool would have a stronger inherent ability to organize collective activities.

CONTRACTUAL CHOICES AND THE NATURE OF THE IRRIGATION SYSTEM

Similar to what Wang et al. (2005) found, our data show that water management reform was implemented in some villages but not others (Table 9.1, rows 1 and 2). If villages that use fixed-wage contracts are categorized as unreformed villages, most of the reformed villages in our sample started to use either a profit-sharing or fixed-wage contract to manage the canals. In our sample, most of the reformed villages (82%) started to use either a profit-sharing or fixed-rent contract to manage their canals in either year 2000 or 2001. Other villages started earlier. For example, one village started to use a fixed-rent contract as early as 1995. Although a substantial share of leaders choose either profit-sharing or fixed-rent contracts (17.5% and 37.5%), fixed-wage contracts are still the dominant form of canal management (45%). Hence there are still many villages in our sample that have yet to reform.

The descriptive statistics are largely consistent with predictions derived from theory. The contractual choices of the leaders seem to vary systematically with the nature of the village's cultivated land, the characteristics of the canal system, the human capital, and the opportunity cost of the leader and managerial candidates.

Table 9.1 Characteristics of Canals, Water Resources, Village Leaders, and Villages Under Different Water Managerial Forms (Ningxia Province, China 2001)

	Managerial Form	Fixed-Wage	Profit-Sharing	Fixed-Rent
1	Number of observations	18	7	15
2	Percentage	(45)	(17.5)	(37.5)
Condition of the Natural Environment				
3	Degree of fragmentation (number of plots/household)	6.76	7.89	9.98***
4	Water abundance (index—4 if abundant; 1 if scarce)	3.61	3.71	3.53
Characteristics of the Canal System				
5	Canal lining (1 if lined, 0 otherwise)	0.28	0.29	0.53*
6	Propensity to silt up (1 if yes, 0 otherwise)	0.61	0.14**	0.60
7	Density of sluice gates (number/km)	41.54	22.60	38.19
Opportunity Cost of the Pool of Managerial Candidates				
8	Share of off-farm income (%)	42.90	27.71**	22.67***
9	Proportion of migrants (%, share of village labor force that out migrate)	21.55	14.23*	14.33*
Human Capital Characteristics (Ability Proxies)				
10	Level of education of the leader (attainment in years)	9.28	8.86	8.47
11	Level of education of the managerial candidates (%, share of labor force with at least high school education)	44.81	44.89	45.41
12	Age of the leader (years)	44.67	45.43	41.93

**denotes the difference between the mean under profit-sharing or fixed-rent contract and under a fixed wage contract is significant at 10%; **significant at 5%; ***significant at 1%.*

When the degree of fragmentation of a village's cultivated land is high, the leader appears to be more likely to use a fixed-rent contract (Table 9.1, row 3). When the canals in a village's irrigation system are lined, villages apparently are less likely to be run as a fixed-wage contract (only 28% of the 18 irrigation systems with lining are run as fixed-wage contracts—row 5). In contrast, irrigation systems that have a higher propensity to silt up appear to be more likely to be run under fixed-wage contracts (row 6). In villages with fixed-wage contracts, the share of off-farm income in a villager's total household income is almost twice that of villages with fixed-rent contracts (42.9% vs 22.7%, row 8). Villages also appear to favor fixed-wage contracts when a greater share of the labor force out-migrates (row 9).

The descriptive statistics, however, also uncover some patterns that are not entirely consistent with the predictions. In examining more closely the relationship between the propensity to silt up and contractual choice, we observe an up-down-up pattern when moving from fixed-wage contracts (highest) to profit-sharing (lowest) to fixed-rent contracts (higher than profit-sharing and lower than fixed-wage). We find another type of nonlinearity in the relationship between the contractual choice and some of the covariates (eg, with water abundance—Table 9.1, row 4; and age of leader—row 12). In these cases, we observe a down-up-down pattern when moving from fixed-wage contracts (lower than profit-sharing but higher than fixed-rent contracts) to profit-sharing contracts (highest) to fixed-rent contracts (lowest). It is possible that in Table 9.1 we are only observing two-way correlations and so we are not able to observe the true underlying relationships. In the next section, we then move to use multivariate analysis to examine more rigorously these relationships.

EXPLAINING CONTRACTUAL CHOICE IN CANAL MANAGEMENT: MULTIVARIATE ANALYSIS

In order to statistically test the predictions from the contractual choice model, we run a series of multinomial logit regressions. The **dependent variable is a discrete outcome variable with three alternatives**: fixed wage contract, profit-sharing contract, or fixed-rent contract. The factors from the descriptive analysis are used as explanatory variables in the regression to test our predictions about the determinants of contractual form. We explain contractual choice as a function of the nature of the village's cultivated land; the design of the canal; the opportunity costs of the pool of managerial candidates and leaders; and the human capital characteristics of the leaders and managerial candidates. Specifically, we use the degree of fragmentation and water abundance to test Prediction 2. We use the canal lining, propensity to silt up, and density of sluice gates to measure the characteristics of the canals in the village to test Prediction 3. We use two measures of the relative opportunity cost of canal managerial candidates and leaders, the share of off-farm income, and the proportion of migrants in the village to test Prediction 4. Finally, we use the human capital characteristics of leaders—the level of education and age of the village leader and the average level of education of those in the pool of managerial candidates—in order to test Prediction 1. The description of the multinomial logit regression and estimation issues are in the section "Multinomial Logit Regression to Explain Contractual Choice" in Appendix to Chapter 9.

Since we are interested in the leader's choice of one type of contract over another, in Table 9.2 we report coefficients that represent the relative probability of choosing one alternative over another. In general, our empirical estimations perform satisfactorily, especially given the fact that our sample is relatively small (Table 9.2). The goodness of fit measure, pseudo R^2, is around 0.9 for the multinomial logit equations with county dummies, which is quite high for analyses that use cross-sectional data. The coefficients are also jointly significant.

Table 9.2 Multinomial Logit Regressions With Fixed Effects at County Level Explaining Contractual Choice by the Village Leader in Ningxia Province

	Base Category	Model 1		Model 2		
		Fixed-Wage		Fixed-Wage		Profit-Sharing
		(1) Profit-Sharing $\log(p_2/p_1)$	(2) Fixed-Rent $\log(p_3/p_1)$	(3) Profit-Sharing $\log(p_2/p_1)$	(4) Fixed-Rent $\log(p_3/p_1)$	(5) Fixed-Rent $\log(p_3/p_2)$
Condition of the Natural Environment						
1	Degree of fragmentation (number of plots/ household)	19.15 (6.31)***	87.48 (13.95)***	−49.326 (3.33)***	23.779 (3.49)***	64.866 (3.61)***
2	Water abundance (index—4 if abundant; 1 if scarce)	141.40 (13.63)***	190.12 (14.46)***	33.924 (2.65)***	55.922 (3.07)***	20.109 (3.19)***
Characteristics of the Canal System						
3	Canal lining (1 if lined, 0 otherwise)	447.38 (14.02)***	420.75 (13.73)***	134.835 (3.73)***	108.493 (3.64)***	−24.137 (4.02)***
4	Propensity to silt up (1 if yes, 0 otherwise)	−148.97 (16.68)***	−9.18 (6.29)***	−237.693 (3.33)***	−2.591 (1.11)	209.026 (3.60)***
5	Density of sluice gates (number/km)	−0.04 (2.77)***	0.09 (6.30)***			
Opportunity Cost of the Pool of Managerial Candidates						
6	Share of off-farm income (%)	−21.53 (14.92)***	−18.60 (13.87)***	−14.913 (3.45)***	−4.999 (3.42)***	8.811 (3.70)***
7	Proportion of migrants (%, share of village labor force that out migrate)	−7.24 (15.80)***	−0.30 (1.45)			

Human Capital Characteristics (Ability Proxies)						
8	Level of education of the leader (attainment in years)	−55.15 (13.05)***	−55.26 (12.64)***	−0.313 (0.14)	−13.201 (3.51)***	−11.9 (3.48)***
9	Level of education of the managerial candidate (%, share of labor force with at least high school education)	18.54 (13.49)***	18.25 (13.62)***	−0.504 (0.79)	4.687 (3.66)***	4.590 (3.48)***
10	Age of the leader (years)	45.54 (14.79)***	32.54 (13.13)***	20.996 (3.54)***	8.316 (3.60)***	−11.266 (3.73)***
	Observations	40		40		
	Log likelihood	−2.945		−15.63		
	Percent correctly predicted (out-of-sample)	43		50		

*Robust z statistics in parentheses; *significant at 10%; **significant at 5%; ***significant at 1%.*

Most importantly, many of our results support the predictions of our model and help us identify factors that induce some villages to run their irrigation systems under fixed-rent contracts while others run them under profit-sharing or fixed-wage contracts. For example, the multivariate analysis is consistent with Prediction 2. The contractual choices vary systematically with the nature of the village's cultivated land. The coefficient in the fixed-rent equation on the variable *Degree of fragmentation* is positive and statistically significant when the fixed-wage contract is treated as the base category (row 1, column 4, Table 9.2). Hence the village leader is more likely to choose a fixed-rent contract relative to a fixed-wage contract. The positive and statistically significant coefficient in the fixed-rent equation when the profit-sharing contract is treated as the base category indicates that the village leader is also more likely to choose a fixed-rent contract relative to a profit-sharing contract (row 1, column 5). In short, when the degree of land fragmentation is higher and the coordination of water delivery is likely to be more complex and require closer supervision, the leader finds fixed-rent contracts more profitable. It should be noted that the coefficients on the degree of fragmentation for the profit-sharing contract have the opposite sign in Model 1 and Model 2. This is the only major difference in the estimation results generated by Model 1 and Model 2. Hence, we did not include the impact of the degree of fragmentation on the choice of profit-sharing contract here.

The relationship between water abundance and the contractual choices is also in line with Prediction 2. The positive and statistically significant coefficients show that the village leader is more likely to choose a fixed-rent contract as opposed to a fixed-wage contract (row 2, column 4) or a profit-sharing contract (row 2, column 5) when water resources become more abundant. Such a relationship may mean that when there is more water, there is likely to be less conflict over water and thus less need for the leader's skill at resolving the water-related conflicts. Moreover, with abundant water resources, there is more scope for the manager to utilize his skill at supervising water usage (eg, managing the sluice gates in a way that avoids excess water loss). Savings on the volume of water as an input in providing irrigation services increase the level of profit, and under such circumstances, the manager is more likely to be willing to take a fixed-rent contract.

The signs on the other sets of variables are somewhat supportive of the other predictions. When looking at the effect of characteristics of the canal system on the decision to choose a fixed-rent contract versus a fixed-wage contract, our data support Prediction 3. The positive sign on the canal lining variable demonstrates that as increasing investment improves the conditions of canals, the leader has a propensity to switch from a fixed-wage contract to a profit-sharing or fixed-rent contract (row 3, column 3 and 4, Table 9.2). On the other hand, the negative sign on the propensity to silt up variable indicates that a canal that often silts up and requires more cleaning makes the leader more reluctant to provide the managers with a fixed-rent contract or a profit-sharing contract (row 4, column 3 and 4, Table 9.2). Likewise, the multivariate analysis shows that in canal systems in which the density of sluice gates is higher, the leader is more likely to use a fixed-rent contract as opposed to a fixed-wage contract or a profit-sharing contract (row 5, column 1 and 2, Table 9.2).

Our estimated effects of variables representing the opportunity costs of the managerial candidates and the leader also are consistent with our theory (Prediction 4). The negative and statistically significant coefficients on the share of off-farm income variable in both the profit-sharing equation and the fixed-rent equation, when the fixed-wage contract is treated as the base category, show that when villagers have more access to off-farm activities, the leader is more inclined to run the canal system under a fixed-wage contract as opposed to a profit-sharing contract or a fixed-rent contract (row 6, column 3 and 4, Table 9.2). Similarly, the coefficient on the proportion of migrants variable indicates that when there are more migration opportunities (ie, it is easier to find an off-farm job outside the village), it is less likely that a managerial candidate will take a profit-sharing or fixed-rent contract as opposed to the case when the village leader runs the canal system under a fixed-wage contract (row 7, column 1 and 2, Table 9.2).

While generally consistent with the basic model, the relative magnitudes (and in one case the sign) of the coefficients on the variables measuring the three characteristics of the canal system and on the coefficient of both the opportunity cost variables show that there are nonlinearities in the relationship. For example, similar to the analysis in the descriptive statistics, as the propensity of the canal to silt up increases, the village leader is less likely to select a profit-sharing contract as opposed to a fixed-rent contract (row 4, column 5, Table 9.2). We also observe the same nonlinearities as the share of off-farm income or the proportion of migrants increases (row 6 and 7, Table 9.2). Under a profit-sharing contract, since the manager and the leader can only claim a fraction of the residual profits, they have weaker incentives. Given costly monitoring in our theoretical framework, it is possible that the manager will shirk his duty, and a moral hazard problem could arise (Marshall, 1956; Otsuka et al., 1992). For example, when the managerial candidates have more access to off-farm employment, they are likely to be able to generate higher earnings by working off-farm. In such a circumstance, after taking a profit-sharing contract, managers have a greater incentive to shirk their water management duties and spend as much time as possible working off the farm. Knowing this, even though the characteristics of the canal and environment are such that profit-sharing may be the optimal contractual arrangements (in the absence of imperfect information), the leader would be less inclined to offer the manager a profit-sharing contract.

Our results also contain nonlinearities that are consistent with the idea that the contractual arrangements between leaders and managers are affected by the role that profit-sharing contracts can play in sharing risk. In our theoretical model, both the leader and the manager are assumed to be risk-neutral. In an environment with uncertainty (reflected by the parameter, q, in Equation 1), whether the leader and/or the manager are risk-neutral will affect the contractual choices. As is well-known in the share contracting literature, with production uncertainty, the choice of contractual form affects not only the share of the responsibilities and the residual profit, but also the sharing of risk between the leader and the manager (Otsuka et al., 1992). Under a fixed-wage contract, the leader is burdened with all the risk. In contrast, the manager shoulders all the risk under a fixed-rent contract. The leader and the manager share the risk under a profit-sharing contract. According to our basic theory, when canals are lined, collective activities play less of a role and so leaders, ceteris paribus, are more inclined to rent the canal system to an individual manager. The positive and statistically significant coefficients on the canal lining variable in the profit-sharing and fixed rent equations (row 3, column 3 and 4, Table 9.2) confirm the idea that the village leader is more likely to use a fixed-rent contract or a profit-sharing contract as opposed to a fixed-wage contract. However, the results also show that the village leader is more likely to use a profit-sharing contract as opposed to a fixed-rent contract (row 3, column 5, Table 9.2). If leaders in such systems are willing to rent out the canal system, but managers, who are risk averse, are less willing to accept such an offer (because of the risk they assume as the contractor), it could be that the leader finds it optimal to engage the services of the manager as a profit-sharing contract. In this case, although the incentives for the manager are less than under a fixed-rent contract, the additional risk sharing offered by the profit-sharing contract is enough to get more managers to accept a managerial role.

Multivariate analysis also provides some support for Prediction 1: the optimal contractual choice depends on the relative abilities of the leader and potential managers. The negative signs on the coefficients of the level of education of the leader indicate that more capable leaders (ie, those with higher education) are less likely to run the canal system under a fixed-rent canal as opposed to running it by themselves under a fixed-wage contract or sharing the responsibilities under a profit-sharing contract (row 8, column 4 and 5, Table 9.2). In contrast, when the pool of managerial candidates is better educated and thus is more capable of running the irrigation system by themselves, leaders are more likely to use a fixed-rent contract as opposed to a fixed-wage contract or a profit-sharing contract (row 9, column 4 and 5, Table 9.2). Older leaders may be less able to manage the canal system (due to need to exert high levels of effort). Hence, older leaders are less likely to use a fixed-wage contract as opposed to a profit-sharing contract or a fixed-rent contract (row 10, column 3 and 4, Table 9.2). As above, however, the basic theory is not complete, since there is some evidence that our results are consistent with the idea that profit-sharing contracts are in fact demanded since they reduce risk.

CONCLUSIONS AND POLICY IMPLICATIONS

The chapter explains the puzzling fact that in pursuing water management reform, leaders have provided incentives to managers in some areas, but not in all. Our findings indicate that one of the reasons that not all leaders provide strong incentives stems from the specific characteristics of the irrigation system. If the conditions of canals do not allow for profitable operation of the canal under a profit-sharing or fixed-rent contract, the leader will not lease out the canal to the manager. In addition, the nature of the village's resources and its economic environment as well as the characteristics of its leaders and the pool of possible managers will affect contract choice. The absolute advantage-based explanation for the determinants of contractual choice explains much of what we observe in rural China. The findings also suggest contractual choice may depend on risk sharing and imperfect information.

Regardless of the exact determinants of contractual choice, our findings do help explain why even though strong incentives promote water savings, they are not used in all villages. The simple answer is that they are not appropriate to all villages. Hence, in China's future design of water management reforms, policy implementation should depend on the local conditions of the villages and it should be recognized that not one reform path fits all villages. Concretely, when designing policies on water management reform, instead of simply requesting leaders to provide incentives, China's policymakers need to take into account the features of the area where the reform is going to take place.

Our results also have implications for the design of China's broad water reform strategy. According to our results, water management reform has the potential to work in some areas. Hence, it should be encouraged. However, in other areas, reforms will not be appropriate. In such areas, pushing water management reform will not only be difficult, but may produce negative results if forced. Losing the leader's active participation could be counterproductive in villages that need collective action to be mobilized. In these other villages, if water is to be saved, upper-level policy officials may have to look beyond water management reform. In general, this means that a more integrated water reform strategy, using water management reform and focusing mostly on complementary policies, may be more successful in the long run.

REFERENCES

Baland, J.-M., Platteau, J.-P., 1999. The ambiguous impact of inequality on local resource management. World Development 27, 773–788.

Dayton-Johnson, J., 2000a. Choosing rules to govern the commons: a model with evidence from Mexico. Journal of Economic Behavior and Organization 42, 19–41.

Dayton-Johnson, J., 2000b. Determinants of collective action on the local commons: a model with evidence from Mexico. Journal of Development Economics 62, 181–208.

Easter, K.W., Dinar, A., Rosegrant, M.W., 1998. Water markets: transaction costs and institutional options. In: Easter, W.K., Rosegrant, M.W., Dinar, A. (Eds.), Markets for Water: Potential and Performance. Kluwer Academic Publishers, Boston, USA.

Eswaran, M., Kotwal, A., 1985. A theory of contractual structure in agriculture. American Economic Review 75, 352–367.

Fujiie, M., Hayami, Y., Kikuchi, M., 2005. The conditions of collective action for local commons management: the case of irrigation in the Philippines. Agricultural Economics 33, 179–189.

Marshall, A., 1956. Principles of Economics. Macmillan, London, UK.

Ostrom, E., 1990. Governing the Commons: The Evolution of Institutions for Collective Action. Cambridge University Press, New York.

Otsuka, K., Chuma, H., Hayami, Y., 1992. Land and labor contracts in agrarian economics: theories and facts. Journal of Economics Literature 30 (4), 1965–2018.

Radosevich, G., 1988. Legal considerations for coping with externalities in irrigated agriculture. In: O'Mara, G.T. (Ed.), Efficiency in Irrigation: The Conjunctive Use of Surface and Groundwater Resources (Chapter 3). The World Bank.

Wang, J., Xu, Z., Huang, J., Rozelle, S., 2005. Incentives in water management reform: assessing the effect on water use, production, and poverty in the Yellow River Basin. Environment and Development Economics 10, 769–799.

CHAPTER

10 Impacts of Surface Water Management Reforms

Since the late 1990s, China's policy makers have promoted surface water management reform. So far, the record seems to be mixed, although most evaluations are based only on anecdotes or case studies (Nian, 2001; Huang, 2001; China Irrigation Association, 2002). Even in those areas in which management reform has been well-designed, effective implementation has been difficult (Ma, 2001; Management Authority of Shaoshan Irrigation District, 2002). Visits to the field can easily uncover cases in which local water management changes were implemented and failed.

Although the literature argues that the success of water management reform depends on both managerial incentives and farmers' participation, individual scholars evaluating the performance of water management reforms typically stress one or the other. In China there is also an implicit debate. Some agencies, such as the World Bank (Reidinger, 2002), and some of China's researchers (Nian, 2001) believe that participation is of primary importance, and they have tried to incorporate participation components into their projects. When China implements reforms in other areas (outside of water management), reformers frequently give high priority to using incentives to encourage behavior that will push forward the new policies (Park and Rozelle, 1998). As water management reform begins to become more widespread, policy makers have to decide whether they should put more emphasis on providing incentives or on encouraging participation, or both.

So far, China's water officials have emphasized the role of managerial incentives in water management reform. In many of the new reform efforts, water managers are provided with monetary rewards if they can meet certain targets, such as achieving water savings. Far fewer efforts have been made to encourage farmer participation in the management of the local irrigation system.

The prominence given to incentives is not new in the context of China's overall economic reform effort. Reformers frequently have relied on incentives to induce agents to exert more effort, allocate resources more efficiently, and enter into new economic activities (Naughton, 1995). The household responsibility system primarily gave incentives to farmers in crop production (Lin, 1992). The fiscal reforms gave local leaders incentives to begin township and village enterprises (Walder, 1995). The grain reforms gave grain bureau personnel the incentive to commercialize commodity trading (Rozelle et al., 2000). Clearly, high-level water officials hope a similar set of reforms can improve the performance of China's water management.

Managing Water on China's Farms. http://dx.doi.org/10.1016/B978-0-12-805164-1.00010-5

Such a strategy may also encourage water managers to improve the efficiency of irrigation systems, although in the case of water management, there are a number of other issues that may create negative externalities. Since the reforms provide financial incentives to the manager to more efficiently manage water, it is possible that the managers act in such a way as to negatively affect production, income, and the poverty status of farmers, while increasing their own income. For example, managers could deliver less water than demanded by farmers or cut off water deliveries to slow-paying, poorer households. Despite the high stakes of the reforms, there has been little or no empirical-based work conducted to understand and judge the effectiveness of water management reform.

The overall goal of this chapter is to evaluate the performance of China's new surface water management institutions by empirically analyzing their effect on water use, agricultural production, and rural incomes. We also examine how these results have been achieved by looking at the role of incentives in water management reform. To pursue these goals, we have three objectives. First, we review the evolution of water management reform and identify the reform mechanisms that might affect rural outcomes, focusing especially on providing incentives to managers and increasing farmer participation. Second, we use multivariate analysis to identify the impact of these reform mechanisms on crop water use, agricultural production, and rural incomes.

REFORM AND THE EVOLUTION OF WATER MANAGEMENT

Based on our field surveys, after upper-level officials began implementing the reforms, surface water is managed in three ways. If the village leadership through the village committee directly takes responsibility for water allocation, canal operation, and maintenance (O&M) and fee collection, the village's irrigation system is said to be run by *collective management*, the system that essentially has allocated water in most of China's villages during the People's Republic period. A *WUA* is theoretically a farmer-based, participatory organization that is set up to manage the village's irrigation water. In WUAs a member-elected board is supposed to be assigned the control rights over the village's water. *Contracting* is a system in which the village leadership establishes a contract with an individual to manage the village's canal networks.

According to our data, since the early 1990s and especially after 1995, reform has successively established WUAs and contracting in place of collective management (Table 10.1). The share of collective management declined from 91% in 1990 to 64% in 2001 (column 5). Across our sample, contracting has developed more rapidly than WUAs. By 2001, 22% of villages managed their water under contracting and 14% through WUAs. During China's economic reforms, many government services have been contracted out to private individuals, including grain procurement, extension, and health services. Assuming the results from our sample reflect the more general trends across some parts of north China, the somewhat more rapid

Table 10.1 Surface Water Management in the Sample Villages in the Sample Irrigation Districts (IDs) in China, 1990–2001

	Ningxia		Henan		
	ID-1	ID-2	ID-1 (Percent)	ID-2	Total
1990					
Collective	100	81	100	100	91
WUA	0	5	0	0	3
Contracting	0	14	0	0	6
1995					
Collective	100	72	100	100	87
WUA	0	10	0	0	6
Contracting	0	18	0	0	7
2001					
Collective	27	51	92	100	64
WUA	50	14	0	0	14
Contracting	23	35	8	0	22

Note: In Ningxia Province, there are 8 villages in the ID-1 and 24 villages in the ID-2. In Henan Province, there are 10 villages in the ID-1 and 7 villages in the ID-2.
Authors' survey.

emergence of contracting may be due to the ease of setting the system up and the similarities of the reforms to the other reforms that have unfolded in rural China (Nyberg and Rozelle, 1999).

While there has been a shift from collective management to WUAs and contracting during the past 5 years, water management reform still varies across the four sample irrigation districts (IDs). WUAs and contracting have developed more rapidly in Ningxia than in Henan (Table 10.1). For example, in 1995 the collective ran 100% of the water management institutions in one of the Ningxia IDs (column 1). By 2001, however, the collective managed water in only 27% of the sample villages. WUAs managed water about 23% of the villages, and contractors managed water in approximately 50% of them. In Ningxia's other sample ID, the share of villages under WUAs and contracting reached 49%, almost the same as those under collective management (column 2). In contrast, significantly less reform occurred in Henan. Only 8% of the villages in one of the sample IDs, and none in the other, had moved to contracting or WUAs by 2001 (columns 3 and 4).

Based on our field survey, although some of the differences in water management among the IDs may be due to the characteristics of local villages and local water management initiatives, the dramatic differences between Ningxia and Henan Provinces suggest that upper-level government policy may be playing an important role. In 2000, to promote water management reform, Ningxia provincial water officials issued several documents that encouraged localities to proceed with water management reform (Wang, 2002). Regional water officials exerted considerable effort to promote water management reform in a number of experimental areas. The sharp shift away from collective management is consistent with an interpretation that these measures were effective in pushing (or at least relaxed the constraints that were holding back) reform. In our field work and during the survey, we spent considerable time discussing with officials how they pushed water management reforms. They told us that, in fact, they used a variety of ways, including issuing policy promotion documents, organizing meetings with local leaders to discuss the reforms, and talking to village leaders in more informal ways. In this way, we believe that the reforms have relaxed constraints that kept village leaders from reforming irrigation management. We believe the main function of the action of the officials has been to send a signal to village leaders that the previous prohibitions against contracting out the village's irrigation system were being relaxed.

The differences among the villages in Ningxia and variations in the way that different regions implemented the reforms (ie, some moved to contracting while others shifted to WUAs), however, show that the reforms are far from universal. In fact, this is what would be expected in China, a nation that often allows local governments considerable room in making their own decisions on the exact form and timing of institutional changes (Jin et al., 2000). In contrast, neither the Henan provincial government nor any of the prefectural governments have issued directives mandating reforms.

VARIATIONS IN REFORM MECHANISMS

While the shift in China's water management institutions demonstrates that the nation's communities are following policy directives that are being developed and issued from upper-level governments, when local leaders set up their organizational frameworks in their villages, practice often varies from theory. For example, in 70% of WUAs, the governing board was the village leadership itself, just as under the old collective model. Of the remaining 30% of WUAs, the village-appointed manager had close ties to the village leadership, with more than half having been a leader under the collective system. In other words, at least in terms of the composition of the management team, most WUAs in our sample differ little from collective management. Furthermore, in reality, farmers had little voice in managing or appointing the management team of their community's irrigation system. Based on our survey data, although 80% of WUAs hold regular meetings, only 30% of them invite farmers to participate. Even in the villages that invited farmers to participate, on average, only 5% of those that attended management meetings were farmer representatives. In this section, we take a closer look at the practical nature of the reformed surface water management institutions, focusing on changes in manager incentives, farmer participation, and overall accountability.

Change in Manager Incentives

An examination of the way that managers are compensated perhaps shows the greatest difference between theory and practice. To show this, however, we need to understand the way farmers pay fees, managers are compensated, and how IDs are paid. In the main text, we only discuss the water fees and payment for managers with incentives. In the collective management system, in villages without reform, there is no excess profit and so they do not face incentives to save water. In collectively managed canal systems, village leaders are told by the IDs how much per mu they need to collect, and they do so (regardless if the water delivered to them is relatively high or low). Farmers also pay on a per mu basis regardless of the amount of water delivered.[1] In these villages, if there is any water savings, it accrues to the ID. Since all of the revenues of the ID go directly to the financial bureau, there is no incentive for ID officials to try to save water under the prereform system. In fact, water management reform has created a complicated system of fees, payments, and charges that embody the primary incentives for the managers to save water. Water fees collected from farmers include two parts: *basic water fees* associated with the fixed quantity of land in the village and *volumetric water fees* loosely associated with the volume of water used. Once the manager collects that total fee from the farmer, he turns the basic fee part to the village accountant who in turn sends it to the township which is supposed to use the funds to maintain the township's canal infrastructure. Set by water bureau officials, the farmer is required to pay the basic water fee (which is based on his land holdings), and part of the basic water fee belongs to the water manager after it is collected. This part of the manager's compensation is paid to him as a *fixed payment* and provides little or no direct incentives to save water.

The calculation of the volumetric fee is somewhat more complex. Prior to the farming year, ID officials determine (on the basis of historic use patterns and other criteria) a targeted amount of water that a village should use (called the *target quantity*). Based on a per cubic meter charge, the total value of the expected water use for the village is then divided by the village's total quantity of land, and this volumetric water fee is added to the basic water fee to create the farmer's total water fee. Neither fee provides the farmer with incentives to save water, since both are fixed amounts based on land holdings. Incentives for water managers, however, stem from the collection of the volumetric fee. While farmers have to pay a fixed amount based on *expected* water use, in implementing water management reform, ID officials agree that the water manager only has to pay the per cubic meter charge for the water that is actually used (*actual quantity*). If the actual quantity of water delivered to the village (at the request of the water manager) is less than the target quantity, the difference between the volumetric fee collected from farmers and the water manager's payment to the ID is his *excess profit*. In communities that give the water manager full incentives, our data found that the share of excess profit over all manager revenue was 56%, indicating that this does represent a significant part of managers' income earnings.

According to our data, there are sharp differences in the way that villages have implemented the incentives part of the reform packages, regardless of whether they

[1]Mu is the unit of land used in China. 1 hectare = 15 mu.

Table 10.2 Incentives Provided to Water Managers in WUA and Contractors in the Sample Irrigation Districts (IDs) in China, 2001

	Percentage of Samples (%)		
	With Incentives	**Without Incentives**	**Total**
Whole Sample			
WUA and contracting	41	59	100
Ningxia Province			
ID-1			
WUA	25	75	100
Contracting	0	100	100
ID-2			
WUA	25	75	100
Contracting	76	24	100
Henan Province			
ID-1			
Contracting	0	100	100

Authors' survey.

are WUAs or contracting (Table 10.2). For example, in 2001, on average, leaders in only 41% of villages offered WUA and contracting (or *noncollective*) managers incentives that could be expected to induce managers to exert effort to save water in order to earn an excess profit (row 1). In the rest of the villages, although there was a nominal shift in the institution type (ie, leaders claimed that they were implementing WUAs or contracting), in fact, from an incentive point of view, the WUA and contracting managers faced no incentives (row 1). In these villages, water managers are like village leaders in a collectively managed system in that they do not have a financial incentive to save water. The incentives offered the managers differ across IDs (rows 2 to 6). Hence, to the extent that the incentives are the most important parts of the reform, the differences across time and space mean that it would not be surprising if in some villages WUAs and contracting were more effective at saving water than in others.

Somewhat ironically, since one of the main goals of water management reform is to provide farmers with better irrigation services, the design of the water management reforms placed little emphasis on the incentives for farmers. In many villages, at most, the water management reforms mandated that water fees paid by farmers should be reduced. This reduction in water fee was part of the initial arrangement between the contracting parties. The reformers, expecting a fairly large reduction in water use, wanted to make sure that farmers received at least some benefit (even though the reform was not supposed to affect their cropping incomes). This benefit was given to the farmers by reducing their water fees. However, the reduction in the water fees in most villages was quite modest. On average, water fees were reduced by only about 9%.

Changes in Farmer Participation

The focus on participation within the field of irrigation management has emerged from a concern about the effectiveness of management. The question that participation addresses is: "Who is best suited to carry out which management functions?" (Vermillion and Sagardoy, 1999). Those in favor of participation believe that farmers should decide on the roles that they would like to perform and the roles that they want managers and their leaders to perform. Participation can involve all aspects of irrigation, from whether or not the irrigation system should use farmer participation to who will head it and how it will run.

In our survey, we attempt to cover several major dimensions of participation. In particular, our definition of participation includes three parts: how farmers participated in the process of the establishment of reform process (eg, the setting up of the WUA); the selection of the managers; and whether or not farmers were invited to attend regular business meetings. These three aspects of decision-making cover most major activities of water management institutions (their creation, leader selection, and input into day-to-day business procedures).

Despite the important role that farmers play in water management in some parts of the world, according to our data, participation is not part of either China's traditional, collectively run water management or contracting. Traditionally, the implementation of many government services in China is carried out from the top down with little consultation with or participation of farmers (Zhang et al., 2002). Although collectively managed services, such as those provided by collectively run water organizations, in theory are supposed to be determined by the entire collective, in fact, village leaders have managed their villages in a large part based on the authority that they have derived from higher-level officials. In our sample villages, we find that farmers participate little (and mostly not at all) in collectively run water management organizations. Similarly, by definition (and according to our survey results), contracting involves transferring control and income rights to an individual and involves almost no participation of farm households.

In contrast, the reforms that led to the creation of WUAs explicitly attempt to encourage farmer participation. Again, however, practice often varies from theory. In our survey areas, farmers have little voice in deciding the establishment of WUAs or appointing the management team of their community's irrigation system. For example, at least in the early stages of the development of WUAs (the only stage of the organizations that we observe since this type of management is so new in our sample villages), our data show that on average only about 13% of WUAs involve farmers in the decision on their establishment (Table 10.3, row 1 and column 3). In fact, most farmers (70%) who are in villages in which the local irrigation system is being nominally managed by WUAs did not even know that they were part of a WUA. Although the meaning of farmer participation in China is similar to that in other international venues, this does not mean that China's farmers are being empowered by WUAs. We are only describing the extent of farmer participation. To the extent that farmer participation does not affect performance, we are unable to say if this is because the extent of participation is insufficient or if it does not inherently (for whatever reason) affect behavior.

Table 10.3 Farmer's Participation and Incentives Provided to Farmers in WUAs in Sample Irrigation Districts (IDs) in Ningxia Province, China, 2001

	Percentage of Samples (%)		
	WID-N (Weining)	QID-N (Qingtongxia)	Whole Sample
Farmer's Participation			
Decision on the establishment of WUA	0	25	12.5
Decision on selecting managers	25	25	25
Regular meetings	0	50	25
Above any activity	25	50	37.5

Authors' survey.

Farmers also are seldom encouraged to participate in other parts of water management. Based on our random sample, *none* of the WUA governing board members are actually elected by farmers. Only 25% of WUAs allow farmers to participate in the process of selecting managers (Table 10.3, row 2). As a result, in most cases (70% of the WUAs), the governing board of the WUA is the village leadership itself. In a minority of the cases (30% of the WUAs), village leaders appointed a chair or manager to carry out the day-to-day duties of the WUAs. In many of these WUAs, however, the managers actually have close ties to the village leadership (eg, the manager frequently is a former village leader or a close relative of a current one). Moreover, although 80% of WUAs hold regular meetings, farmers are invited to participate only in 25% of them.

Accountability

Compared with collective management and contracting, WUAs are more accountable to farmers. As discussed above, we assume that a relatively high degree of transparency, at least in part, reflects a relatively high degree of accountability. According to the field survey, we found that the degree of transparency for WUAs is higher than other management forms. In fact, all WUAs have some degree of transparency. Nearly 40% of WUAs shared all three types of information about the irrigation system with farmers (in other words, the WUA told farmers how water fees are generated, what volume of water was actually delivered by the ID to the village, and the actual area that was irrigated). About 50% of WUAs shared two of the three types of information. The rest (about 10%) shared at least one type with farmers. In contrast, neither collective managers nor contractors shared any of this information with farmers.

WATER MANAGEMENT AND CROP WATER USE

In the search for water savings, although it is possible that water managers may use certain methods of water management that would save water at the expense of farmers, in fact, most managers (perhaps under the scrutiny of village leaders and

ID officials) attempted to develop new ways of managing water that increased water use efficiency without having a systematic negative effect on production or incomes. In particular, based on our field survey, irrigation managers took actions to save water in a number of ways. They both improved the operation of the system (by supervising water delivery more intensively and using new techniques) and increased canal maintenance. For example, in the study regions, rice is one of the main crops. In local irrigation systems using traditional leader-run management regimes, leaders often used a practice in which there was continuous flooding of the fields for long periods of time during the season. Obviously, such a system is less supervision-intensive, as the only time the official needs to spend on management is the few minutes that it takes to open a few gates in the canal network when there is water in the main canals from the irrigation district. They can then forget about their management duties. In such a system, a lot of water flows through the village's canal system and directly into the outflow ditches.

However, in some reformed regions (in many cases under the direction of local extension agents), local canal managers adopted a system of irrigation called "alternate wetting and drying irrigation." According to this technology practice, after the irrigation canals are used to flood the fields, they are closed for a period of time right up to the time that the soil begins to dry out. At this point, the fields are then flooded again. Of course, to do this properly, it takes a lot more supervision time as the water deliveries are on-again and off-again and need more precise timing. According to a joint study by the International Water Management Institute, International Rice Research Institute, and Wuhan University's Hydrology Department, alternate wetting and drying irrigation can save up to 30% of the delivery of water to the fields without affecting yields (Barker et al., 2001). In some of the experiments, yields rose.

Of course, according to self-reported management efforts, we cannot rule out that part of the water savings did not come from unilateral reductions in water deliveries by canal managers (despite the fact that canal managers were not supposed to do so, by the terms of the contract). To examine this issue, we asked farmers about the timing of deliveries of irrigation water during the sample year and the impact that delayed deliveries (if any) had on their crop output. In response to these questions, we found that, although in our sample farmers did experience delays in irrigation that caused reduction in yields, such timing problems were not significantly related to whether an irrigation system's manager faced incentives (the correlation coefficient between incentives and delays is only 0.09 and insignificant from zero). In addition, although we found that the share of irrigation delay on plots managed by those with incentives is somewhat higher than those without incentives, it is not significantly so. The reasons for the delays are also different among plots run by managers facing different types of incentives. For example, most irrigation delays under management without incentives are related to poor management and low rates of payment of the village's water fee to the ID (Table 10.4). In contrast, most delays on plots managed by irrigation managers that do face incentives are due to water scarcities that kept water from being delivered to the ID.

Table 10.4 Irrigation Delays and Reasons for Delays by Various Types of Incentive Mechanisms in China's Surface Water Irrigation Management Institutions, 2001

	Nature of Managerial Incentives	
	With Incentive	Without Incentive
Irrigation Delays[a]		
All sample plots	74	294
No irrigation delays (%)[b]	72	81
Yield reductions due to irrigation delays (%)[c]	14	14
Reasons for Irrigation Delays[d]		
No water in canals due to dry weather (%)	35	9
Poor management (%)	65	73
Late payment of water fees, ID does not allocate water to village's canals (%)	0	18

[a] *Irrigation delays are reports by farmers that because irrigation water was not available at times when it was needed, the crop suffered damage.*
[b] *No irrigation delays represents the share of plots of farmers in the sample that did not suffer any irrigation delay.*
[c] *Yield reductions due to irrigation delay measures for plots that suffered irrigation delays an estimated of delay-related yield falls.*
[d] *Rows 4 to 6 in each column add to 100%.*
Authors' survey.

Much of the water savings we observed, then, came from management-induced efficiency improvements. Villages used "water rotation" irrigation (instead of flooding the entire village through a single outlet); "timed released" irrigation (a system that more carefully times the opening and closing of irrigation inlets and outlets); and improved canal maintenance that is implemented by lateral desilting and keeping the canal network inside the village free of debris and plant matter.

Although the major objective of water management reform is to save water, descriptive statistics using our data show that water use in some areas that have established WUAs and contracting is lower than those areas still under collective management, and higher in others (Table 10.5). For example, in the second ID in Ningxia (ID2), the water use per hectare in areas that have reformed (WUAs and contracting) is lower than those areas in which the collective still manages the water (rows 5 and 6 versus row 4). However, in Ningxia's other ID (ID1) and in Henan, water use per hectare is higher in those villages that have shifted to WUAs or contracting (rows 1 to 3; 7 and 8).

While the effectiveness of changing from collective to noncollective management in terms of water saving is not clear, descriptive statistics from our data show the importance of policy implementation. In particular, the potential importance of incentives in making the reforms work is shown when comparing water use in those villages that provide their water managers with incentives to those

Table 10.5 Relationship Between Incentives to Managers, Farmer's Participation, and Crop Water Use in Sample Irrigation Districts (IDs) in Ningxia Province, 2001

	Crop Water Use		
	(1) With Participation	**(2) Without Participation**	**(3) Test of Difference in Means (Column 1 vs. Column 2)[a]**
Measure of farmer's participation	(cubic meters per hectare)		
(1) decision on the establishment of WUA			
QID-N (Qingtongxia)	22,668	15,207	*0.61*
(2) decision on selection of managers			
WID-N (weining)	24,641	14,955	*0.39*
QID-N (Qingtongxia)	26,614	25,133	*0.61*
(3) Attendance at regular meetings			
QID-N (Qingtongxia)	16,108	15,337	*0.36*
(4) any of above activities in (1), (2), or (3)			
WID-N (weining)	19,928	15,083	*0.39*
QID-N (Qingtongxia)	26,614	25,133	*0.36*

[a] *Column (3) is the t-statistic for the difference between columns 1 and 2. None are significantly different at standard levels of confidence.*
Authors' survey.

that do not (Table 10.4). After reform, when managers face incentives to earn profits by saving water, water use per hectare is lower by nearly 10% when compared to collectively managed systems across our Ningxia sample (row 1, columns 1 and 3). In contrast, when leaders implement water management reform without providing incentives, water use is actually higher (column 2). When examining the individual IDs in Ningxia, we also find that in both IDs, water use is lower (or perhaps it does not rise as much) when incentives are provided during reform than when they are not. In ID2, for example, water use in lower in both noncollective systems with and without incentives, but it is even lower for those with incentives (row 3). In ID1, although water use in the both noncollective systems is higher, it is less high for those with incentives (row 2). We also find the same patterns occur when examining individual crops (rows 4 to 6).

Although the positive relationship between incentives and water use is supported by the descriptive statistics, the story is different when examining the relationship between farmer participation and water use. The data show that in villages in which farmers participate in water management, water use is not lower (Table 10.5, columns 1 and 2, rows 7 and 8). Specifically, if the management of WUAs allows farmers to participate in some way, the point estimate of crop water use is actually higher than the point estimate of those villages in which farmers do not participate.

Statistical tests of the difference between the means of villages with and without participation demonstrate that there is no significant difference (column 3). Our data also show that there are no statistical differences between villages that allow farmers to participate in different types of activities (eg, the decision to establish the WUA—rows 1 and 2; the decision to select managers—rows 3 and 4; and the encouragement by managers for farmers to attend regular meetings—rows 5 and 6).

ESTIMATION RESULTS OF ECONOMETRIC MODELS

Based on the above discussion, the link between water use per hectare and its determinants has been specified by econometric models. The description of the models are included in Methodological Appendix to Chapter 10.

After holding constant other factors, our results show that the mere act of shifting management from the collective to either a WUA system or contracting by itself does not lead to water savings (Table 10.6, row 1). The signs on the coefficients of the WUA and contracting variables are negative, suggesting that water use is lower in villages that have moved to noncollective management. However, the standard errors are all large relative to the magnitude of the coefficients, which implies that nominal institutional reform has no significant impact on saving water.

When officials provide water managers with incentives, without regard to whether they shifted to WUA or contract management, managers reduce water deliveries in the village (Table 10.6, row 3). Econometric results show that the coefficient on the incentive indicator variable is negative and significant (at the 10% level), when compared to the collective management, the omitted institutional type. In other words, without regard to the form of the water management institution, if managers face positive incentives, water use per hectare can be reduced by more than 6000 cubic meter per hectare, about 40% of average water use (row 3, column 4).

In contrast, our results show that the participation of farmers in water management does not reduce water use, as the corresponding coefficient is not significant in the water use equation (Table 10.6, row 5). These results imply that even when farmers are invited to participate in water management activities and the decision-making process, water use does not fall. While perhaps surprising, it should be remembered that although we randomly selected our villages, the sample only covers villages in old irrigation districts that have been established for many years. It could be that in new irrigation systems, such as those in countries which the World Bank has supported, the role of participation could be more important (Reidinger, 2002). Also, it could be that in our sample, village participation is not being encouraged in a way that is conducive to improving performance. In other words, if more active participation were encouraged, perhaps water management could be improved to the point that water could be saved.

Table 10.6 Regression Analysis of Determinants of Crop Water Use at Household Level

	Water Use per Hectare			
	OLS	**OLS**	**IV**	**IV**
Water Management Institutions				
Type of institution				
Share of WUA	−1311 (0.70)		−1920 (1.00)	
Share of contracting	−704 (0.49)		−2469 (1.34)	
Incentive				
Share of noncollective with incentives to managers[a]		−2637 (1.76)*		−6166 (1.82)*
Share of noncollective without incentives to managers[a]		955 (0.53)		1557 (0.45)
Participation				
Participate in any activity of water management (1 = yes 0 = no)		1169 (0.44)		13,429 (0.55)
Transparency				
Transparency in degree of water management		−2419 (0.95)		−10,081 (0.63)
Production Environment				
Share of village irrigated area serviced by surface water	2391 (0.99)	2029 (0.84)	2560 (1.08)	2520 (1.07)
Village water scarcity indicator variable (1 = yes 0 = no)	−3574 (3.13)***	−3703 (3.15)***	−3463 (3.03)***	−3549 (3.13)***
Value per hectare of accumulated investment into village irrigation infrastructure	−0.107 (1.01)	−0.038 (0.36)	−0.114 (1.11)	0.103 (0.61)
Cropping Structure				
Share of sown area in rice in 1995	10,592 (4.18)***	10,590 (4.20)***	10,655 (4.23)***	10,740 (4.21)***
Household Characteristics				
Age of household head	519 (1.17)	419 (0.94)	552 (1.25)	528 (1.20)
Age of household head, squared	−6.282 (1.28)	−5.217 (1.06)	−6.705 (1.37)	−6.328 (1.29)
Education of household head	−82 (0.50)	−66.9 (0.41)	−79 (0.48)	−46.8 (0.28)
Farm size	−10,487 (2.23)**	−8629 (1.73)*	−8964 (1.89)*	−6338 (1.25)
Irrigation District Indicator Variables				
QID-N (Qingtongxia)	−9888 (6.50)***	−8987 (5.95)***	−9968 (6.69)***	−8943 (4.55)***
PID-H (People's victory)	−11,151 (4.94)***	−10,743 (4.87)***	−11,588 (5.12)***	−10,584 (4.12)***
LID-H (Liuyuankou)	−15,752 (5.68)***	−15,334 (5.64)***	−16,105 (5.82)***	−15,193 (5.05)***
Constant	14,261 (1.43)	15,725 (1.57)	13,822 (1.39)	12,172 (1.22)
Observations	189	189	189	189
Adjusted R-square	0.44	0.45	0.45	0.45

** Absolute value of t statistics in parentheses.* Significant at the 10% level; ** significant at the 5% level; *** significant at the 1% level.*

[a] Noncollective institutions include WUAs and contracting.

Our calculations with survey data.

WATER MANAGEMENT, PRODUCTION, INCOME, AND POVERTY

According to our results, water management reform, at least when implemented as designed, leads to water saving and meets the primary goal of water sector officials. However, it is possible that the success from such a policy comes at a cost, either of falling production or falling income. In this section, we examine how water management reform affects agricultural production, rural incomes, and poverty.

Descriptive statistics from our data show that water management reform negatively influences agricultural production (Table 10.7, rows 1 to 3). Compared with villages that continue to operate irrigation by collective management, in the villages that provide incentives to managers to save water, wheat yields decline by nearly 10%. Maize and rice yields also decline by 9 and 12%, respectively. The negative effect of incentives on production is even clearer when comparing the yields between villages that nominally implement reforms but do not provide incentives to water managers with those that do provide incentives (rows 1 to 3, column 1 versus 2). In the case of wheat and maize yields, while production in villages with managers that have positive incentives fall, those in villages that have moved to WUAs and contracting, but have not provided incentives, actually rise marginally. In the case of rice, yields fall for villages that only reform nominally, but not as far as for villages that provide incentives to their managers. Since the pattern in production is consistent with, though in the opposite direction of, the correlations between water management and water use, the descriptive data suggest that water savings through management reform may only be able to come at a cost of lower yields.

In contrast, descriptive statistics show no evidence of a negative impact of incentives on farmer income (Table 10.7). Evidence from our survey reveals in the villages in which leaders reformed their water management system and provided incentives to managers, farmers actually earn higher income (row 4). Surprisingly,

Table 10.7 Incentives, Production, Income, and Poverty in the Sample Irrigation Districts (IDs), Ningxia and Henan Provinces, 2001

	Income (Dollars)	Cropping Income (Dollars)	Poverty Incidence (%)
Incentives to Managers			
Noncollective with incentives[a]	302.0	138.8	11.1
Noncollective without incentives[a]	254.4	101.4	6.5
Collective	210.0	93.9	7.5
Participation of Farmers			
With participation	293.3	120.5	0
Without participation	236.7	103.4	8.4

[a] *Noncollective institutions include both WUAs and contracting.*
Authors' survey.

crop income also is higher in villages that have provided managers with incentives (row 5). Part of the explanation for the difference between yields and income may be due to the fact that water fees also fall in villages that have reformed. It also may be that farmers are shifting their production decisions and allocating labor to other enterprises in villages that provide water managers with incentives. Econometric analysis is needed to isolate the effect of reform on cropping income and to distinguish between water management reform and poverty effects. In contrast to the case of income, our descriptive data show that poverty is worse in those villages that provide managers with incentives (row 6). Farmer participation does not appear to have a negative influence on either farmer income or poverty (rows 4 and 5).

ESTIMATION RESULTS OF ECONOMETRIC MODELS

In addition to water management reform, other socioeconomic factors influence agricultural production, income, and poverty. In order to answer the question of whether water management reform affects outcomes, it is necessary to control for these other factors. Therefore, we have specified several econometric models to analyze the effects of water management reform on agricultural production, income, and poverty. The description on the models are show in Methodology Appendix to Chapter 10.

Our results show that reforming water management reduces wheat yield but has no significant impact on the yields of maize and rice. From the wheat water use model, when villages provide water managers with incentives, managers reduce water use per hectare about 3800 cubic meters, a decline of about 50%. At the same time, the coefficient on the predicted water use variable in the wheat yield equation is positive and statistically significant (Table 10.7, column 1, row 1). The estimated water use elasticity for wheat yield is 0.226. Overall, our estimates of the size of the decline in water use and the responsiveness of wheat yields to water use imply that water management reform reduces wheat yields by about 11%. In contrast, although we find that incentives have a negative association on water use, the estimated water use elasticities for maize and rice are indistinguishable from zero (columns 2 and 3).

If our plot level analysis of water management and production are correct, then this would mean that in our sample areas the main tradeoff between the water savings from management reform and production occurs for wheat and is less severe or absent for maize and rice. The conclusion is plausible and, although its validity may only be true for our sample region, it is consistent with many of the observations we made in the field. Wheat is the crop that depends more than any other on irrigation because its growth period occurs almost entirely during the dry season. Water cutbacks should be expected to reduce yields. Maize, in contrast, is grown during the wet season, and water managers that have an incentive to save water may be able to time their use of irrigation water with the rains while those that have no interest in saving water might adhere to a predetermined water delivery schedule, no matter what the weather. In the case of rice, although the crop is dependent on large volumes of irrigation water, experiments by domestic and international water scientists have shown that there are many new ways of managing rice irrigation (eg, alternative wetting and drying—see Barker et al., 2001) that can lead to water savings but do not have significant yield effects in many cases. New water management technologies, however, require effort to learn and implement. Our results, then, may demonstrate that it is managers with incentives who have been able and willing to use new technologies, which bring water savings without large yield declines.

Our research results also demonstrate that water management reform has no statistically significant impact on farmer incomes (Table 10.8). When we use either an Ordinary Least Squares (OLS) or Two Stage Least Squares (2SLS) approach, the coefficients on the incentive variables in the both total and cropping income models are not statistically significant. Consistent with the descriptive statistics (which find no obvious fall in income in those villages that give water managers incentives), our results may suggest that whatever negative income effect there is from falling wheat production, it is being offset partially by reductions in water fees (though, as seen above, the reductions in water prices were fairly small). It could also be that the average reduction in income due to lower wheat yields is small enough, only $11.41 (11% of average wheat yield, 4740 kg per hectare, times average wheat sown area per household, 0.17 ha, times the price of wheat, $0.13 per kg) that they cannot be detected statistically. Moreover, since the fall in household income is less than 1.2%, the losses in cropping likely are being offset by other actions taken by households (eg, because water management is better, it is possible that farmers can focus more on other economic activities).

Similar to the results for incentives, farmers will not earn less money due to their participation in water management. Since participation of farmers has no effect on water use, it is not hard to understand why it has no negative influence on farmer income. In some areas of China, observers have noted arrangements in which village governments agree to pay for a targeted amount of water that is supplied to the village by the ID, but can make adjustments to the amount of fees that are ultimately charged to the farmers based on the actual amount of water provided. In other words, savings are passed on to farmers and are not captured by the canal managers. While this is interesting, this is not what is happening in our study villages.

Similar results also can be found in the poverty model. Since we measure poverty status as "under the poverty line or not," our results say that there is no effect on household poverty status of a village's decision to provide water managers with incentives or involve farmers in surface water management. If universally true, such a finding would be important, since critics of water management reform often point out that one possible adverse consequence of using incentives to induce water savings is that managers may cut back on water deliveries to marginal users, who may also be those on the poorest land with the lowest income. Our results here, however, should be interpreted with caution. First, we have not identified what may be behind this result. In many villages, leaders have specified strict rules in their agreements with water managers that they cannot exclude households from water allocation schedules. Second, as seen by examining the estimated equations in Table 10.9, only a few of the coefficients are significant, a sign that our sample may be too small

Table 10.8 Regression Analysis of Determinants of Farmer Income

	Total Income per Capita		Cropping Income per Capita	
	OLS	**IV**	**OLS**	**IV**
Water Management Institutions				
Incentives				
Share of noncollective with incentives to managers[a]	203 (0.61)	838 (1.24)	−118 (0.90)	91.5 (0.35)
Share of noncollective without incentives to managers[a]	−15.7 (0.05)	−383 (0.57)	38.5 (0.28)	−44.5 (0.17)
Participation				
Participate in any activity of water management (1 = yes 0 = no)	232 (0.45)	2496 (0.53)	27.7 (0.14)	1700 (0.92)
Transparency				
Transparency in degree of water management	−161 (0.33)	−1344 (0.44)	−256 (1.34)	−1263 (1.06)
Production Environment				
Share of village irrigated area serviced by surface water	345 (0.74)	311 (0.68)	−132 (0.72)	−123 (0.68)
Village water scarcity indicator variable (1 = yes 0 = no)	182 (0.79)	142 (0.64)	−12.3 (0.14)	−6.238 (0.07)
Value per hectare of accumulated investment in village irrigation infrastructure	0.069 (3.35)***	0.060 (1.83)*	0.014 (1.76)*	0.016 (1.22)
Cropping Structure				
Share of village rice area in 1995	215 (0.44)	232 (0.47)	−43.1 (0.23)	−21.9 (0.11)
Household Characteristics				
Age of household head	179 (2.05)**	176 (2.04)**	49.5 (1.46)	52.6 (1.55)
Age of household head, squared	−1.701 (1.76)*	−1.689 (1.76)*	−0.575 (1.53)	−0.606 (1.61)
Education of household head	23.0 (0.73)	21.7 (0.68)	−6.673 (0.54)	−6.671 (0.54)
Farm size	3205 (2.86)***	2990 (2.65)***	3267 (7.36)***	3149 (7.00)***
Total productive asset per capita	0.112 (3.48)***	0.110 (3.45)***		
Productive assets used in agricultural production per capita[b]			0.077 (1.66)*	0.080 (1.72)*
Number of plots per household	−125 (3.68)***	−129 (3.80)***	−4.577 (0.35)	−6.509 (0.49)
Production Shocks				
Production shocks	−233 (1.24)	−220 (1.18)	−189 (2.59)**	−182 (2.49)**
Irrigation Districts Indicator Variables				
QID-N (Qingtongxia)	−302 (1.01)	−457 (1.17)	−145 (1.23)	−234 (1.50)
PID-H (People's victory)	−1065 (2.38)**	−1177 (2.25)**	−69.9 (0.40)	−115 (0.56)
LID-H (Liuyuankou)	−938 (1.75)*	−1059 (1.77)*	−63.1 (0.30)	−108 (0.46)
Constant	−2599 (1.34)	−2332 (1.21)	−465 (0.62)	−492 (0.65)
Observations	189	189	189	189
Adjusted R-square	0.24	0.24	0.35	0.35

** Absolute value of t statistics in parentheses.* Significant at the 10% level; ** significant at the 5% level; *** significant at the 1% level.*

[a] Noncollective institutions include both WUAs and contracting.

[b] Productive assets include assets used for agricultural and nonagricultural production activities.

Authors' calculations with survey data.

Table 10.9 Regression Analysis of Determinants of Poverty

	Dummy of Poverty[b]	
	OLS	IV
Water Management Institutions		
Incentives to managers		
Share of noncollective with incentives to managers[a]	0.069 (0.96)	0.063 (0.43)
Share of noncollective without incentives to managers[a]	0.033 (0.45)	−0.072 (0.49)
Participation		
Participate in any activity of water management (1 = yes 0 = no)	−0.146 (1.33)	−0.166 (0.16)
Transparency		
Transparency in degree of water management	0.095 (0.89)	0.163 (0.24)
Production Environment		
Share of village irrigated area serviced by surface water	−0.180 (1.79)*	−0.167 (1.67)*
Village water scarcity indicator variable (1 = yes 0 = no)	−0.014 (0.29)	−0.007 (0.15)
Value per hectare of accumulated investment in village irrigation infrastructure	−0.000 (1.16)	−0.000 (0.92)
Cropping Structure		
Share of village rice area in 1995	0.002 (0.02)	0.013 (0.12)
Household Characteristics		
Age of household head	0.010 (0.56)	0.008 (0.43)
Age of household head, squared	−0.000 (0.70)	−0.000 (0.59)
Education of household head	−0.010 (1.47)	−0.010 (1.41)
Arable land per hectare of household	−0.242 (1.00)	−0.260 (1.06)
Total productive asset per capita	−0.000 (0.79)	−0.000 (0.72)
Number of plots per household	0.013 (1.78)*	0.013 (1.78)*
Production Shocks		
Dummy of production shocks (1 = yes 0 = no)	0.097 (2.40)**	0.095 (2.32)**
Irrigation District Indicator Variables		
QID-N (Qingtongxia)	0.027 (0.42)	−0.001 (0.01)
PID-H (People's victory)	0.044 (0.46)	0.006 (0.05)
LID-H (Liuyuankou)	−0.048 (0.41)	−0.084 (0.64)
Constant	0.014 (0.03)	0.103 (0.25)
Observations	189	189
Adjusted R-square	0.01	0.001

** Absolute value of t statistics in parentheses.* Significant at the 10% level; ** significant at the 5% level; *** significant at the 1% level.*

[a] Noncollective institutions include WUA and contracting.

[b] If per capita net income of household is lower than the national poverty line ($80.87 per capita per year in 2001), the dummy of poverty is one; otherwise, the dummy of poverty is 0.

Authors' calculations with survey data.

to identify poverty effects. In short, while interesting, we believe our current results may be more important as a tool that raises awareness of possible associations rather than providing definitive answers. Future research should try to pinpoint the source of this effect and use larger data sets to strengthen our understanding of these issues.

CONCLUSIONS AND POLICY IMPLICATIONS

In this chapter, we have sought to understand the reform of China's surface water management systems and its effect on water use, output, income, and poverty. Research results show that since 1990, collective water management has been replaced by WUAs and contracting in many locations. In some regions, the reform institutions have become the dominated form of management. Spread mostly by the efforts of water officials, we have shown that implementation has often deviated from theory. Participation by farmers has played only a minor role in most villages. In some villages, reform has been only nominally implemented, and there are few apparent differences when comparing the "reform" institutions to traditional management forms. In part because of these implementation problems, our analysis has shown that nominal reform has had little effect on water use.

The absence of a systematic relationship between nominal reform and water use, however, does not mean that the entire reform process has failed. Indeed, one of the main features of China's water management reforms, the provision of incentives to water managers, appears to have succeeded in achieving large water savings while having only a small or no effect on agricultural production, rural incomes, or poverty. Our findings demonstrate that in villages that provided water managers with strong incentives, water use fell sharply. The incentives also must have improved the efficiency of the irrigation systems since the output of major crops, such as rice and maize, did not fall, and rural incomes and poverty remained statistically unchanged. Only wheat production fell. Although our study needs to be undertaken in other areas in the future before the results can be generalized to the rest of China, at least in the sample sites that provided their manager incentives, water management reform has been nearly a win—win policy. We note that we can find little if any effect of participation by farmers on water use in our sample sites.

Overall, we believe that our findings support the conclusion that the government should continue to support water management reform. Officials that want the reforms to succeed should make an effort to ensure that more emphasis be put on the effective implementation. Although the negative impacts on production and farmer income were not found, in the longer run, as water management reform reaches into more water scarce areas and seeks to continue to achieve water savings in areas that have already cut back on use, there may be sharper tradeoffs between water use and production and income. When the tradeoffs are larger, officials still may choose to opt for pushing reforms that save water. In these cases, since the farmers that lose access to water could also suffer production and income falls, policies to mitigate the adverse consequences should be developed.

Although the literature emphasizes the importance of participation for water management reform, we find little if any effect of participation by farmers on water use in our sample sites. Perhaps the degree of participation in our sample sites is so low that it does not play an important role. Our survey shows that, in fact, farmers have not participated actively in many important activities of water management. In any event, the results suggest that further analysis of the determinants and effect of participation is needed.

It should be noted that it is not automatic that field level delivery savings (the type of savings that we are talking about here) lead to true basin-wide savings. It is possible that water lost during delivery and on the field could help recharge underground aquifers. However, since there is so little groundwater used in our IDs, this is not really an issue. It also is possible that if all of the water that was used on the fields prior to reform went into the groundwater and flowed back into the river, there would be really no true savings. While the share of groundwater that is "wasted" during delivery to the field in some parts of the Yellow River system may flow back into the river, in other parts (eg, in Inner Mongolia, Shaanxi, and Henan) there are sinks and areas in which the river bed is above the area being irrigated. In China's dry, hot, and windy areas along most parts of the Yellow River, overirrigation could lead to significant amounts of evaporation. Hence, although we do not know for sure (and cannot find any hydrologists that know for certain), it seems that in this case, reducing field deliveries could be leading to true water savings.

REFERENCES

Barker, R., Loeve, R., Tuong, T.P., 2001. Water saving irrigation for rice. In: Proceedings of International Workshop, Held by International Water Management Institute, International Rice Research Institute, Wuhan University of Technology and Zhejiang University in Wuhan, 23–24 March 2001, Wuhan, China.

China Irrigation District Association, 2002, Participatory irrigation management: management pattern reform of state-owned irrigation district. Paper Presented at the Sixth International Forum of Participatory Irrigation Management, Held by the Ministry of Water Resources and the World Bank, 21–26 April 2002, Beijing, China.

Huang, W., 2001. Reform irrigation management system, realizing economic independency of irrigation district. In: Nian, L. (Ed.), Participatory Irrigation Management: Innovation and Development of Irrigation System. China Water Resources and Hydropower Publishing House, Beijing, China.

Jin, H., Qian, Y., Weingast, B., 2000. Regional Decentralization and Fiscal Incentives: Federalism, Chinese Style. Hoover Institution, Stanford University. Working Paper.

Lin, J., March 1992. Rural reforms and agricultural growth in China. American Economic Review 82, 34–51.

Ma, Z., 2001. Deepening reform of farmer managed irrigation system, promoting sustainable development of irrigation district. In: Nian, L. (Ed.), Participatory Irrigation Management: Innovation and Development of Irrigation System. China Water Resources and Hydropower Publishing House, Beijing, China.

Management Authority of Shaoshan Irrigation District, 2002. Positively promoting reform based on practices of irrigation district, obtaining achievement of both management and efficiency. Paper Presented at the Sixth International Forum of Participatory Irrigation Management, Held by the Ministry of Water Resources and World Bank, 21–26 April 2002, Beijing, China.

Naughton, B., 1995. Growing Out of the Plan: Chinese Economic Reform, 1978–1993. Cambridge University Press, New York, USA.

Nian, L., 2001. Participatory Irrigation Management: Innovation and Development of Irrigation System. China Water Resources and Hydropower Publishing House, Beijing, China.

Nyberg, A., Rozelle, S., 1999. Accelerating China's Rural Transformation. Published by the World Bank.

Park, A., Rozelle, S., 1998. Reforming state-market relations in rural China. Economics of Transition 6, 461–480.

Reidinger, R., 2002. Participatory irrigation management: self-financing independent irrigation and drainage district in China. Paper presented at the Sixth International Forum of Participatory Irrigation Management, Held by the Ministry of Water Resources and the World Bank, 21–26 April, Beijing, China.

Rozelle, S., Park, A., Huang, J., Jin, H., 2000. Bureaucrat to entrepreneur: the changing role of the state in China's grain economy. Economic Development and Cultural Change 48, 227–252.

Vermillion, D., Sagardoy, J.A., 1999. Transfer of irrigation management services: Guidelines, 58. FAO Irrigation and Drainage Paper.

Wang, J., 2002. Field Survey Note in Ningxia Province. Center for Chinese Agricultural Policy, Chinese Academy of Sciences (unpublished).

Walder, A.G., 1995. Local governments as industrial firms: an organizational analysis of China's transitional economy. American Journal of Sociology 101 (September), 263–301.

Zhang, L., Huang, J., Rozelle, S., 2002. Growth or Policy? Which Is Winning China's War on Poverty. Center for Chinese Agricultural Policy, Chinese Academy of Sciences. Working Paper.

CHAPTER

11 Evaluation of Water User Associations

Since the late 1990s China's policy makers have promoted surface water management reform. So far, the record seems to be mixed, although most evaluations are based only on anecdotes or case studies (Nian, 2001; Huang, 2001; China Irrigation Association, 2002). Even in those areas in which management reform has been well-designed, effective implementation has been difficult (Ma, 2001; Management Authority of Shaoshan Irrigation District, 2002). Visits to the field can easily uncover cases in which local water management changes were implemented and failed.

Perhaps because of this mixed performance, shortly after the launching of the Water User Association (WUA) movement in China, those that believed that WUAs were a way to effectively manage water began to formalize the basic precepts that were needed to run a successful WUA. Specifically, in the mid-1990s World Bank project managers delineated a set of five principles that they believed were necessary for the successful implementation of a decentralized water management reforms based on WUAs. Principles 2 to 5 were first explicitly spelled out in the mid-1990s (World Bank, 2003). At a somewhat later date (a couple of years later, in the late 1990s), project managers added one more principle, which we have included as Principle 1 (Xie, 2007). The Five Principles are:

Principle 1—adequate and reliable water supply: A WUA is organized only where an adequate and reliable water supply is available and where on-farm delivery infrastructure are in good condition and can be properly maintained by WUA members.

Principle 2—legal status and participation: A WUA should be the farmers' own organization, a legal entity and has a leadership elected by its members.

Principle 3—WUAs organized within hydraulic boundaries: The jurisdiction of a WUA should be the hydraulic boundaries of the delivery system.

Principle 4—water deliveries can be measured volumetrically: A WUA should be able to receive its water under contract from its water suppliers and water should be able to be measured volumetrically.

Principle 5—WUA equitably collects water charges from members: A WUA should equitably assess and collect water charges from its members and make payment for the cost of water.

Managing Water on China's Farms. http://dx.doi.org/10.1016/B978-0-12-805164-1.00011-7

At least since the late 1990s, then, World Bank water project teams in China and some of their government collaborators have insisted that any successful WUA program would necessarily be based on these Five Principles. Surprisingly, given the high profile that the World Bank's WUA projects have assumed, in fact, there has never been a rigorous evaluation conducted by an independent research team. Since their introduction, no one knows the answer to the question: How have WUAs spread across China? In the areas that have been implemented by the World Bank, how have they been implemented? Have project managers followed the Five Principles? When the World Bank implements WUAs, is there any evidence that China's own water officials have learned from the efforts? Do WUAs matter? Is there evidence that when WUAs are implemented that water use falls, crop yields rise, and efficiency increases?

The overall goal of this chapter is to continue the evaluation of China's WUAs. We examine the prevalence of WUAs organized based on the Five Principles, and assess their effectiveness compared with WUAs not organized according to the Five Principles.

Four types of water management institutions will be a part of our analysis: Bank-supported WUAs; WUAs in areas adjacent to the World Bank project sites (non-Bank WUAs); WUAs in non-World Bank areas (in particular, in our Ningxia sample, or Ningxia WUAs); and collective water management institutions (ie, villages that manage water the traditional way—or villages in China where the village leadership (or, the leaders of the collective))—manages the water. After documenting the differences in the ways that villages manage water, we then examine differences in the performance of villages. In other words, our unit of analysis becomes the type of institution (Bank WUAs; non-Bank WUAs; Ningxia WUAs; and collectively managed villages).

To meet these objectives, the rest of the chapter is organized as follows. First, we discuss the data and use them to examine how WUAs are organized. In the discussion of the nature of WUA organization, differences among the different types of WUAs are examined. Next, we look at differences in the ways that different types of WUAs have affected performance. In the final section we conclude.

ORGANIZING CHINA'S WUAs: THE FIVE PRINCIPLES?

We use data from the China Water Institutions and Management (CWIM) and Bank surveys to document the nature of China's WUAs. Because of the hypothesized importance of the Five Principles, we use these to organize the analysis. We also examine ways in which different types of WUAs are treated differently—in ways unrelated to the implementation of the five principles.

THE UNDERLYING LOGIC OF THE WORLD BANK PRINCIPLES

In two of Elinor Ostrom's most well-known papers on the determinants of success of collective action (Ostrom, 2000, 2009), she discusses the elements that are needed for institutions that require collective action to succeed and survive. In fact, these treatises can help identify the more general, underlying logic of the Five Principles of the World Bank's WUA development efforts. First, Ostrom emphasizes that the nature of a resource system's current productivity is an important determinant of whether or not collective action is feasible (Ostrom, 2009). According to Ostrom (2000), if a water resource is already exhausted, local users will not see a need to manage the resource. Hence, in this way Ostrom's logic can help explain why the World Bank made "access to an adequate and reliable water supply" their first principle for organizing WUAs. If there is not a reliable water supply, or if the water system's delivery infrastructure is in poor condition, the high cost and low potential benefit from the effort needed to organize an active WUA will discourage participation.

The second principle proposed by the World Bank is also consistent with the ideas of Ostrom. In order for WUAs to succeed, Principle 2 states that the WUA should be the farmers' own organization and it should have a leadership elected by its members. In defining a successful model of collective action, Ostrom (2000) states that most individuals in the resource command system should be a part of the system that makes and modifies the rules. When users have full autonomy to craft and enforce their own rules, the collective body will face lower transaction costs in managing the resource (as well as lower costs in defending a resource against invasion by others). In addition, according to Ostrom, most long-surviving resource regimes select their own monitors (or leaders), who are accountable to the users or are users themselves and who keep an eye on resource conditions as well as on user behavior (Ostrom, 2000). This, too, is a transaction cost-based argument since when leaders are respected, and themselves are local users and are known to the group and have legitimacy, self-organization is more likely to be sustained.

Ostrom's work can also be used to explain the inclusion of Principle 3 in the World Bank's WUA prescription (which states that WUAs should be organized within hydraulic boundaries of the delivery system). In fact, when discussing the principles of long-surviving, self-organized resource regimes, the presence of clear boundary rules has been listed as the first principle (Ostrom, 2000). Under this principle, participants can know clearly who is in and who is out of a defined set of relationships and thus with whom the members should cooperate. In the case of a surface water irrigation system, if a WUA manages all of the water within a single, well-defined hydrological boundary, it will be easier for the members of the WUA to realize who is inside and outside, and why they need to integrate decisions about water allocation and water delivery coordination. When all of the water that enters a single catchment is the extent of the resource that needs to be managed, the information problems are lessened and management decisions are made clearer.

Finally, Principles 4 and 5 are also consistent with the ideas of Ostrom's collective action theory. The World Bank suggests that WUAs should be able to receive its water under contract from its water suppliers and that the water should be able to be measured volumetrically when they receive water from water suppliers (Principle 4). It also suggests that the WUA should equitably assess and collect water charges from its members in a way that the amount of the water payments is sufficient to cover the costs of the resource management and delivery (Principle 5). In her theory of successful collective action, Ostrom believes that it is important that local rules in use should be able to restrict the amount, timing, and technology of harvesting the resource, allocate benefits proportional to required inputs and that rules of access to the resource (and payment for the resource) are crafted in such a way that they take local conditions into account. When this happens, there is a sustainable balance of inputs and outputs that relieves stress on the organization. In many ways this is what is accomplished when WUAs sign the formal contracts with the Irrigation District for delivering a certain amount of water (that can be measured volumetrically). The payment scheme of farmers for the water is also important for these reasons. When the amount of water that is to be delivered is known and the payment for the water is born by the members, better decision making can be made (since savings are shared; and expansions are jointly financed and enjoyed).

PRINCIPLE 1: ADEQUATE AND RELIABLE WATER SUPPLY

The data that we collected in the Bank WUA, non-Bank WUA, collectively managed, and Ningxia WUA villages demonstrate that the Bank villages were not randomly chosen in terms of access to reliable supplies of irrigation water. According to our findings, Bank villages had a number of characteristics that endowed them with more reliable supplies of water (Fig. 11.1, Panel A). For example, in a number of ways the infrastructure of Bank village was better. While 90% of the cultivated land in Bank villages was irrigated, only 63% of the cultivated land was irrigated in collectively managed villages. The canal system itself was also better. In 93% of the villages, the branch canals had lining. In contrast only 79% of non-Bank, 50% of collectively managed, and 22% of Ningxia villages had lined branch canals. Moreover, more than 85% of the length of the branch canals was lined; while the share was lower in all of the other villages and less than half in most of the other villages.

The nature of the scarcity of water also differed among the types of villages (Fig. 11.1, Panels B and C). According to our data, the farmers in Bank villages faced less water shortages and had access to more abundant and flexible water resources. While farmers in 93% of non-Bank villages claimed water was scarce, those in 60–67% of Bank and collective villages said water was scarce. When a follow-up question was asked about the perception of the share of years between 2000 and 2005 that water was in sufficient supply, the responses by farmers in Bank villages suggest that water was more abundant in these villages (80%) than when compared to the non-Bank, collective, and Ningxia villages (40–60%).

Although for space purposes, we do not show the data here, the perception of village leaders were consistent with the perceptions of farmers. While in 46% of collectively managed villages, 53% of Ningxia villages, and 60% of non-Bank villages leaders and WUA managers stated that there were years of water shortages

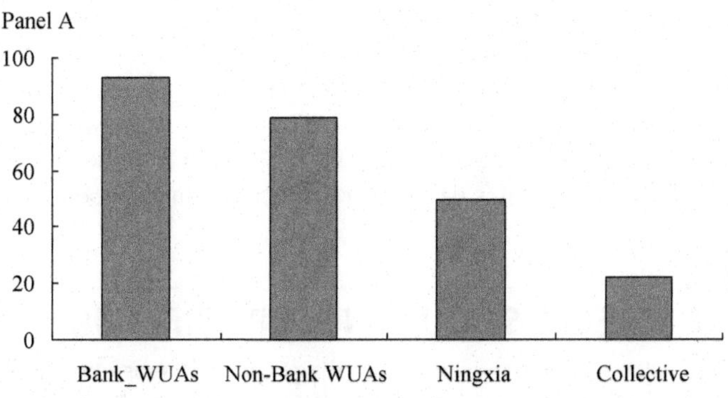

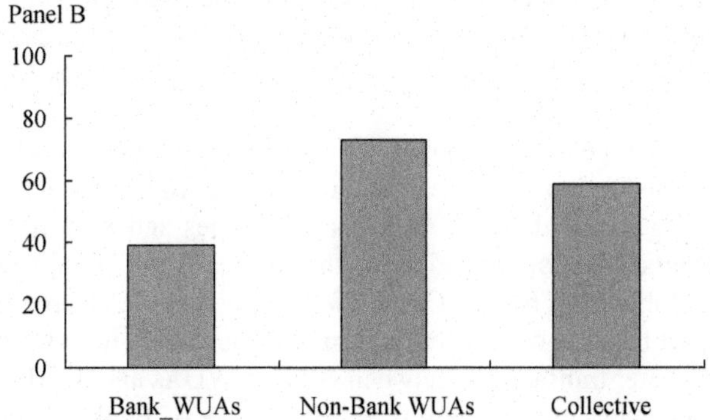

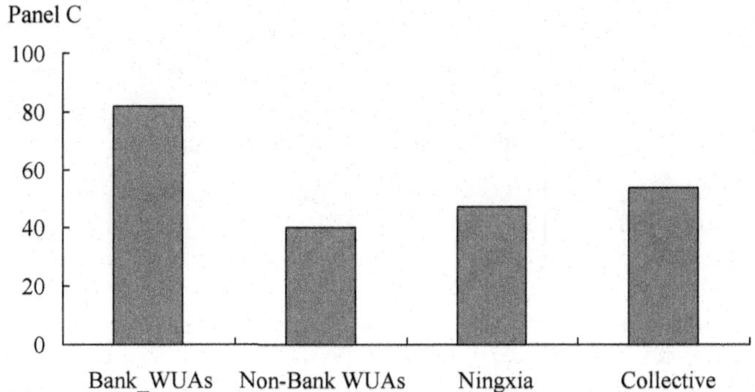

FIGURE 11.1

Local infrastructure and water supply conditions. Panel A: local infrastructure: villages with lined branch canals, Unit: share of villages with lined branch canals (%); Panel B: water supply conditions: whether the supply of water is short? Unit: share of villages with short water supply (%); Panel C: water supply conditions: share of years with sufficient water between 2000 and 2005, Unit: share of years with sufficient water (%).

Source of data: 2004 CWIM survey and 2006 Bank survey.

in the early 2000s (between 2002 and 2005), village leaders and WUA managers in only 19% of Bank villages stated that there were years without enough water. Moreover, according to village leaders, there was better access to conjunctive water use (ie, there were groundwater resources available even if there were shortages in deliveries of surface water) in Bank villages (47%) than in collectively managed villages in the Bank survey sites (0%) or the Ningxia villages (3%).

PRINCIPLE 2: LEGAL STATUS AND PARTICIPATION

The legal status also differs between Bank villages and villages outside of the Bank site (ie, in Ningxia; Table 11.1). For example, WUAs inside Bank sites are somewhat more formal than other WUAs. In 100% of Bank and non-Bank villages, the WUA has a constitution (row 1). In contrast, fewer Ningxia WUA villages (80%) have constitutions.

Greater differences appear when looking beyond the relatively simple formality of having a written constitution (which can often be copied from a publicly available document). According to our data, 100% of Bank villages and 80% of non-Bank villages have registered their organizations with the local Civil Affairs Bureau as an "official WUA" (Table 11.1, row 2). Only 30% of Ningxia WUAs have registered. Moreover (and in part because of their more formal legal status), in terms of written contracts that guide water transactions between villages/WUAs and the ID, the Bank villages are also somewhat more formal (row 3). In 86% of Bank villages, WUA

Table 11.1 The Difference of WUAs in Different Types of Villages

	Bank Villages WUA	Non-Bank Villages WUA	Ningxia Villages WUA
Share of WUAs having written constitution (%)	100	100	90
Share of WUAs that registered (%)	100	80	30
Share of WUAs having written contract with irrigation districts (%)	86	67	20
Share of WUAs in which farmers participate in meetings (%)	96	93	19
Share of villages' farmers participating in meetings (%)	25	11	2
Share of WUAs in which the chair is village leader (%)	19	20	85

Source of data: 2004 CWIM survey and 2006 Bank survey.

managers have a written contract with the ID. This is true of 67% of non-Bank villages. However, the share of villages in Ningxia with water delivery contracts with the ID is much lower (only 20%).

In the same way that there are differences between Bank WUAs and WUAs in non-Bank sites in procedures that are involved with setting up WUAs, there also are sharp differences in participation once the WUAs are in operation (Table 11.1, rows 4 and 5). In both Bank and non-Bank villages (96% and 93%, respectively), farmer-respondents confirmed that WUA management and operation meetings were open to the participation of farmers. In only 19% of villages in Ningxia, however, did farmer-respondents say meetings were open. The rules on meeting openness translated into higher farmer attendance in management meetings—especially in the case of Bank villages. The rate of participation by farmers was 25% in Bank villages (ie, one out of four farmers in the village attended at least one meeting during the year). The share of participation was 11% in non-Bank villages. It was only 2% in Ningxia villages.

Finally, the differences in participation rules and rates between villages in the Bank site and villages in Ningxia also are correlated with differences in governance. In 85% of the villages in Ningxia, the village leader is also the formal head of the WUA. During many interviews with farmers we discovered that when the village leader was the head of the WUA many farmers believed that there was only a change in the name of the water management institution and that there was no substantive change. In the case of the Bank (non-Bank) villages, however, only 19–20% of WUAs were headed by the village leader. Clearly, it appears as if there were true governance differences between villages inside and outside of Bank project sites.

PRINCIPLE 3: WUAs ORGANIZED WITHIN HYDRAULIC BOUNDARIES

We also can use our data to show that there are differences in the principle used to set up WUAs with regards to the hydraulic boundaries of the delivery system. In the Bank villages and in the non-Bank Villages (within the Bank project site) all (100%) of the villages are in WUAs that are created within the hydraulic boundaries of the irrigation delivery system. The Ningxia non-Bank WUAs, however, violate this principle. In more than 50% of Ningxia WUAs, the hydraulic boundaries do not completely align with WUA boundaries. In a majority of the villages, the WUA is super imposed on a single village while the village's water delivery system is shared with another village. Of course, in some sense, this difference in the execution of Principle 3 may be responsible for the observed differences in governance. When more than one village is involved in a single WUA, there are many reasons why a village leader (who might be perceived as favoring his/her villagers over others) would not be elected WUA chair.

PRINCIPLE 4: WATER DELIVERIES CAN BE MEASURED VOLUMETRICALLY

Similar to the other three principles, the organizers of WUAs in the Bank sites (in the case of WUAs in both the Bank and non-Bank villages) have also been successful in implementing Principle 4. According to our survey, water managers in Bank (non-Bank) villages pay for water by water use in 93–100% of the villages (Fig. 11.3). This means that in most or even all villages, water can be measured (and charged for) volumetrically. Because of this, of course, in these villages there should be large incentive to save water. In the Ningxia villages, however, this number is lower. Only 65% of WUA villages can pay the Irrigation District (ID) for water according to how much water was used, and 35% of villages still pay for water deliveries by area.

PRINCIPLE 5: NATURE OF WAY IN WHICH WUA COLLECTS WATER CHARGES FROM MEMBERS

In assessing differences in the ways that WUAs and other water management organizations manage the collection of water fees there are a number of fundamental differences. One of them, however, is not "if water fees are collected." In all villages in our sample—100% of Bank WUAs, non-Bank WUAs, Ningxia WUAs, and collectively managed villages—collect water fees from farmers (Fig. 11.2).

Beyond the fact that water fees are collected, however, villages differ in the ways that fees are managed. Bank and non-Bank WUAs in Bank sites exerted effort to make the size and basis of the water fee charge (from the WUA to the farmer) transparent (Fig. 11.3, Panel A). More than 90% of farmers in the WUAs in Bank sites said they knew that the nature of the water charge from the WUA to them was well publicized. Only 5% of Ningxia WUA farmers, and no farmers in the collectively

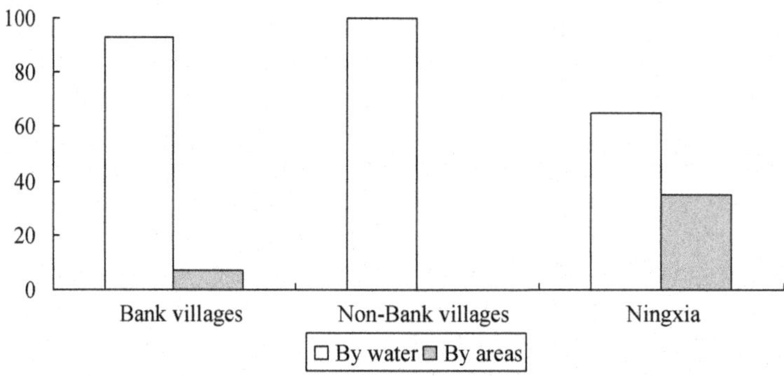

FIGURE 11.2

Types of charging water fees from ID to village/WUA. Unit: share of villages (%).

Source of data: 2004 CWIM survey and 2006 Bank survey.

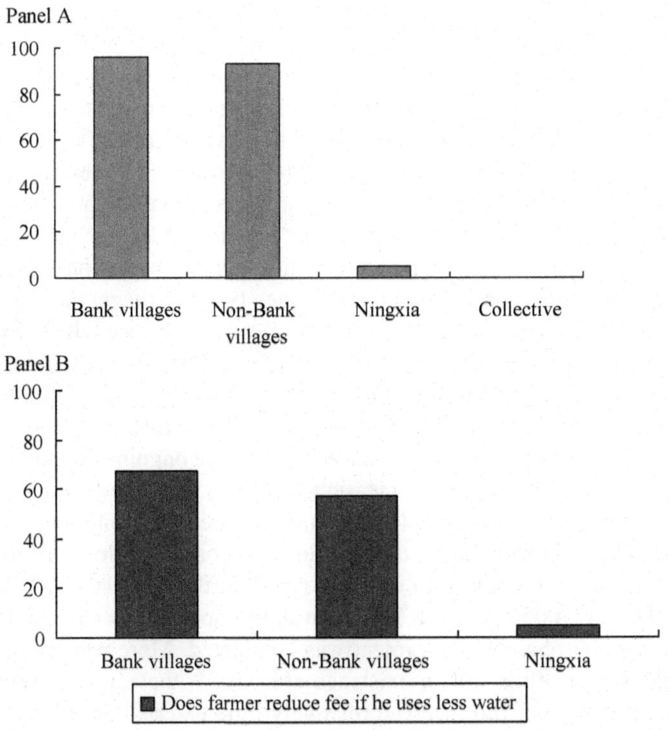

FIGURE 11.3

Nature of way WUAs collect water charges from members. Panel A: transparency of WUA management, Unit: share of villages that make water fees open to farmers (%); Panel B: water fee-based incentives for farmers to participate, Unit: share of villages (%).

Source of data: 2004 CWIM survey and 2006 Bank survey.

managed villages (in Bank project sites), were aware that information on water fees was being displayed in the village.

Likewise, farmers in most Bank WUA villages (67%) and non-Bank WUA villages (57%) had their water fee reduced if they used less water (Fig. 11.3, Panel B). This occurred in only 5% of Ningxia WUAs. In other words, WUA officials in the Bank sites provided an incentive (or at least compensation) to farmers when they used less water.

A SUMMARY: THE FIVE PRINCIPLES AND BEST PRACTICE

When summarizing the findings from the previous two subsections we, in fact, find that the Bank WUAs (as well as non-Bank WUAs in the Bank sites) are being implemented largely—though not fully—in ways that are consistent with best practice or the Five Principles (Table 11.1). According to the Five Principles, WUAs have access to adequate and reliable water supply (row 1); they should have clear legal status and encourage farmer participation (row 2); they should be organized within hydraulic boundaries (row 3); they should be able to receive water from the ID that is measured volumetrically (row 4); and they should have clear and reasonable water fee systems—between the WUA and its members (row 5). As seen from our summary table, the WUAs in the Bank villages—when compared to non-Bank villages and especially Ningxia villages—are using best practice (at least in a relative sense). There is room for improvement, particularly in the implementation of Principle 4. However, our data makes clear the ranking of each type of village: the Bank villages in the Bank sites are implementing WUAs in closest compliance with the Five Principles. Remarkably, perhaps, the non-Bank WUAs in the Bank sites are a close second. Clearly, there is either learning or extension efforts that are spilling over from the Bank to the non-Bank villages. The Ningxia villages rank far lower.

In the literature, there is evidence that WUAs require ongoing support to survive. Therefore, one concern is that the measured differences in performance between Bank WUAs and other institutional forms may be due to the ongoing support and not to differences from applying the Five Principles (or the spillover from applying the Five Principles). According to our survey, while the earliest WUA was established in 1998, by 2005 all of the WUAs in the sample had been established and the implementation phase of the project was completed. Moreover, although there was a Bank project office still in existence after 2005, these were mostly staffed by local water officials, who received relatively little external assistance and were charged with carrying on all of their other responsibilities. Since our survey was organized in 2006, we believe that most of the marginal differences among Bank, non-Bank, and collectively managed villages are explained by the spread of ideas about the Five Principles. There may be a bit of residual effect from special, ongoing extension support, but it is most likely only minimal.

PERFORMANCE

In this section we have two distinct parts. The first part looks at how the WUAs in the different types of villages have performed in terms of "implementing their procedures and changing their management approaches." Because of the obvious possibility that WUA managers may exaggerate the extent of the changes that they have made, in this section we rely on the observations and opinions of farmers only. In the second part, we examine the impact that different types of WUAs have had on how water is managed and, in turn, the impact upon water use, yields, income per capita, and cropping patterns. Although we include the descriptive analysis here, in footnotes we refer to multivariate analysis that for the most part is consistent with our descriptive findings.

It is possible, of course, that the differences we measure between Bank WUAs and non-Bank WUAs, and between Bank WUAs/non-Bank WUAs and villages managed by traditional collective management institutions, are a result of more than the adoption of the Five Principles of WUA Management alone. It may be that there is something unobservable creating the measured difference (ie, there may be a problem of endogeneity due to unobserved heterogeneity). We recognize this danger. And, as a result in this chapter, we believe that we did as much as possible, given that the Bank and non-Bank WUAs were not randomly assigned. We used multivariate analysis, holding constant a number of observable factors before examining the effect of the Five Principles (and/or of WUAs as a management entity). We chose our sample carefully in order to minimize unobserved heterogeneity (at least that which is due to the location of the villages) by choosing the non-Bank WUA villages and collectively managed villages that were geographically close to Bank WUA villages. We also compare the Bank WUAs with WUAs in another environment (those in Ningxia). All of these efforts were made to minimize the problems of endogeneity due to the existence of unobserved heterogeneity. But it is also important to remember that while the results may still be affected by some unobserved heterogeneity, when comparing our results against the previous literature on WUAs (where there is almost no effort to consider these problems), our results are likely one of the strongest efforts to appear in the literature to date.

IMPACT ON MANAGEMENT PRACTICES AND OUTCOMES

Our findings indicate that the promotion of Bank WUAs, in fact, has made a significant and material difference in the way that WUAs execute management practices. This is seen in two ways. First, by comparing Bank WUAs with non-Bank WUAs in terms of how they use each individual management practice. When doing this, we see that the differences are significant. Specifically, in the case of 30% of the Bank villages, farmers reported that there were new management procedures, which resulted in improved coordination of irrigation deliveries. While low, it is true that this number is only 17% for non-Bank WUAs. The difference is about 43%. Similar differences can be found in comparing the other management practices, such as new canal maintenance activities (100% difference), new water saving practices (64% difference), and "concrete" changes on management (42% difference). In relative terms, our study found consistently measured differences in management practices.

Second, we also see differences when we create a more comprehensive measure for the change of any management practice. In total, at least one change (new ways of coordinating deliveries; canal maintenance; new water saving practices; or change management procedures) was made in 93% of Bank villages and 43% of non-Bank villages. When examined in this way, the difference between Bank and non-Bank villages is both significant in relative terms (54%) and in absolute terms (ie, 50 percentage points). Hence, based on this empirical evidence, we do not believe that it is erroneous to conclude that the implementation of Bank WUAs has led to better management practices.

While specific changes in water management practices were not perceived by farmers in all villages, there was an overwhelming perception by farmers that some of the ways of doing business improved. In 96% of Bank villages and 93% of non-Bank villages farmers reported that information on the volume of water use was communicated to the members of the WUA in a transparent manner. Farmers in most villages—both Bank villages (77%) and non-Bank villages (100%)—also told enumerators that the WUA leadership shared with the membership, in an open way, information regarding the total area that was irrigated (a number that is important in some villages to keep track of and calculate water charges). Hence, while some villages may not have changed the exact technical activities by which they managed water, according to our data, almost all villages began to change the way that they do business with, and relate information to, farmers in the village.

The perception by farmers about the degree to which they participated in irrigation management (as opposed to the physical act of attending meetings, which was reported in Table 11.1) also shifted, especially in Bank villages (Fig. 11.4, Panel A). Farmers in more than half of the Bank villages reported that most farmers in the villages (any that wanted to) actually participated in (and could in part influence) all of the management activities. Farmers in another 11% of villages said that most farmers participated in what they believe were the important management activities. While fewer farmers in non-Bank villages told enumerators that most farmers participated in all management activities (only 29%), the number of villages in which farmers claimed to participate in all, or all important, activities was the same for non-Bank villages ($65 = 29 + 36$) as for Bank villages ($63 = 52 + 11$). Only 33% of farmers in Bank villages, and 21% of farmers in non-Bank villages, said that there was no participation in management. This number is very small relative to the findings in the Ningxia villages, in which more than 90% of villages reported little or no participation by farmers (Wang et al., 2006).

Perhaps the most compelling evidence of the success that the Bank villages achieved is in improving water management—even relative to the non-Bank villages. During the survey, the enumerators asked farmers if they believed that water management had improved, relative to the traditional collectively managed form of water control, after the adoption of their village's WUA. One-hundred percent of Bank-villages (versus only 50% of non-Bank villages) said that, overall, water was better managed by the WUA. Higher percentages of farmer focus groups also stated that there was more effort centered on saving water (93%); more timely delivery of water (96%); a reduction in water charges to farmers—in either an absolute or relative sense (89%); and less conflict when water fees were being collected (81%). The comparable percentage figures are much lower for the non-Bank villages. These data show that the some combination of factors—the way the WUA was set up, governed, and/or managed—have greatly improved at least the perception of farmers about the way Bank WUAs are managing water. From this point of view, the Bank WUA projects can be considered a success.

In addition, there is a perception by farmers that WUAs—especially those in the Bank villages—were able to better resolve conflicts (Fig. 11.4, Panel B). To elicit the opinions of farmers on this, enumerators first asked the respondents whether or not, before the establishment of WUAs, there was any conflict within the village or

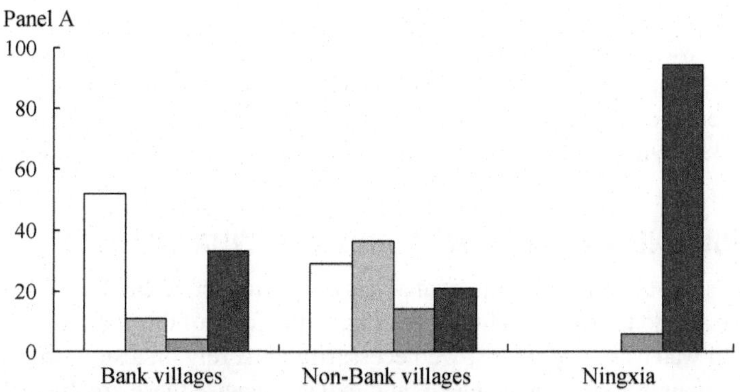

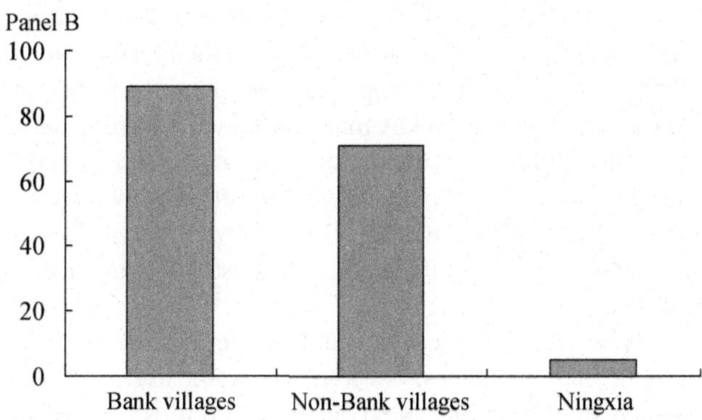

FIGURE 11.4

Performance of WUAs. Panel A: nature of actual farmer participation, Unit: share of villages (%); Panel B: reduction in conflicts among farmers after WUA, Unit: share of villages (%).

Source of data: 2004 CWIM survey and 2006 Bank survey.

between villages over water allocation (when water supply were scarce) or over the order in which water was delivered (to both farmers with the villages and among villages along the canal). In both cases (ie, either regarding conflicts among farmers within villages; or among villages within the ID), the Bank WUAs were more effective in reducing conflict. In fact, while in 89% of Bank villages conflict among farmers within villages were reduced (versus 71% in non-Bank villages), there were no Bank villages in which conflicts rose (while conflicts rose in 14% of non-Bank WUAs). While we do not know the exact reason why, it likely has something to do with the participatory nature of WUAs and the way WUAs give farmers a platform from which they can address and resolve problems.

The relative success of Bank villages in resolving conflicts among villages within the ID was even greater. In 73% of Bank villages, farmers told enumerators that conflicts among villages that shared the same canal reduced after initiating the WUA. Farmers did not report rises in conflicts in any of the Bank villages. Conflicts fell in non-Bank villages also, but in a much lower share of villages (only 21%).

WATER USE AND YIELDS IN WUA AND NON-WUA VILLAGES

Fig. 11.5 presents the initial results of analysis of the impact of the Bank's WUAs on performance. To do the analysis for the impact of the form of the water management institution on water use and yields, we necessarily must rely on a sub-sample of the villages. We mainly look at rice, wheat, and maize, because these are the only crops that were grown in more than 50% of all villages. They are also still China's three main crops in terms of sown area and output. In addition, when we asked each village leader about water use and yields, we only collected information on the three major grain crops grown in the village. If rice, wheat, and corn were not one of the three major grain crops, we did not collect information on them. This rule was chosen to avoid collecting information on a crop that was of minor importance in a village and on which the village leader/WUA manager may not have good information. Finally, we could only collect information on water use if the village actually had sown area for the crop that was irrigated. For a subset of the villages, farmers cultivated wheat or maize on nonirrigated land. Although we have yield information for these crops, we did not use them in Fig. 11.5 because we wanted the sample coverage to be the same in both tables.

Relying on estimates by village leaders of water use per hectare for three major crops grown in the Bank survey villages, we find that water use in the Bank and non-Bank villages (which are all in Bank sites) on all major crops is far below water use in collectively managed villages. For example, in the case of rice, farmers in the Bank (non-Bank) villages used 6585 (5985) cubic meters per hectare, while those in the collectively managed villages used 8145 cubic meters per hectare. This means that rice farmers in the Bank and non-Bank villages used around 20% less water. Although it might seem counterintuitive that the use of irrigation water on rice should be less than the use of irrigation water on wheat, it should be remembered that the crops (as well as other crops in the study) are grown at different times of the year (Table 11.2). Therefore, these figures are plausible since rice is grown during the rainy season and, according to our data, requires less irrigation delivery. In the rest of the chapter, although we only say "water use," we mean "use of irrigation water."

Although we do not systematically study the impact of water savings on ecosystem sustainability, it is worth thinking through the possible implications. In many cases, when there are water savings from the adoption of new management practices in the agricultural sector (as we have found in this case), it can also contribute to more effective integrated water resources management and ecosystem sustainability. Integrated water resources management is a set of systematic procedures that can contribute to the sustainable development, allocation, and monitoring of water

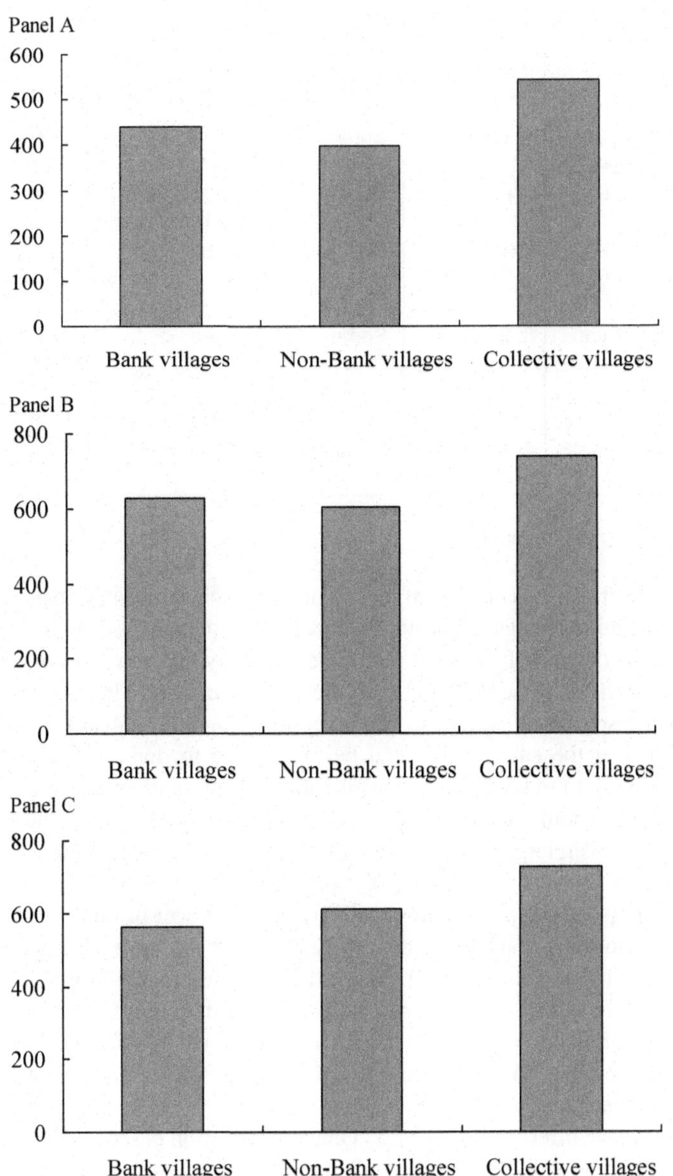

FIGURE 11.5

Water use for rice, wheat, and maize under alternative water management institutions. Panel A: water use for rice under alternative water management institutions, Unit: m^3/mu; Panel B: water use for wheat under alternative water management institutions, Unit: m^3/mu; Panel C: water use for maize under alternative water management institutions, Unit: m^3/mu.

Source of data: 2004 CWIM survey and 2006 Bank survey.

Table 11.2 Examination of Implementation of WUA Five Principles

Principles	Bank WUAs	Non-Bank WUAs	Ningxia's WUAs	Summary
Adequate and reliable water supply	Best	Some yes/ some no	Some yes/ some no	Best practice: Bank WUAs
Legal status and participation	Best	Good	Poor ... little participation	Best practice: Bank WUAs
WUAs organized within hydraulic boundaries	Most	Only some	None	Best practice: Bank WUAs
Water deliveries can be measured volumetrically	Less	Some	Most	NO (Ningxia is best)
Nature of way WUAs collect water charges from members	All/ transparent	All/ transparent	All/non transparent	Best practice: Bank WUAs

resource use in the context of social, economic, and environmental objectives. In its simplest incarnation, integrated water resources management is a logical and intuitively appealing concept. Its basis is that there are many different uses of finite water resources. Moreover, these different uses are interdependent. Therefore, faced with limited water resources, if water use efficiency in the agricultural sector is improved, it is possible that the saved water can be reallocated to domestic, industrial, and ecosystem sectors. In many cases, the availability of new sources of water can reduce pressures in the system and can lead to greater stability. However, it is also possible that if water has been largely consumed in the agricultural sector and the saved water is discharged into a system that is not set up to capture the saved resource, that there likely is not material impact on the sustainable development of other sectors.

The data also show that similar differences in water use appear for the other major crops (Fig. 11.5, Panels B and C). Wheat farmers in the Bank (non-Bank) villages used 9420 (9075) cubic meters per hectare compared to more than 11,115 cubic meters per hectare in the collective villages, a level about 15% lower. Maize farmers in Bank and non-Bank villages also used from 16 to 23% less water than those in collective villages.

So what can we infer from Fig. 11.5? One interpretation based on Fig. 11.5 might be that Bank and non-Bank WUAs are managed in such a way that they save water (as well as being managed better) because when local WUAs are managed according to the Five Principles they can manage water more efficiently. However, interpreting these results requires caution, because they need to be examined in conjunction with information on yields.

To illustrate one of the problems of coming to a conclusion about the efficiency of a single form of water management based on water use alone, we also present yields by water management type in Fig. 11.6. According to our data, while water

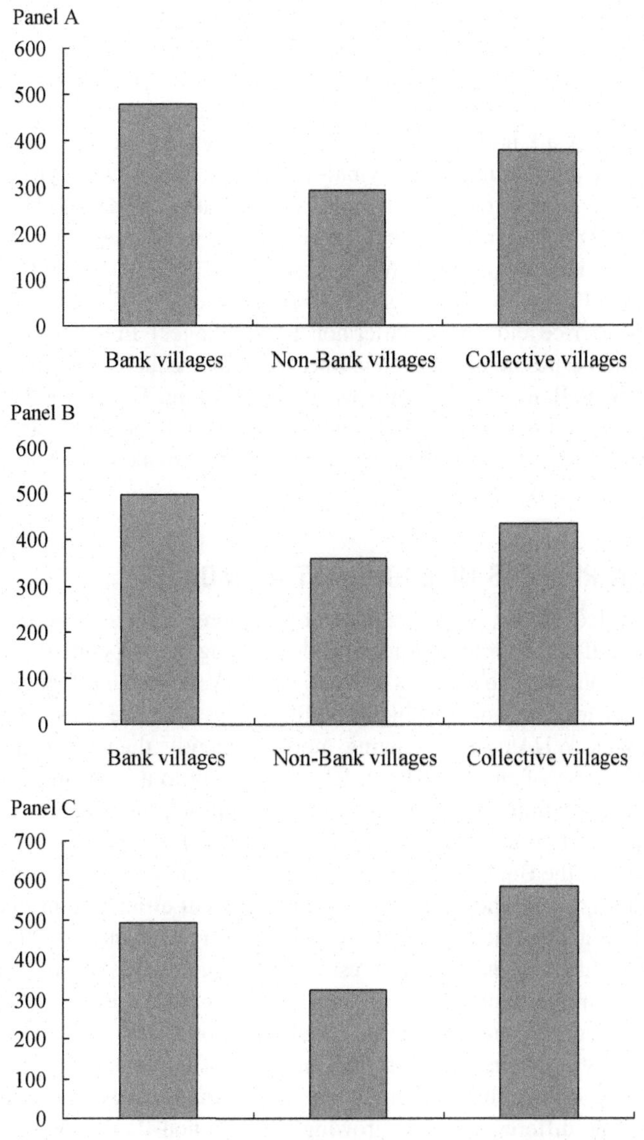

FIGURE 11.6

Crop yields for rice, wheat, and maize under alternative water management institutions. Panel A: crop yields for rice under alternative water management institutions, Unit: kg/mu; Panel B: crop yields for wheat under alternative water management institutions, Unit: kg/mu; Panel C: crop yields for maize under alternative water management institutions, Unit: kg/mu.

Source of data: 2004 CWIM survey and 2006 Bank survey.

use is lowest in the non-Bank villages, yields are also lower. In fact, yields are lower in the non-Bank villages for all crops (rice, wheat, and maize) when comparing them to crops cultivated in both Bank and collective villages. Although the level of crop yields in non-bank villages are lower than bank villages in the descriptive statistics, remember that they are not statistically different using standard statistical tests. It is plausible that the distributions of the yields of farmers in these two types of villages are statistically the same, even when using multivariate analysis. In the case of the Bank results, it is only for maize that we be relatively more confident that WUAs lead to more efficient use of water. While the yields of the crops in the Bank villages are highest for all crops, water use is only lowest in the case of maize (and it is only second lowest for rice and wheat—after non-Bank villages). In fact, there is an additional problem, it is possible that even in the case of maize there are other factors that make yields rise in Bank villages while water use is falling. Therefore, it is important to conduct more rigorous multivariate analysis. We also found that water productivity for the Bank WUAs is higher than that in both non-Bank WUAs and collective villages (Fig. 11.7).

ESTIMATION RESULTS OF ECONOMETRIC MODELS

For the interested reader, we have included a more detailed description of the approach and full report of the findings in Methodological Appendix to Chapter 11.

In fact, the multivariate analysis—that is, when we hold the effects of a number of other covariates and non-varying ID effects constant—we have evidence that farmers in Bank WUAs are producing more efficiently than those in collective IDs, and in the case of rice, non-Bank IDs. According to a joint analysis of water use and yields, we find that when compared to the use of water in collectively managed villages (the base set of villages), farmers in Bank WUAs use less water. The coefficient on the Bank WUA variable is negative and statistically significant. In the rice yield regression, however, there is no statistical difference in rice yields in the Bank WUAs and the rice yields in the base collectively managed villages. Based on these results (and a joint statistical test of the two coefficients), we can conclude the rice farmers in the Bank WUAs are more efficient since they use less water and get the same yields. In short, it can be concluded from the analysis that the Bank WUAs, those that are using best practices in terms of implementing the Five Principles, are more efficiently managing their water. Interestingly, according to our results there is no difference in rice growing between non-Bank WUAs and collectively managed villages (despite the fact that the mean of the water use in non-Bank villages was lower than that in collective villages, but not statistically so).

When looking at the results of all crops, the conclusion that Bank villages are more efficient in their water use stands (relative to collectively managed villages), and we find that non-Bank WUAs are also more efficient than collectively managed villages. In both the cases of Bank WUAs and non-Bank WUAs, the coefficients on their variables show that water use on all crops is lower than that in collectively managed villages. Also, there is not statistical difference in yields. Hence, while providing more

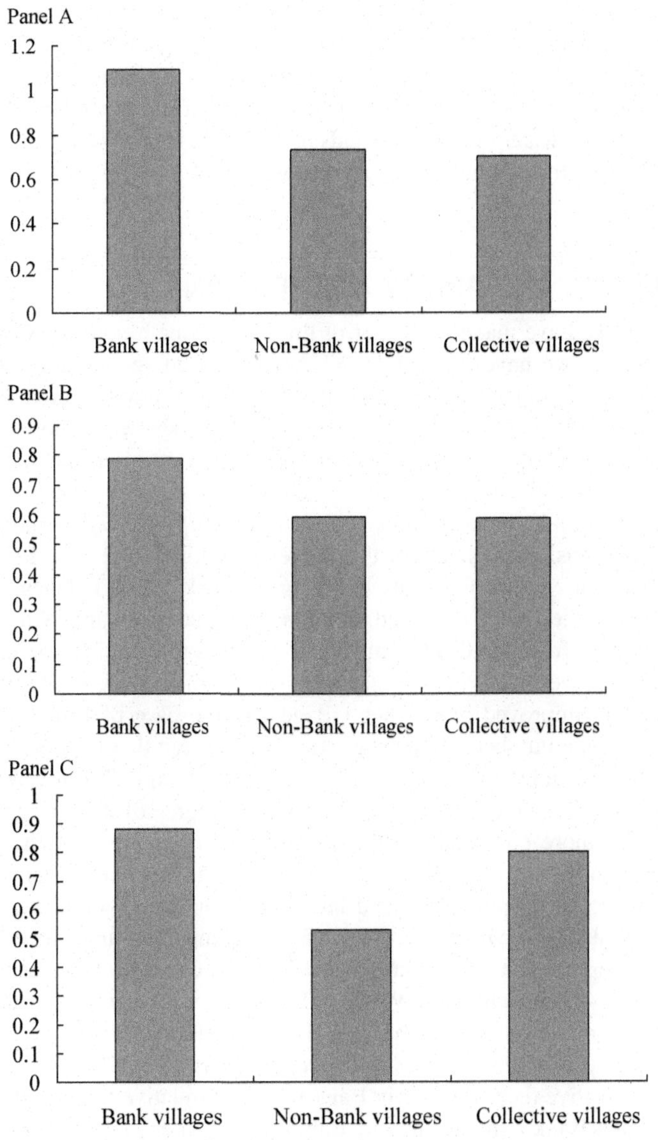

FIGURE 11.7

Water productivity for rice, wheat and maize under alternative water management institutions. Panel A: water productivity for rice under alternative water management institutions, Unit: kg/m^3; Panel B: water productivity for wheat under alternative water management institutions, Unit: kg/m^3; Panel C: water productivity for maize under alternative water management institutions, Unit: kg/m^3.

Source of data: 2004 CWIM survey and 2006 Bank survey.

evidence that the Bank WUAs are more efficient than collectively managed villages, it also suggests that there may have been a positive effect of having Bank WUAs in the region (and the learning that came from Bank WUAs and non-Bank WUAs). The fact there may be spillovers is supported by statistical tests that show there are no statistical differences between water use and/or yields between the Bank WUAs and non-Bank WUAs when using the aggregate crop measures.

CONCLUSIONS AND POLICY IMPLICATIONS

In this chapter we have made a number of findings. Using a northern China representative data set, we have found that WUAs are indeed spreading across China. According to our estimates, about 10% of villages have now adopted WUAs. Of course, this does not mean all WUAs work well and manage water more efficiently; in the rest of the chapter, we address the nature of implementation, and the impact, of WUAs.

Perhaps most fundamentally we have found that the Bank WUAs have excelled in many dimensions, particularly in the implementation of the Five Principles of WUA management. Using four sets of villages (Bank WUA villages, non-Bank WUA villages, collectively managed villages in Bank sites, and Ningxia WUA villages) with data from the CWIM and Bank data sets to examine implementation and performance, we find evidence that many of the management practices of the Bank WUAs are improved over other villages. In the case of Principle 1, Bank WUA villages had a number of characteristics that showed that they were endowed them with more reliable water supply. In the case of Principle 2, Bank WUAs have been set up and are operating with a relatively high degree of farmer participation. The leaders are more consultative. The procedures are clearer and the processes more formal. Bank WUA villages are also set up in a way that makes them consistent with Principle 3; WUAs are organized largely within their hydraulic boundaries. Finally, the Bank WUA villages are successful in implementing the Principles 4 and 5. For example, most of the Bank WUA villages can deliver water volumetrically (Principle 4); and all Bank WUA villages collect water fees from farmers (Principle 5). Hence, from this analysis, WUAs in the Bank villages can be thought of as operating according to the best practices in terms of the Five Principles.

While the positive record is true in Bank villages—relative to all of the comparison cases—non-Bank villages in the Bank survey sites; collectively managed villages in the Bank survey sites; and in the Ningxia (and other) villages outside of the Bank survey sites, it is also true that there is evidence that the Bank's effort to promote WUAs extends beyond their own project villages. The openness, consultative nature, and transparency that are found in the Bank WUAs are nearly matched by the non-Bank WUAs. Non-Bank WUAs clearly follow more formal procedures and adopt better practices than the other two comparison groups (collectively managed villages and Ningxia WUAs). That good practices can spillover into adjacent areas is an important finding. We do not know why this diffusion is occurring, nor do we currently possess the data needed for it to be understood, but we suggest that this process be studied to draw lessons for further improvement. If specific activities promote learning across boundaries, they should be adopted in other projects; also, managers may want to be intensify these activities to make the diffusion effect stronger and more widespread.

So, what are the results of efforts to implement WUAs according to the Five Principles? In our analysis of water efficiency use, we find convincing results that water is being used more efficiently in Bank WUA villages. This is true in rice growing villages and in all villages that cultivate rice, wheat, and maize. To a lesser extent (but significantly), the same is true for non-Bank WUAs. According to the criteria of improving the efficiency of water use, Bank WUAs, and the other WUAs in the same regions are successful.

In this chapter, we have shown almost unambiguous evidence that the procedures put in place have resulted in a system that is operating according to design. There is an overwhelming perception that the new institutional form is making real differences in particular aspects of water management, and that water management is improving in general. There is a perception that there is less conflict—both within the village and among villages. In general, WUAs can truly be said to be contributing to China's "new harmonious society." From this analysis we conclude that there is a direct causal link between WUA water management reform with World Bank characteristics and the positive feeling the farmers have about their village's water management activities. This is an important contribution and a success to be proud of.

Despite the great number of strengths, there are still puzzles in our results and potential weaknesses in the ways that the establishment and implementation of WUAs are approached. Beyond water efficiency, the impact on actual economic performance (income and cropping structure) is more difficult to document. The point estimates of yields in Bank WUA villages are lower. The point estimates of the effect of WUAs on yields are negative (though insignificant). There is no measured absolute rise in income due to WUAs in the Bank villages. In other words, using the strictest statistical methods, it is possible to reject the hypothesis that WUAs increase yields or increase incomes. Somewhat surprisingly, grain area is higher in Bank WUA villages. If the intent of the policy is to promote cash crops and specialization into nongrain crops, there is no evidence that it is happening in Bank WUA villages. According to this strict interpretation, while the Bank WUAs have brought nominal changes, they have not generated the more fundamental changes that can lead to improved economic welfare or structural change.

REFERENCES

China Irrigation District Association, 2002. Participatory irrigation management: management pattern reform of state-owned irrigation district. In: Paper Presented at the Sixth International Forum of Participatory Irrigation Management, Held by the Ministry of Water Resources and the World Bank. April 21–26, 2002, Beijing, China.

Huang, W., 2001. Reform irrigation management system, realizing economic independency of irrigation district. In: Nian, L. (Ed.), Participatory Irrigation Management: Innovation and Development of Irrigation System. China Water Resources and Hydropower Publishing House, Beijing, China.

Ma, Z., 2001. Deepening reform of farmer managed irrigation system, promoting sustainable development of irrigation district. In: Nian, L. (Ed.), Participatory Irrigation Management: Innovation and Development of Irrigation System. China Water Resources and Hydropower Publishing House, Beijing, China.

Management Authority of Shaoshan Irrigation District, 2002. Positively promoting reform based on practices of irrigation district, obtaining achievement of both management and efficiency. In: Paper Presented at the Sixth International Forum of Participatory Irrigation Management, Held by the Ministry of Water Resources and World Bank. April 21–26, 2002, Beijing, China.

Nian, L., 2001. Participatory Irrigation Management: Innovation and Development of Irrigation System. China Water Resources and Hydropower Publishing House, Beijing, China.

Ostrom, E., Summer 2000. Collective action and the evolution of social norms. The Journal of Economic Perspectives 14 (3), 137–158.

Ostrom, E., July 2009. A general framework for analyzing sustainability of socio-ecological systems. Science 325 (24), 419–422.

World Bank, June 2003. Water User Association development in China: participatory management practice under bank-supported projects and beyond. Social Development Notes 83. http://siteresources.worldbank.org/INTRANETSOCIALDEVELOPMENT/873467-1111666620939/20502171/SD+Note+83+22-Jul-03.pdf (accessed 29.01.09.).

Wang, J., Xu, Z., Huang, J., Rozelle, S., 2006. Incentives to managers and participation of farmers in China's irrigation systems: which matters for water savings, farmer income and poverty? Agricultural Economics 34, 1–16.

Xie, M., 2007. "Global development of Water User Associations (WUAs): lessons from South-East Asia". Water User Association development in Southeastern European Countries. In: Hussain, I., Naseer, Z. (Eds.), Proceedings of the Regional Workshop on WUAs Development, June 4–7, 2007 – Bucharest, Romania.

SECTION

IV

Future Options

CHAPTER

Irrigation Water-Pricing Policy

12

We have shown in previous chapters that government efforts to encourage households to conserve water have achieved only mixed results. Adoption rates of water saving technologies are startlingly low. Restrictions on groundwater drilling go unheeded. Many surface water management reforms have not been effective so far.

Under these circumstances, China's water officials and scholars have begun to consider reforming the pricing of irrigation water as an important policy instrument for dealing with the water scarcity problem (eg, Wang, 1997; Wei, 2001). Throughout this book, when trying to identify why past policy efforts have not been effective, one theme has continuously come up: the absence of economic incentives facing water users. Other researchers have reached similar conclusions (Lohmar et al., 2003; Yang et al., 2003). Similar to many places around the world, the cost of water is low in China. Groundwater users only need to pay for the cost of energy to pump water out. No extraction fees are charged. Surface water is also priced much lower than its engineering cost (Zheng, 2002). Despite the fact that the agriculture is the main water-using sector in China (68% in 2001, Ministry of Water Resources, 2002), charges for irrigation water have not been raised much. Furthermore, inside most irrigation districts, water fees are assessed based on the size of irrigated area. When the cost of water is low, or not related to the quantities demanded, the benefit from saving water is also low. As a result, the current water-pricing policy in the agricultural sector (as oppose to the industrial and residential sectors) has not been effective in providing water users with incentives to save water. China's government has raised the price of water that is charged for residential use and industrial use. For example, the price of tap water in Beijing has increased nine times since 1991, from a level of $0.014/m^3$ to $0.44/m^3$ (Chen et al., 2005). Reform to make the price of water reflect its true value is critical to providing agricultural users an incentive to save water.

While there is increasing consensus that reforming water pricing is necessary, two basic issues need to be addressed before any new policies can be made. The first issue is the effectiveness of increasing the cost of irrigation. Previous economic studies in a number of developed countries have shown that demand for irrigation water is inelastic (eg, Moore et al., 1994; Ogg and Gollehon, 1989). If water users in China are not responsive either, raising the price of water will not be an effective mechanism to reduce demand. If water users do respond to price changes, it is important for policy makers to learn about the nature of the responses when planning

Managing Water on China's Farms. http://dx.doi.org/10.1016/B978-0-12-805164-1.00012-9

price interventions, because water use reductions could reduce crop production and thus affect China's food security.

The second issue is the welfare impact of higher irrigation costs on producers. In the political economy that dominates policymaking in China, it is imperative to assess how much harm producers would experience should pricing policies be effectively implemented. The current government is intent on reducing farmer burdens and raising incomes even if other long-term problems (eg, the unsustainable tapping of groundwater resources) are undermined (Lohmar et al., 2003).

Despite the fact that dealing with water scarcity is among the most critical issues on the government agenda and that good policies require that officials understand the nature of water demand, only a few studies have analyzed water demand in rural China (eg, Chen et al., 2005; Yang et al., 2003). Many of the previous studies are largely qualitative. To our knowledge, there are rarely any rigorous quantitative analyses of household water demand that can be used to advise China's policy makers.

The goal of this chapter is to analyze the potential of reforming groundwater pricing as a way to encourage water conservation and to assess its impacts on crop production and producer welfare in rural China. To meet this goal, we develop an approach that can inform policy makers about the effectiveness of water-pricing policy as well as how water-pricing policy should be implemented. The first step in the approach involves estimating a set of crop-specific production frontiers as well as household level technical inefficiency parameters. The estimation results aid us in measuring the relationship between water and crop output as well as the value of water to households. Our results show that in general there is a large gap between the cost of water and the value of water to producers. Using the estimation results in a series of simulation analyses, we examine the effects of water-pricing policy on water savings. In particular, we examine two different water-pricing policies, one that takes into account the gap between the cost of water and water value (henceforth, the *informed policy*); the other ignores this gap (henceforth, the *uninformed* policy). Finally, we analyze the impacts of water-pricing policy on crop production, especially that of grain crops and producer welfare.

This chapter is meant as a starting point of quantitative research on irrigation water-pricing reform in China. Due to the broad nature of the topic, we must limit the scope of this study. Several important issues are left to future work. In particular, partly due to the lack of data, we are not able to include the implementation cost of water-pricing policy. New institutional arrangements have emerged in both groundwater-using areas and surface water-using areas (Huang et al., 2008; Wang et al., 2005a,b). An important issue to examine is how water-pricing policy would perform under these new institutional arrangements. A good discussion on these issues, that uses examples outside China, can be found in Dinar and Saleth (2005). We only focus on the agricultural sector in this study for two reasons. First, agriculture is the major water user (more than 60% of the nation's water supply). Second, it is not likely that there will be large-scale water transfers between agriculture and other sectors in the future. This is because the central route of the South–North water transfer project passes through Hebei province (our study site) and will supply water to urban and industrial sectors. With this alternative, when the price of water is raised, the transfer of water from agriculture to other sectors, as depicted in many other studies (eg, Weinberg, 2002), is less likely.

This chapter uses only the household-level data from 24 communities in Hebei province in the 2004 China Water Institutions and Management (CWIM) Survey. About 12% of China's grain is produced in Hebei province (Ministry of Agriculture of China, 2004). The major crops in Hebei province are wheat, maize, and cotton. Wheat is planted in the previous winter and harvested in the spring. Maize and cotton are planted in the summer and harvested in the fall. To construct a measure of the volume of water applied, we asked respondents in the CWIM survey to report, for each crop, the length of irrigating time, the total number of irrigations during the entire growing season, and the volume of water applied per irrigation. In addition, we asked households to report the amount they paid for irrigation water for each crop in order to calculate the cost of water. The price of water is defined as the average cost per cubic meter of water. In the rest of the chapter, the cost of water and the price of water are used interchangeably.

NATURE OF IRRIGATION WATER DEMAND IN NORTHERN CHINA

The characteristics of our sample data allow us to study the nature of irrigation water demand in northern China. Most importantly, we observe large variations in the prices of water paid by households in our data. Communities in our sample data rely on groundwater for more than 80% of their irrigation water. Most variations in water prices come from the differences in the depth to water in wells across space. Strong positive correlations between the price of water and the depth to water for wheat, maize, and cotton illustrate this difference (Table 12.1, column 1 and 2). Households that paid more for per unit of water are usually those that faced greater depth to water, because it costs more to pump water when it is located deep within the earth. Moreover, since the depth to water varies significantly across space, from less than 20 m to more than 100 m (column 1), households in different quartiles of depth to water paid very different prices for water. For example, maize-growing households in the fourth quartile (the farmers pumping from the deepest wells) paid as much as $0.062/m^3 for water while those households in the first quartile paid as little as $0.006/m^3 (column 2, row 7 and 10).

With the large variations in the prices of water across space, we observe several differences in water-using patterns. First, we observe a strong inverse relationship

Table 12.1 The Cost of Water, Depth to Water and Water Use in Hebei Province, China, 2004

	Percentile of the Cost of Water	(1) Depth to Water (m)	(2) Average Cost of Water (Dollars/m^3)	(3) Volume of Water Use per Unit of Land (m^3/ha)
Wheat				
1	Average	38.4	0.029	4455
2	0–25%	15.9	0.008	5321
3	26–50%	19.4	0.019	4956
4	51–75%	51.9	0.031	4628
5	76–100%	69.0	0.174	2276
Maize				
6	Average	44.7	0.029	2022
7	0–25%	15.0	0.006	2640
8	26–50%	35.4	0.018	2534
9	51–75%	62.2	0.029	1730
10	76–100%	65.3	0.062	1184
Cotton				
11	Average	59.1	0.035	1477
12	0–25%	41.3	0.017	2322
13	26–50%	45.7	0.028	1950
14	51–75%	47.3	0.041	1394
15	76–100%	108.0	0.062	978

Data source: Authors' survey in 2004 (CWIM data).

between the price of water and the level of water use among households that irrigated their crops. As the price of water rises, households adjust water use by decreasing their water use per unit of irrigated area. In this chapter, we define this as *adjustments at the intensive margin* or stress irrigation. For example, wheat-growing households that face a price of \$0.0085/m^3 for water applied 5321 m^3/ha; other wheat-producing households that pay \$0.06/m^3 used only 2276 m^3/ha (Table 12.1, column 2 and 3, row 2 and 5). There also are large adjustments and monotonically decreasing water use among households that grew maize or cotton in response to rising water prices.

It should also be noted that, on average, crop water use calculated from the survey data is consistent with findings from agronomy studies in China. The estimated crop irrigation water requirement (the difference between evapotranspiration and effective precipitation) in Hebei province, conditional on the average rainfall level between 1952 and 1998, was shown to be 2620 m^3/ha for wheat, 1340 m^3/ha for maize, and 1260 m^3/ha for cotton (Chen et al., 1995). Taking into account irrigation efficiencies (which have been estimated to be between 0.6 and 0.7 in Hebei province, Chen et al., 1995), the levels of crop water use from our data (in the column 3 of Table 12.1) are close to these estimates. Because the growing season of maize and cotton (late July or August to October) coincides with the rainy season in Hebei province, they require much less irrigation water than wheat.

In addition to the adjustments at the intensive margin, households also respond to price increases in several other ways. In particular, households may choose not to irrigate some of their crops or change their crop mix. We defined these responses as *adjustments at the extensive margin.* On average, households in the first quartile of depth to water left 12% of their sown area to be rainfed (Table 12.2, row 1, column 2). The share of sown area that is rainfed increases to 37% among households in the fourth quartile of depth to water (row 4, column 2).

Households also tend to allocate greater shares of sown area to nongrain crops as the depth to water increases. In our data, nongrain crops include cotton, vegetables, fruits, flowers, and peanuts. On average, the households in the first quartile allocate 15% of their sown area to nongrain crops (Table 12.2, row 1, column 3). The share more than doubles for households in the third and fourth quartiles (rows 3 and 4, column 3). Such patterns are consistent with findings of other studies in the United States (eg, Gardener, 1983). Relative to grain crops such as wheat and maize, nongrain crops usually generate higher per-hectare net return but also impose higher nonwater costs since they are more labor- and capital-intensive. Changes in relative prices caused by increases in water prices result in more use of labor and capital. This could bring on a crop mix change away from grain crops toward nongrain crops.

Table 12.2 The Depth to Water and Crop Mix in Hebei Province, China, 2004

	(1)	(2)	(3)
Percentile of the Depth to Water	**Average Depth to Water (m)**	**Percentage of Rainfed Sown Area (%)**	**Average Share of Household Sown Area That Cultivates Nongrain Crop[a] (%)**
0–25%	6	12	15
26–50%	21	15	25
51–75%	58	28	33
76–100%	91	37	31

[a] *Nongrain crops include cotton, vegetables, fruits, trees, and peanuts.*
Data source: Authors' survey in 2004 (CWIM data).

EFFECTIVENESS AND IMPACTS OF WATER-PRICING POLICIES IN RURAL CHINA

Our data show that households respond to changes in water prices through intensive margins as well as through extensive margins. Given this nature of water demand, it is more meaningful to analyze water use at the household level instead of the crop level. The latter cannot capture the extensive margin adjustment. This is particularly true for Hebei province where most sample households are engaged in producing multiple crops including wheat and maize (Table 12.3, row 1). Some households also grow cotton. To estimate the household-level water demand, we focus on the household maximization model from which household water demand is derived and estimate the production frontier and technical inefficiency parameters, the key elements in measuring the relationship between water and output. The methods used to estimate these parameters are presented in Methodological Appendix to Chapter 12.

We then parameterize the household maximization problem with estimation results and use simulations to analyze the effects of water-pricing policy on water use, crop production, and household income. Moore et al. (1994) used rigorous econometric analyses to calculate the intensive and extensive margins. In our case, since large price changes, not marginal changes in water prices, are more relevant, simulations are more appropriate. In the simulations, treating the current costs of water as the baseline water prices, we first run a baseline model by solving the profit maximization problem for each household. We then increase the price of water to several different levels but hold prices of other inputs and output constant. We solve the profit maximization problem at each of these new water price levels. Simulation results form the basis of our policy analyses. By comparing the changes in household water uses, we can predict the extent of water savings that occur when water prices are raised to different levels. Since the simulations also generate the

Table 12.3 Number of Sample Households That Grew Wheat, Maize, or Cotton, Hebei Province, China, 2004

		(1) Total Number of Households	(2) Number of Households That Grew Wheat	(3) Number of Households That Grew Maize	(4) Number of Households That Grew Cotton
1	**Total**	**88**	**63**	**86**	**18**
2	Xian County	30	26	28	8
3	Tang County	29	19	29	1
4	Ci County	29	18	29	9

Data source: Authors' survey in 2004 (CWIM data).

level of crop outputs and household profits, we can also predict the impact on crop production and income.

Since wheat, maize, and cotton account for 80% of total sown area in our sample, we only include these three crops in the simulations. If more crops, such as nongrain crops (eg, vegetables and fruits), were included, households would be able to adjust more at extensive margins by switching to these crops. If opportunities exist for households to switch to grow these high-value crops in large sown areas (eg, improved market access), households may start to invest in more efficient irrigation technology. The increase in water use efficiency might lead to a rise in water demand (Dinar and Zilberman, 1991). We also keep the same rotations. In the summer season, only wheat is grown while in the fall season, either cotton or maize (or both) is grown.

EFFECTIVENESS OF WATER-PRICING POLICY: INFORMED POLICY AND UNINFORMED POLICY

The effectiveness of water-pricing policy depends crucially on the responsiveness of households.

If the water price is lower than the value of water to households, households will not change water uses at all in response to small changes in water prices. Thus a necessary task in designing a water-pricing policy is to determine whether the current price of water reflects the value of water to households. This is our first step in this subsection. We then compare the effects of two types of water policies that differ in their treatment of the value of water.

The value of water is measured by the increment in household profit due to one more unit of water available to households. This change in net income method has shown to generate better estimates of water values than other approaches, especially in the presence of fixed allocable inputs (Young, 2005). We calculate the value of water to households in two steps. In the first step, we solve the profit maximization problem using baseline levels of prices and resource constraints. In the second step, we relax the water constraint by one unit but hold everything else constant (eg, prices of inputs and output, constraints on land and labor). We then calculate the change in household profit. The increment in profit is the value of water to household after netting out the cost of obtaining water. Equivalently, it is the gap between the cost of water and the value of water to households. We can also call it the shadow value of water since it gives households' willingness to pay to relax the water constraints by one unit. The shadow value can also be interpreted as the scarcity value of water. This is because, if households were not constrained water availability (in other words, if water were not scarce), then relaxing the water constraints would not increase household profit and the shadow value would be zero.

Our results show that there is a large gap between the cost of water and the value of water in most households in Xian and Ci County (Fig. 12.1, Panel A). Since resource constraints are season-specific, we have calculated the value of water to households for both the summer and fall seasons. For most households in Tang

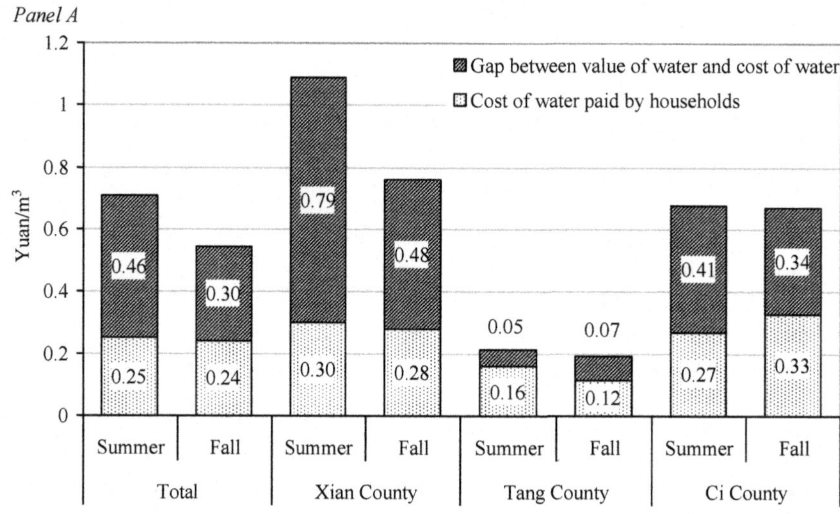

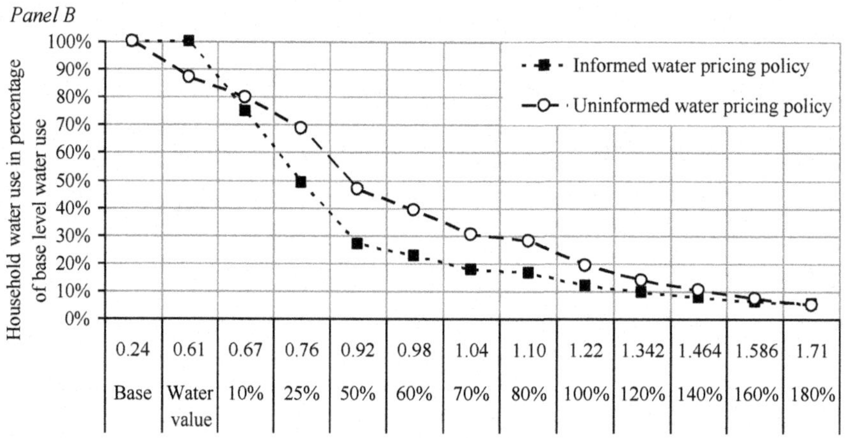

FIGURE 12.1

Effects of higher water prices on household water use. Panel A: value of water and cost of water (Yuan/m^3). Panel B: comparison of informed and uninformed water-pricing policy.

County, the cost of water they paid is the same as the value of water. In Xian County, however, the gap is almost double the irrigation cost in both seasons. In Ci County, the gap is also large in both seasons. The same finding has been observed in many other countries: water is usually underpriced, and its price usually does not reflect the scarcity value (Dinar and Saleth, 2005). From our results, it is clear that, at least in Xian and Ci counties, households will not change their water use much in response to small increases in water prices. This is because the new price level after small increases would still be lower than the household's actual value of the water. As a result, households still maximize their profits by using the same level of water as before the price change.

The large gaps in Xian and Ci counties are consistent with findings from our community leader survey forms. In response to a question about whether or not there was sufficient water in the wells to meet demand during the irrigation seasons for 2002, 2003, and 2004, 13 out of 24 community leaders said there was not. Most of these communities are located in Xian and Ci counties. The constraints on groundwater arise from both the increasing water demand and diminishing supply of water (ie, the rapidly declining groundwater levels in these communities). Since water resources are scarce in many communities in Xian and Ci counties and prices do not reflect the scarcity value of water, it is not surprising to find large gaps between the current costs of water and the value of water.

Given the large gap between the price of water and the value of water, policy makers can design two types of water policies: an *uninformed* policy and an *informed* policy. When implementing an uninformed policy, we assume policy makers are not aware that there is a gap between the current water cost and the true value of water to households. As a result, officials consider the current price of water as the starting point and simply raise the price of water from there. In contrast, policy makers can first find out whether or not the current cost of water reflects the household's value of water. They could do so by collecting information and generating estimates of households' actual value of water. With such information, an informed policy could be implemented in a two-step way. In the first step, the price of water of each household could be increased from its current level to a level that equals their value of water. With this step, the price of water would reflect exactly the true value of water. In the second step, the price of water could then be increased to a point at which users would begin to cut back water use enough to meet the water-saving target.

In order to make uninformed and informed policies comparable, we make sure the changes in the average prices of water under the two policy regimes are the same. For example, under the informed policy scenario, when the price of water each household faces is increased from the base level to their value of water, the average price of water increases from \$0.03/m^3 to \$0.074/m^3, which is a \$0.045/m^3 increment (Fig. 12.1, Panel B). Then under the uninformed policy, we increase the price of water each household faces uniformly by a level that is close to \$0.045/m^3, so that the after-change average price is also \$0.074/m^3—the same as under the informed policy. Similarly, in another change under the informed policy, when the price is first increased to the level of water value and then increased further by 50% of the water value (a two-step procedure), the average price is raised by \$0.082/m^3 and reaches \$0.11/m^3. Then we also raise the price under the uninformed policy by \$0.082/m^3 to \$0.11/m^3 (in one step). Since the average price of water before and after the changes under the informed policy and the uninformed policy are the same, we can put changes in household water use under these two policies on the same graph and plot them against the average water prices.

The simulation results show that the informed policy has the potential to induce sizable water saving. In the first step of the informed policy, which makes sure all households are at the point in which the cost of water is equal to the value of water, households do not change water uses by construction. Once the cost of water has hit the level of water value, however, households are highly responsive to price changes. Suppose policy makers plan to reduce water by 20%. Using the units of the vertical axis of Panel B in Fig. 12.1, this means households need to reduce their water use to 80% of the base level. In order to meet the 20% water-savings target, after the price is increased to $0.074/m^3 in the first step, the price only needs to be further raised by 10% (from $0.074 to $0.081) to move households from zero water reduction to 20% water reduction. In order to achieve a 50% water-savings target, the price of water only needs to be raised to $0.092/m^3, only $0.012/m^3 higher than the level that was needed to hit a 20% target. Therefore, when the price of water reflects the value of water, water-pricing policy can be an effective tool in dealing with the water scarcity problem.

Our results also show the price of water needs to increase greatly. For example, in order to meet a 20% water-saving target, the average price is increased to a level close to $0.081/m^3, leading to a 180% increase in the average price of water. It is important to note, however, that most of the rise in the price of water is in the first step of the informed policy. Of the total rise of price ($0.052/m^3), 87% ($0.045/m^3) is needed just to get all households to the point that the cost of their water is equal to the value of water in production. Such large price rises may be conflicts with other policy goals that aim at keeping food production high and lifting rural incomes. This issue is addressed in later subsections.

Comparisons of the informed and uninformed policies indicate that water-pricing policy can be implemented more effectively when policy makers recognize that the current cost of water is far below the level of the value of water. For example, if water officials set a water-saving target of a 50% reduction in household water use (ie, households reduce water use to 80% of the base level use), under the informed policy, policy makers would increase the price to $0.092/m^3. However, to achieve a similar saving under the uniformed policy, the price would need to be raised to $0.11/m^3. This is because under the uninformed policy, policy makers increase the price by an amount that is the same for all of the households, regardless of whether the household has a high or low value for water. Because of the large gaps between the cost of water and the value of water, especially in Xian County and Ci County, if the price were increased only to $0.092/m^3 (the average water price under the informed policy), it would still be below the true value of water to some of the households. These households would not respond to price changes at all, and so the 50% water-saving target would not be achieved. As a result, policy makers have to raise the price to $0.11/m^3 to insure that price exceeds the level of water values for enough households that the water savings reach the target. Although the change in the average price is the same under both policies, the informed policy increases prices in a more targeted way, guaranteeing that the water price each household faces reflects the value of water. Since all households are responsive under the informed policy, the same amount of water price increment is much more effective. In this case, following the uninformed policy would force policy makers to increase the price of water to a higher level than necessary to meet the same water-savings targets. This higher price would not only result in higher costs for farmers, but also increase the financial burdens of the water-pricing policy if policy makers planned to compensate farmers for their higher costs.

The informed policy, however, does not always outperform the uninformed policy. When the water-saving target is small (eg, less than a 20% reduction in our case, as marked by the intersection of the informed policy and uninformed policy in the upper left corner of Panel B in Fig. 12.1), the uninformed policy works more effectively in reaching the target. This is because only a small proportion of households need to be responsive in order to reach a small water-saving target. Therefore, the uniform increase in water prices under the uninformed policy is sufficient. The uninformed policy works better because it does not require the large increment in water prices to get all households to the point that they are facing their actual water value, as is needed in the first step of the informed policy. When the water-saving target is ambitious (eg, more than 90% in our case), there also is not much difference between the performance of the informed and uniformed pricing policies. This is because a large price increment would be needed to meet such a target under either policy (about 120% of the water value even under the informed policy in our case, Panel B). When the water price is increased greatly, it is likely the price level reaches or exceeds the value of water to most households. Consequently, most households would be responsive under either type of policy, which results in little difference between informed pricing policy and the uniformed pricing policy.

Whether an informed policy or an uninformed policy should be pursued depends on the specific water-saving target as well as the implementation cost. In general, the cost of implementing an uniformed policy is lower since it does not involve collecting information that is needed to estimate household-level water demand. So an uninformed policy is appropriate when the water-saving target is small (less than 20% reduction in our case), since the informed policy does not outperform the uninformed policy. If the water-saving target is more than a 20% reduction, it is more difficult to make a choice. Raising water prices is more effective under the informed policy. However, the cost of collecting information to estimate the value of water may be high. Policy makers need to determine whether the benefit outweighs the increment in the implementation cost if the informed policy, instead of the uninformed policy, is used. Although we do not have data on implementation cost, our analysis serves as a starting point in that it provides the benefit of implementing an informed policy.

IMPACTS OF WATER-PRICING POLICY ON CROP PRODUCTION

Although increasing the price of water has been shown to be effective in reducing water use, when making policies, leaders must also take into account other impacts of higher irrigation costs. We examine how increasing the price of water will affect *crop production* in this subsection and *producer welfare* in the next subsection. In the rest of our analysis, we focus only on the informed policy scenario. We run four different simulations. In each simulation, we first raise the price of water each household faces to their value of water, and then increase the price of water further by percentages of the level of water values. The price increments in the second step of these four simulations are 10%, 25%, 50%, and 100% of the level of water values, respectively (Fig. 12.1, Panel B). Under the informed policy, production does not change at all during the first step. As seen in Panel B of Fig. 12.1, water use does not change when the price is raised to each household's value of water. It follows that production also does not change. Because there is no change in water use or crop production, the effect of this step is not graphed in Fig. 12.2.

Consistent with findings from the descriptive analyses, when the price of water is raised above the level of the value of water, households indeed will adjust their use of water (seen in Fig. 12.1 above), and these changes occur at both the intensive and extensive margins (Fig. 12.2, Panel A). Importantly, when the price increment is small, most of the adjustments come from intensive margins. For example, when the price of water is increased by 10% after being increased to the level of water value, about 80% of the total reduction in water use comes from adjustments at intensive margins. In the case of wheat, on average, households reduce their water use per hectare from 4436 to 3637 m^3. Water use per hectare in the base run is obtained from simulations. These figures will be slightly different from the observed data in Table 12.1. Maize and cotton producers also cut back water use per hectare from 2150 to 1516 m^3 and from 1653 to 1244 m^3, respectively.

At the same time, households adjust at extensive margins as well. For example, when the price of water increases by 10%, adjustments in the extensive margin account for 20% of the total adjustments (Fig. 12.2, Panel A). About 3% of the total change comes from shifting from irrigated to nonirrigated agriculture and 17% comes from shifting the crop mix. In our case, the shift in the crop mix mainly comes the shift from maize to cotton, which requires less water relative to maize.

While most of the adjustments occur at the intensive margins when price rises are relatively small, as the price rise gets higher to target higher reductions in water use, more of the adjustments come from the extensive margins. For example, when the water price is double the level of water value (ie, a 100% increment in the price), almost 75% of the total water reduction comes from adjustments at extensive margins (versus 20% when the price was increased by 10%, Fig. 12.2, Panel A). Most changes at the extensive margin occur when farmers choose to stop irrigating their crops (69%). The remaining 6% comes from changing crop mix. In contrast, only 25% of the fall comes from adjustments at the intensive margins.

Simulations show that adjustments at the intensive and extensive margins affect crop production in two ways. First, stress irrigation reduces the yields of all three crops. For example, when the price increment is 25% of the water value, the average reduction in yields are 23.4% for irrigated wheat, 11% for irrigated maize, and 4.8% for irrigated cotton. With lower yields, the level of crop production is, of course, lower for all crops, ceteris paribus. Yield changes due to adjustments at the intensive

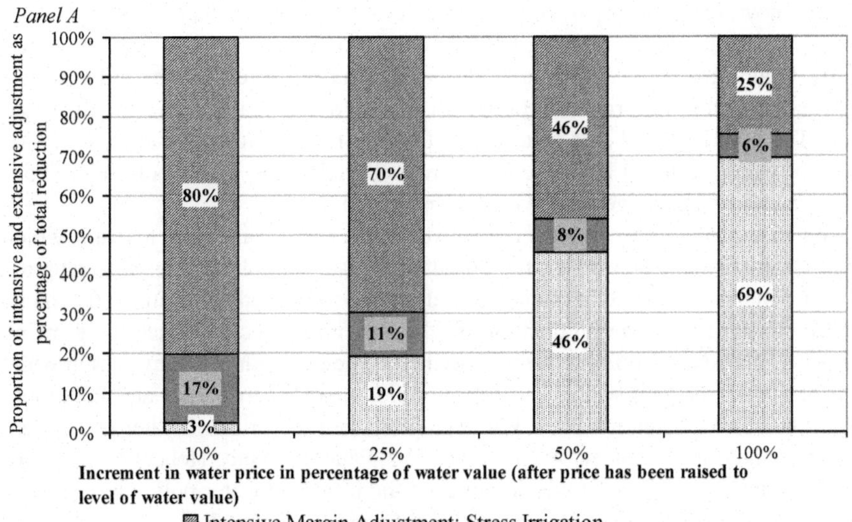

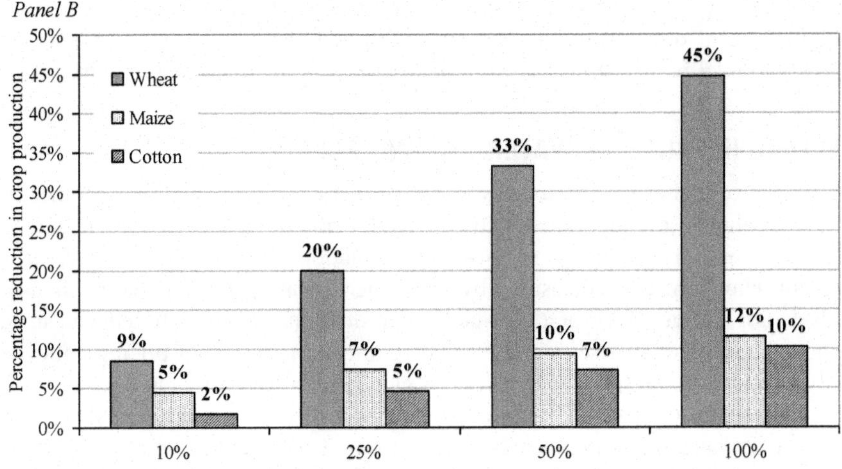

FIGURE 12.2

Effects of higher water prices on crop production under informed water-pricing policy. Panel A: composition of water-use adjustments in response to higher water prices. Panel B: crop production reduction in response to higher water prices.

margin, however, is only part of the explanation for grain production changes. Adjustments at the extensive margins shift crop production from grain to nongrain crops. Grain area (the sown area of wheat and maize) is reduced by 4.3% with a 25% increase in water prices. Farmers also switch from irrigated area to nonirrigated area. The total irrigated area of all crops falls by 15.6%. Hence, in total, when the price of water is raised by 25%, grain production falls by 14.3%, of which 3.5% points came from changes at the extensive margin.

When accounting for both the both lower yields and less irrigated acreage, which result from rising water costs, the simulation results imply that a wide-ranging, pan-provincial water-pricing policy would reduce food production in China significantly. In particular, the production of wheat is most affected. Since the growing season of maize and cotton in Hebei province coincides with the rainy season while that of wheat does not, wheat production relies more on irrigation and falls more when the cost of irrigation rises. For example, when the price of water is doubled, wheat production is reduced by 45% (Fig. 12.2, Panel B). Since Hebei province produces about 12% of China's wheat output, if the informed water policy were implemented only in Hebei, the fall in wheat output would be equivalent to more than 5% reduction in China's total production of wheat. Furthermore, as policy makers choose to continue increasing water prices, the amount of crop production reduction that is due to adjustments at the extensive margin is greater. For example, when the price of water is doubled, of the 45% reduction in wheat output, 32.4% points (or 72% of the total reduction) come from adjustments at the extensive margin.

WELFARE IMPACTS OF WATER-PRICING POLICY

The impact of higher irrigation costs is not limited to crop production. Incomes of rural households are also lower if the water-pricing policy is implemented (Fig. 12.3, Panel A). In the first step of each simulation, since the real price of water each household faces (as measured by the value of water) did not change, households do not change their water uses or crop production. Incomes are reduced since the actual cost of water rises and the negative effect on income of pricing policy is solely attributed to higher water prices. As can be seen from moving from bar 1 to bar 2 in Panel A, on average, crop income drops by $32.41 per household.

When irrigation costs are increased during step 2, although income continued to decline due to higher water price, the rate of decline slows. For example, when policy makers increase the price of water by 10% (after the initial increment in the first step), on average, crop income decreases from $234.34 to $197.58 (Fig. 12.3, Panel A). A 10% increase in the water price only drops crop income further by $3.63. This is because farmers respond to changes in water price since it reflects the value of water to them. Since farmers respond to an increase in water price by reducing levels of water use, the impact on crop income is smaller than that in the initial step. Crop income drops from $234.34 to $183.56 when the price of water is doubled. It should be noted that in our analysis, we do not consider any general equilibrium effects. If water-pricing policies were implemented over large areas of China and millions of

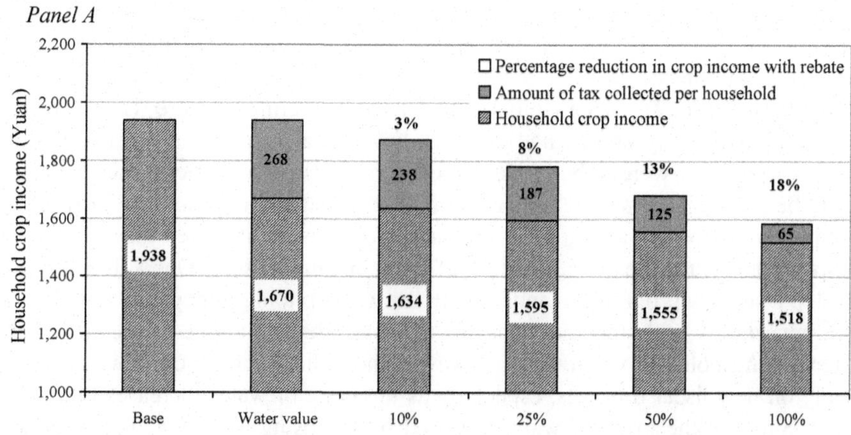

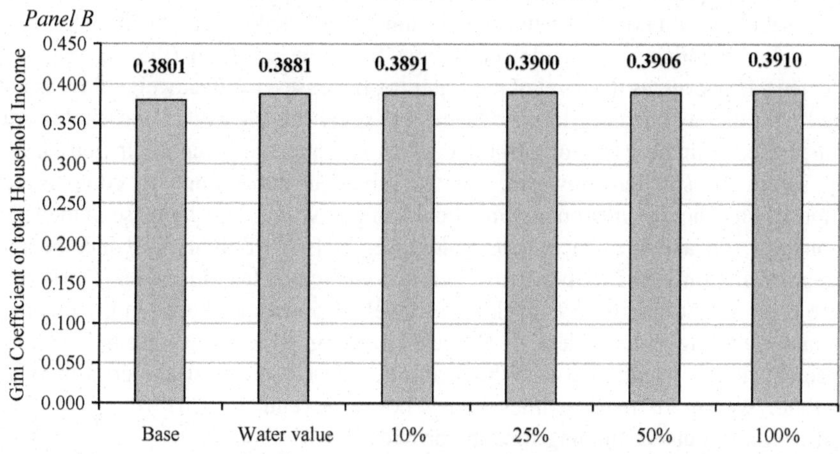

FIGURE 12.3

Effects of higher water prices on produce welfare under informed water-pricing policy. Panel A: effects of higher water prices on household crop income. Panel B: effects of higher water prices on distribution of household total income.

farmers changed their crop mix, the price of grain crops might rise. If this effect were considered, the income impact of higher irrigation costs would be lower.

Hence, while water policy has great potential in saving water, the impact of water-pricing policy on producer income poses a major challenge to China's policy makers in today's political-economic environment. China has made remarkable progress in alleviating poverty, reducing poverty from 53% of the population in the 1980s to 8% in the 2000s (Rozelle et al., 2003; Ravallion and Chen, 2007). The country's leaders are intent on continuing to eliminate extreme poverty in rural China. A set of tax reforms that targets an eventual elimination of taxation on rural households has been implemented (Brandt et al., 2005). For example, agricultural tax was completely eliminated in pilot provinces in 2004 and nationwide by the end of 2006 (Oi et al., 2012). Given China's current political climate, there will be strong resistance against any policy that results in lower rural incomes. Almost certainly, if any water policy were to be implemented in rural China, complementary policies would be needed to offset the impacts of higher irrigation costs on rural income.

Since rural households shoulder the burden of conserving water, they should be compensated with at least the amount of their incomes losses. One solution is to develop a subsidy program in tandem with the water-pricing policy that would provide households with income transfers to offset the reduction in income from water-pricing policies. Our results, however, show that such a policy would have to rely on new fiscal transfers, especially as the price of water increases to higher levels. Suppose the price of water is raised by imposing a tax per unit of water use. When the price of water is raised from its initial actual cost to the level of the household's value of water, most of the amount needed to fund the transfer program (administrative costs aside) can come from the program (the tax revenue collected). However, as the level of the water tax increases, the deadweight loss associated with the tax becomes larger. Our results show this clearly. If households are compensated with the rebate of the collected tax revenue, the reduction in household crop income is smaller. However, the tax rebate is not enough to compensate completely for the loss in crop income. For example, with a 25% increase in the irrigation cost, on average, each household loses \$41.48 of their crop income on average, while only \$22.61 per household is collected as tax (Fig. 12.3, Panel A). There is an \$18.86 gap (or 8% of the base level crop income) between the income loss and the tax revenue collected. The level of the gap increases with the level of increment in the irrigation cost. When irrigation cost is doubled, the crop income loss (\$50.79) is more than six times the level of tax revenue (\$7.86). To compensate for these gaps, outside funding must be provided.

Despite its significant negative effects on average income, our simulation results show that water-pricing policy does not deteriorate the distribution of income. For example, doubling irrigation cost only increases the Gini coefficient of household total income from 0.3881 to 0.391, which is only a 0.7% increase (Fig. 12.3, Panel B). This is consistent with findings in Dinar and Tsur (1995). In our case, one important reason for this small impact is that in rural China, land is equally allocated to households both in terms of land size and soil quality.

CONCLUSIONS AND POLICY IMPLICATIONS

Tackling the growing water scarcity problem has become one of the most important tasks that faces China's leaders. Past water policies, including the policies that increase water supplies alone and those that promote the adoption of water saving technologies, have not been effective. Relying on a set of household-level data, this chapter examined the potential for conserving water through water-pricing reform. We also examined the impacts on production and producer welfare.

Our results show that the current cost of water is far below the true value of water in many of our sample areas. Since water is severely underpriced, water users are not likely to respond to small increases in water prices. Therefore, one of our main findings is that a necessary step in establishing an effective water-pricing policy is to increase the price of water up to a point that it equals the value that water has to the household. Increases in water prices once they are set at the value of water can lead to significant water savings. In short, unlike past water research, our study shows that water-pricing policy, by directly giving users incentives, has the potential of resolving the water scarcity problem in China.

Our analysis also shows that higher water prices affect other aspects of the rural economy. Higher irrigation costs will lower the production of all crops, in general, and that of grain crops, in particular. This may hurt the nation's food security goal of achieving 95% self-sufficiency for all major grains in the short run. Furthermore, when facing higher irrigation costs, households suffer income losses, although income distribution does not deteriorate. As a result, it is imperative that complementary policies be used to offset these negative impacts. For example, a comprehensive set of subsidy policies are needed to offset the loss in income. To be effective in reducing water, of course, subsidies must be decoupled from production decisions.

This chapter provides both good news and bad news to policy makers. On the one hand, water-pricing policies have great potential for curbing demand and helping policy makers address the emerging water crisis. Irrigation is central for China to maintain food security in the long run and will continue to be one investment that enables China to lift its future production of food and meet its food grain security goals (Huang et al., 1999). The goal of water-pricing policy, which is to manage water resources in a sustainable way, does not conflict the long-run goal of the nation's food security policy. On the other hand, dealing with the negative production and income impacts of higher irrigation costs will pose a number of challenges to policy makers, at least in the short run. If China's leaders plan to increase water prices to address the nation's water crisis, an integrated package of policies will be needed to achieve water savings without hurting rural incomes or national food security.

REFERENCES

Brandt, L., Rozelle, S., Zhang, L., 2005. Tax-for-Fee Reform, Village Operating Budgets and Public Goods Investment. World Bank Beijing Office.

Chen, Y., Guo, G., Wang, G., Kang, S., Luo, B., Zhang, D., 1995. Water Requirement of China's Major Crops. Hydropower Publishing House, Beijing, China.

Chen, Y., Zhang, D., Sun, Y., Liu, X., Wang, N., Savenije, H.H.G., 2005. Water demand management: a case study of the Heihe river basin in China. Physics and Chemistry of the Earth 30 (6–7), 408–419.

Dinar, A., Saleth, R.M., 2005. Issues in water pricing reforms: from getting correct prices to setting appropriate institutions. In: Folmer, H., Tietenberg, T. (Eds.), International Yearbook of Environmental and Resource Economics 2005/2006: A Survey of Current Issues. Edward Elgar, Cheltenham, UK.

Dinar, A., Tsur, Y., 1995. Efficiency and Equity Considerations in Pricing and Allocating Irrigation Water. The World Bank Policy Research Working Paper No. 1460.

Dinar, A., Zilberman, D., 1991. The economics of resource-conservation, pollution-reduction technology selection: the case of irrigation water. Resource and Energy Economics 13 (4), 323–348.

Gardener, B.D., 1983. Water pricing and rent seeking in California agriculture. In: Anderson, T. (Ed.), Water Rights: Scarce Resource Allocation, Bureaucracy, and the Environment. Ballinger Publish Company, Cambridge, Massachusetts, USA, pp. 83–116.

Huang, J., Rozelle, S., Rosegrant, M.W., 1999. China's food economy to the twenty-first Century: supply, demand, and trade. Economic Development and Cultural Change 47 (4), 737–766.

Huang, Q., Rozelle, S., Msangi, S., Wang, J., Huang, J., 2008. Water management reform and the choice of contractual form in China. Environment and Development Economics 13 (2), 171–200.

Lohmar, B., Wang, J., Rozelle, S., Huang, J., Dawe, D., 2003. China's Agricultural Water Policy Reforms: Increasing Investment, Resolving Conflicts, and Revising Incentives. Market and Trade Economics Division, Economic Research Service, U.S. Department of Agriculture. Agriculture Information Bulletin No. 782.

Ministry of Agriculture of China, 2004. China Agriculture Yearbook. China Agriculture Publishing House, Beijing, China.

Ministry of Water Resources, 2002. China Water Resources Bulletin. China, Beijing.

Moore, M.R., Gollehon, N.R., Carey, M.B., 1994. Multicrop production decisions in western irrigated agriculture: the role of water price. American Journal of Agricultural Economics 76 (4), 859–874.

Ogg, C.W., Gollehon, N.R., 1989. Western irrigation response to pumping costs: a water demand analysis using climatic regions. Water Resources Research 25 (5), 767–773.

Oi, J.C., Babiarz, K.S., Zhang, L., Luo, R., Rozelle, S., 2012. Shifting fiscal control to limit cadre power in China's townships and villages. The China Quarterly 211, 649–675.

Ravallion, M., Chen, S., 2007. China's (Uneven) Progress against Poverty. Journal of Development Economics 82 (1), 1–42.

Rozelle, S., Zhang, L., Huang, J., 2003. China's war on poverty. In: Hope, N.C., Yang, D.T., Li, M.Y. (Eds.), How Far across the River? Chinese Policy Reform at the Millennium. Stanford University Press, Stanford, CA, USA.

Wang, H., 1997. "Distorted Water Prices". Chinese Water Conservation Yearbook. China Water and Electricity Publishing House, Beijing, China.

Wang, J., Huang, J., Rozelle, S., 2005a. Evolution of tubewell ownership and production in the North China plain. The Australian Journal of Agricultural and Resource Economics 49 (2), 177–195.

Wang, J., Xu, Z., Huang, J., Rozelle, S., 2005b. Incentives in water management reform: assessing the effect on water use, production, and poverty in the Yellow river basin. Environment and Development Economics 10 (6), 769–799.

Wei, B., 2001. Suggestions on the Reform of Water Supply Pricing Systems. The Water Rights and Water Market Forum, Beihai, China.

Weinberg, M., 2002. Assessing a policy grab bag: federal water policy reform. American Journal of Agricultural Economics 84 (3), 541–556.

Yang, H., Zhang, X., Zehnder, A.J.B., 2003. Water scarcity, pricing mechanism and institutional reform in northern China irrigated agriculture. Agricultural Water Management 61 (2), 143–161.

Young, R.A., 2005. Determining the Economic Value of Water: Concepts and Methods. Resources for the Future, Washington DC, USA.

Zheng, T., 2002. Several Issues on the Reform of the Pricing of Water Supplied in Irrigation Projects. China Water-Pricing Reform Workshop, Haikou, Hainan Province, China.

CHAPTER

Water Allocation Through Water Rights Institution

13

In chapter "Water Scarcity in the Northern China," we showed how rapid economic and population growth has driven increasing demand for water from industrial and urban sectors and placed pressure on resources available for agricultural production. Historically, the transfer of water from agriculture to urban and industrial use has caused social unrest with disputes and clashes erupting as farmers struggle to retain access to water resources. Further, the potential to improve farm incomes by moving from staple agricultural commodities, such as wheat and corn, into higher value crops such as horticulture has also placed pressure on developing mechanisms to reallocate water used for irrigation within the agricultural sector, including interregionally.

In chapter "China's Agricultural Water Policy Reforms: Increasing Investment, Resolving Conflicts, and Revising Incentives," we described and evaluated a number of government policies and new institutions designed to ration and allocate existing water resources more efficiently. The issue of water rights is a silent backdrop to these reallocation mechanisms. The definition and establishment of water property rights will be an important component of both water and agricultural policy reform.

This chapter explores the institutional framework for water management in northern China. This region was selected because it contains around two-thirds of China's cultivated land but less than a quarter of the nation's water resources. Almost 50% of the nation's gross domestic product is generated in northern China, a region that is a significant producer of wheat and maize (Ministry of Water Resources, 2000). Preliminary results from farm household and water manager surveys are used to estimate the benefits of water reallocation in the Yellow River Basin (YRB). Using information from this case study, future directions for water property rights and policy are discussed in light of China's commitment to a more efficient allocation of water resources both within and between irrigation districts (IDs), and between competing uses.

THE YELLOW RIVER BASIN (YRB)

The Yellow River is the second longest in China with a total length of almost 5500 km (Map 13.1). The YRB covers an area of 795,000 km^2 and is characterized by varying climate and relief. The basin can be divided into three distinct reaches: the upper, middle, and lower, according to the following characteristics. The upper reaches are mountainous, with deep valleys until the river flows across the alluvial

Managing Water on China's Farms. http://dx.doi.org/10.1016/B978-0-12-805164-1.00013-0

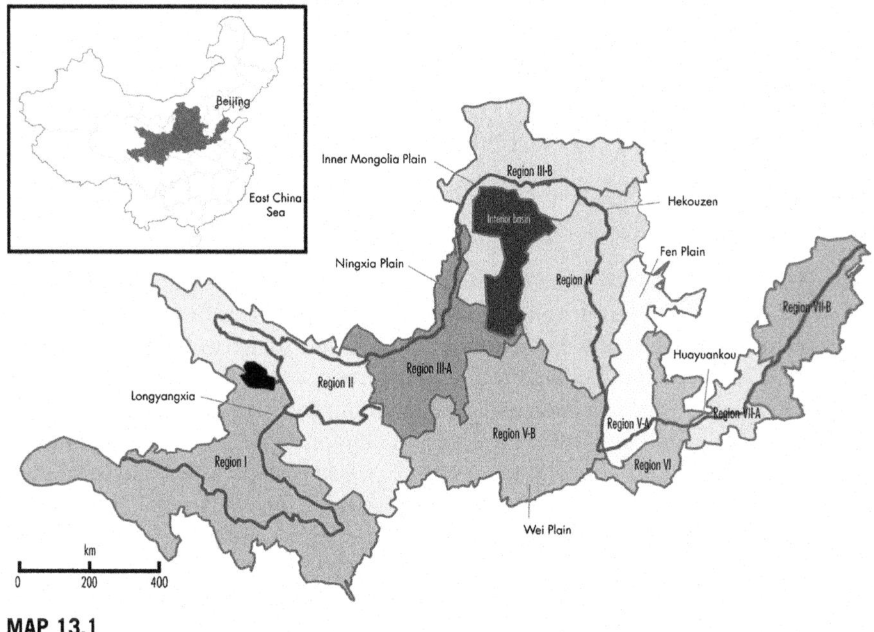

MAP 13.1

China: Yellow River Basin water resource regions.

Source: World Bank, 1993. Water Resources Management: A World Bank Policy Paper. World Bank, Washington, DC.

plains of Ningxia and Inner Mongolia. The reach from Lanzhou to the Mongolian steplands receives minimal rainfall but faces large irrigation demands. The middle reaches, between Hekouzhen and Huayuankou, encompass the major irrigation areas of Shanxi and Shaanxi that are fed by the Yellow River's two major tributaries, the Fen River and the Wei River. Massive amounts of loess soil enter the main stem and tributaries, resulting in sediment loads that are unprecedented in the world's major waterways. The lower reaches stretch from the Taihang Mountains to the Bohai Sea. Much of the alluvial plain of the lower Yellow River area is below sea level where the river is "suspended" (IWMI and YRCC, 2002; YRCC, 2005).

The annual precipitation distribution declines from 600 to 200 mm, more or less progressively, from southeast to northwest. Annual average runoff is around 58 billion m^3 (58,000 ML). Several large dams control flooding as well as run-off from melting ice, and also provide sediment mitigation, hydropower, and water supply services. Water consumption is estimated to be around 31 billion m^3, with agricultural uses accounting for around 80% of total water consumption in 2000. The average per capita share of water resources in the YRB is less than 20% of China's average, and considerably less when compared to the rest of the world (IWMI and YRCC, 2002; YRCC, 2005).

The YRB is regarded as the "cradle of Chinese civilization" and irrigated agriculture has been practiced here for thousands of years. Massive government investment in irrigation infrastructure increased rural livelihoods and agricultural output in the

Table 13.1 Agricultural Water Use and Farmer Income

Province	Agricultural Water Use (10^9 m^3)	Irrigated Area (1000 ha)	Crop Water Use per ha (m^3/ha)	Farmer Income (Dollars/Year)
Sichuan	122	2501	48,888	272
Gansu	97	988	98,401	205
Qinghai	20	194	105,220	215
Ningxia	76	410	185,439	247
Inner Mongolia	449	2538	176,880	269
Shanxi	36	1104	32,165	277
Shaanxi	55	1315	41,546	206
Henan	146	4802	30,347	286
Shandong	188	4797	39,244	380

Note: Agricultural water use is given in cubic meters (m^3).
Data source: National Statistical Bureau, 2003. China Statistic Yearbook (2003). National Statistical Bureau, Beijing, China.

1960s and 1970s. Irrigated agriculture has expanded significantly over the past four decades and new irrigation and agronomic technologies have increased yields considerably in some areas. Key agricultural data for nine provinces in the YRB in 2002 are given in Table 13.1. Irrigation requirements vary considerably because of the large variation in soil and climatic conditions across the basin. This is, in part, reflected in the variation in crop water use between provinces. Based on provincial level data, per hectare crop water use in upstream provinces (such as Qinghai, Ningxia, and Inner Mongolia) are higher than other provinces. This may also be due to these upstream regions having greater access to water resources. Climatic conditions in the lower reaches, such as in Henan and Shandong, mean that crop water requirements are lower than in other parts of the basin. Due to variation in irrigation conditions and other factors, land use intensity and cropping structures in these regions also varies. More favorable agronomic and climatic conditions mean that farmer incomes are higher in the downstream provinces (IWMI and YRCC, 2002; Huang et al., 2005). Unsurprisingly, this has led to many conflicts between upstream and downstream users, which will be discussed in more detail below.

For most of the last 30 years the Yellow River has run dry for some period before reaching the ocean—another consequence of increasing demand for water in the basin. The flow cutoff events brought a growing awareness of the need for environmental water as these events have serious repercussions for sediment transportation, shipping, and the ecology of the river delta and coastal fisheries.

While there are still opportunities for water savings, utilizable water resources are currently fully exploited in the YRB so meeting new demands will almost certainly be met by reducing supplies from other sectors. As agriculture is a large water user, further reductions in supply seem inevitable although this will have implications for rural livelihoods and the long-standing policy of food self-sufficiency. One of the key policy challenges is how to reallocate water supplies while maintaining rural incomes and agricultural output. The focus of the remainder of this chapter is to assess the role that water property rights could play in facilitating water reallocation, and to estimate the impacts of reallocating water on farm household income in the YRB.

INSTITUTIONAL ARRANGEMENTS FOR WATER ALLOCATION

WATER ALLOCATION AMONG PROVINCES ALONG THE YRB

As an agency of the Ministry of Water Resources (MWR), the Yellow River Conservancy Commission (YRCC) has no actual water allocation power in the river basin and has limited power in upstream and midstream water management (Wang et al., 2003). Water allocation in the upstream and midstream sections of the Yellow River is controlled by upstream and midstream control commissions of the Yellow River. Differing functions and purposes of the two departments have rendered cooperation on water allocation difficult.

Before the implementation of the Yellow River Water Allocation Program, water users across the basin could freely draw water from this "open-access resource" (Ma et al., 2007; Wang et al., 2007). Regional water use was simply determined by water withdrawal capacities. Open-access entails a high risk of resources degradation. To satisfy rapidly growing water demands, provinces along the Yellow River constructed a large number of water withdrawal facilities. From the 1950s to the 1990s, water consumption of the Yellow River increased by 150%. As a result of poor management and low water-use efficiency, river waters trickled to a standstill in the lower end of the basin since the early 1990s (Wang et al., 2003). Water use per capita and as a share of GDP in the four upstream provinces of Qinghai, Gansu, Ningxia, and Inner Mongolia are much higher than in the downstream provinces of Shaanxi, Shanxi, Henan, and Shandong. Due to geographical constraints, it is particularly difficult to get water in Shanxi, where per capita water use and water use as a share of GDP was less than 200 and 400 m^3, respectively. In Ningxia these figures are more than 1500 and 400 m^3, respectively.

Increasing depletion of the Yellow River has attracted serious attention from the Central Government and the State Council. In December 1998, the State Council, the State Planning Commission, and the MWR jointly issued the Annual Allocation and Main River Water Quantity Controlling Program of the Yellow River, authorizing the YRCC to exert integrated control of Yellow River waters (mainly for surface water) (Ma et al., 2007; Wang et al., 2007). Since March 1, 1999, the YRCC controls the nonflood period waters from Liujiaxia reservoir to Toudaoguai and from Sanmenxia reservoir to the Lijin gauge. Based on previous experience, the new program emphasizes the importance of enforcement and incentive policies to induce water authorities in the upstream regions to participate in the program and adhere to the national government's directives and their own agreements. Under the high attention and leadership of the MWR and close cooperation among the riparian provinces, autonomous regions, and relevant departments along the Yellow River, integrated water control has obtained significant achievements. Compared with an average of 126 days in which the river flow to the Sea was cut off during 1995–1998, flow at Lijin Station was cut off only for 8 days between 1999 and 2000, and since 2000, river flow has not been interrupted.

However, the experience of the new Annual Allocation program has received some criticism, and there are concerns about the sustainability of the process in the long run (Wang et al., 2003). First, as the authority (YRCC) started enforcing cuts in water withdrawal for upstream users, who began to object. Compared to 1999, total water withdrawals from the Yellow River were cut by 0.53 billion m^3 (BCM) in 2000. To date, the downstream regions have not had to pay compensation for the incremental units of water that they have received as a result of water cuts to upstream users. Moreover, because water sales in the upstream regions have fallen, upstream irrigation districts (IDs) are beginning to experience declines in revenues. In interviews during 2001, we were told that a significant amount of maintenance would have to be delayed or canceled because of the revenue shortfall. If such practices continue and become more severe, the entire irrigation system could suffer serious damage.

In response, upstream water suppliers have demanded to increase prices for water to maintain revenues. Although initially there was some opposition by the central government, after the introduction of the YRB water allocation program, water prices in the upstream areas were allowed to rise. For example, Ningxia doubled the price of water in agriculture from \$0.0009 to \$0.002/m^3 in 2000 (Wang et al., 2003). As a result of higher water fees, the water authorities have regained some of their lost revenues. However, as soon as the revenues began to balance, there were more cuts in water that began to cause concern from Ningxia water officials. The concerns increased in 2001 when officials kept water prices constant. Any visitor to Ningxia (or any other upstream province) will quickly hear the complaints about the unfair transfers of water to downstream users for which they are not receiving compensation.

Uncompensated water transfers between poorer provinces to more well-off provinces can have important impacts on the poor (Wang et al., 2003). Reallocations to date have occurred between the relatively worse-off upstream provinces of Ningxia and Gansu to richer provinces, such as Shandong. Most likely, the shift of water access away from upstream provinces will have a negative effect on some producers in the ID and (as discussed earlier) could adversely affect the operation and future effectiveness of the ID itself. To mitigate these negative effects, central government leaders should estimate the value of compensation to be paid and implement compensation.

Given the scarcity of water resources in the YRB and the lack of compensation for transferring water to better-off downstream users, tensions and conflicts among water stakeholders have risen. Although YRCC is authorized by the State Council to control the Yellow River water resources, YRCC is not a first-tier government agency. The water law does not clearly define the powers of YRCC for administrative enforcement, supervision, and punishment, and the relationship between the YRCC and local governments remains unclear. Integrated management of the

Yellow River is made more difficult by the fact that water-control policies are relatively new. Further, the Yellow River Water Quantity Control Management Method does not have detailed implementing regulations. In addition, the water control plan has hardly been implemented; detailed plans for implementation of the Yellow River Water Quantity Control are therefore urgently needed. Currently, the YRCC has only conducted integrated control for two main river reaches: from Liujiaxia Reservoir to Toudaoguai and from Sanmenxia Reservoir to Lijin gauge; other sections remain without integrated management.

WATER ALLOCATION WITHIN PROVINCES

Generally, the upper-level water resources bureaus, at the provincial and prefecture levels, prepare annual water allocation plans for lower-level regions and then submit them for approval to the local governments (Wang et al., 2003). Before formulating water allocation plans, water resources bureaus at upper levels will solicit opinions on water requirements from the water management agencies at the lower levels. Opinions on water allocation from lower-level water management agencies are the foundation of water allocation plans and are then adjusted and coordinated by upper-level water management agencies according to the overall water supply and demand situation. After the local government approves the water allocation plan, it is sent to lower-level governments for implementation. Last year's water supply and demand situations and the current year's water demand are important parameters for the formulation of water allocation plans. Because water supply in the current year cannot be predicted with great certainty and some water use sectors will not be satisfied with the water allocation plan, conflicts among various water use departments tend to arise and plans tend to be adjusted (Fig. 13.1).

While water resources bureaus at the upper level coordinate and decide on water allocation between upstream and downstream water users within their administrative boundaries, actual water allocation decisions are made by the upstream local government, and downstream water users have little power to influence these allocation decisions. Furthermore, higher-level government agencies have limited influence on lower-level agencies in actually implementing water allocation plans, as relationships are mainly technical, not administrative. When water conflicts occur, the water resources agency at the upper level will try to mediate. If the problem cannot be resolved through mediation, the local government in charge of all bureau's management at the local level will intervene. For serious conflicts, local or higher-level courts will be called upon.

Allocation Among and Within the Main and Branch Canals: Case of Ningxia

The issues surrounding water allocation among the provinces are similar to those among the main canal and branch canals within an ID (Wang et al., 2003). Water allocation at the main or first-level canals and the second-level canals within a province's ID are controlled by one of the following government agencies—the provincial, prefectural, or county's water resources bureau—depending on the size

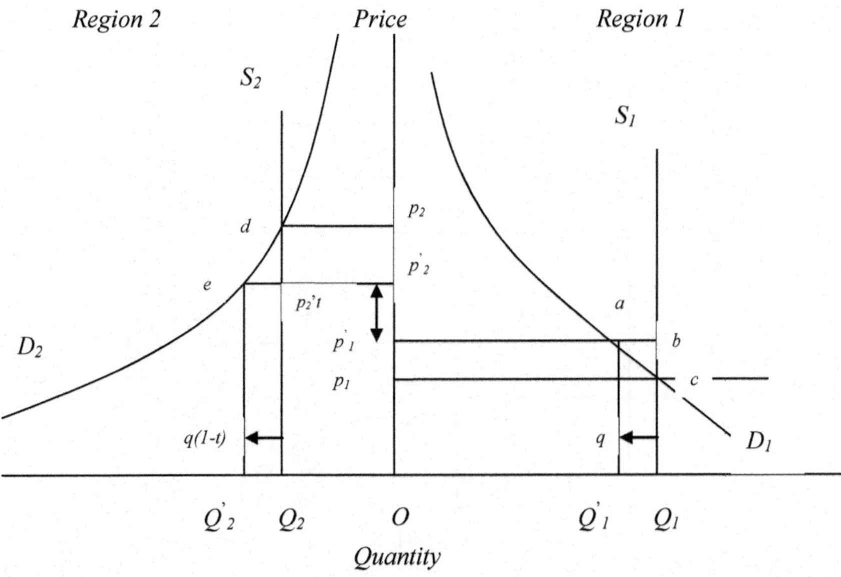

FIGURE 13.1

Water trade between two regions.

and reach of the ID (Fig. 13.2). By regulation, the general principle for water allocation is to give the first priority to downstream users, lift irrigation (vs gravity irrigation), the canal's high off-takes (vs the lower off-takes), agriculture (vs forestry), more vulnerable groups, and large consolidated tracts of land (vs fragmented land).

In Ningxia Province the regulations on water allocation fully acknowledge the primacy of efficiency, but they also attempt to encourage equity (Wang et al., 2003). Unfortunately, due to lack of data, we cannot conduct a systematic assessment of the implementation of allocation regulations. However, based on field interviews and survey data, we can provide partial and fragmentary evidence on the actual implementation of the water allocation principles, to see which ones are pursued and which ones are relatively ignored.

The basic water allocation rule is to allocate water considering water quotas while abiding by the allocation principles. The water quota is critical to guide water allocation practices. In China, a water quota is an engineering standard developed by the water resources bureau that indicates the amount of water that is needed for crop production. The water quota differs by crop type, the availability of water supply, delivery efficiency of the surface water system, the location in the canal, and rainfall, climatic, soil, and other local characteristics that influence the crop water demand. The purpose of the water quota is to provide IDs with a way to allocate water and limit the overexploitation of an area's water resources.

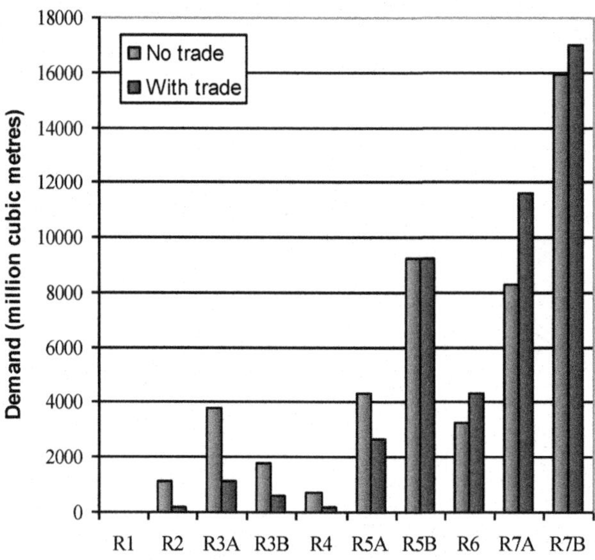

FIGURE 13.2

Water demand, by region.

Although the quota-based water allocation rules have been designed so that all areas should be able to irrigate their crops with the volume of water specified by the quota, some areas exceed their quotas. Given that exceedance of quotas is tolerated by ID officials, the pattern of water distribution may indicate that upstream IDs are granted priority in water allocation over downstream IDs. Unlike the differences that occur in the execution of water allocation regulations among IDs, in some cases water managers allocate water more evenly among upstream and downstream areas within IDs taking into account provincial regulations that downstream areas have priority.

Thus, actual water allocations are affected by policies and principles, but also vested interests of ID officials. To the extent that poorer people live in the downstream areas of IDs, there is mixed evidence that official regulations promote poverty alleviation. The regulations stipulate that officials are supposed to send more water to downstream regions and give priority to vulnerable groups, such as the poor. The actual record shows, however, that implementation of these regulations is spotty. Officials in some areas follow the downstream first rule, while those in other areas do not. There is no evidence, on a district-wide scale, that water officials purposely give priority to the poor.

Allocation Among and Within the Tertiary Canals (Within the Village)

At the tertiary canal level, local water managers are responsible for water allocation. In the following sections we systematically examine the equity and poverty implications of different water management mechanisms (Wang et al., 2003). Our field survey found various allocation rules used by mangers in the course of their water allocation duties. During our survey, we identified three kinds of water allocation approaches: equity, efficiency, and payment capacity. The way in which each village decides to allocate water has most likely evolved as a result of a complex set of characteristics of the village, the nature of its water resources, and local cropping patterns. Explaining why a certain village allocates water in a certain way is beyond the scope of this report. Instead, what we are able to do is describe the fundamental characteristics of the allocation rules and examine how many villages have adopted different approaches.

The equity approach means that water resources are equally allocated to all water users along the canal. Such a rule allows the poor and other vulnerable groups to get access to water. In practice, rules are often promulgated to provide water to those farmers at the end of the canals first, and those nearest last. In our sample, we find that 13% of sample villages within our survey use this method of water allocation.

According to the efficiency approach, village water managers irrigate as the water flows into the canal. Irrigating the nearest fields first is physically the most efficient way to reduce delivery induced water loss. Interestingly, despite the emphasis of IDs on equity, a much greater number of villages (70%) claim they use the efficiency method of water allocation. Finally, some villages use other methods based on the payment order. A mere 2% of villages provide water on a first pay, first serve basis. There are no established rules in the rest of the villages.

EFFECTS OF WATER REALLOCATION IN THE YRB

For the purpose of this study the YRB has been divided into 10 regions based on hydrologic, agroclimatic, and soil conditions. This is consistent with previous work undertaken by the World Bank (1993). The important administrative boundaries and provinces included in each region are provided in Table 13.2 and shown in Map 13.1. There are three basic principles underlying this classification: First, within regions, natural geographic conditions, water resource development and use, and water conservancy development and objectives are sufficiently similar or sufficiently dissimilar; second, distinctions important to different main-stem river sections or tributaries are maintained; third, if possible, the administrative boundaries and catchment areas corresponding to major works on the main stem or tributaries are preserved.

The methodology used to model water reallocation in the Yellow River Basin involved four steps. First, a flexible production function was estimated to characterize the agricultural production technologies in the basin for each region and for seven agricultural crops; wheat, maize, rice, vegetables, cotton, soybean, and potatoes. Second, a nonlinear profit maximization problem was formulated for each region. The objective function embeds the flexible crop production functions and input costs. The constraints reflect regional resource limits on the availability

Table 13.2 Regions of the Yellow River Basin (YRB)

Region	Provinces Included in Each Region
1	Sichuan (3), Gansu (1), Qinghai (11)
2	Gansu (23), Qinghai (16)
3A	Gansu (7), Ningxia (16), Shaanxi (1)
3B	Inner Mongolia (28)
4	Shanxi (21), Inner Mongolia (2), and Shaanxi (10)
5A	Shanxi (50) and Henan (6)
5B	Gansu (23), Ningxia (4), and Shaanxi (68)
6	Shanxi (6), Shaanxi (1), and Henan (20)
7A	Henan (37)
7B	Shandong (65)

Note: For column (2), data in parenthesis is number of counties in each province that should be in the region.
Data source: Authors' calculation based on county level data.

of land and water as well as any policy restrictions. Third, the optimization model was solved over a range of water prices to estimate the parameters of a regional water demand function. Fourth, these water demand functions were incorporated into a spatial equilibrium model of regional water markets for the YRB to estimate the potential gains from water transfers within the basin. The detail descriptions of these models are included in Methodological Appendix to Chapter 13.

Based on the model estimation, the total value of agricultural production in the YRB is estimated to be around $7.4 billion. Total cultivated land in the YRB is almost 16 million hectares (Table 13.4). Across the 10 regions, cultivated land

Table 13.3 Water Use and Price and Estimated Elasticities by Region, 2002

Region	Base Year Water Use (million m^3)	Scarcity Value of Water[a] (Dollars/1000 m^{3b})	Demand Elasticity
1	18	0.77	−0.76
2	1129	0.77	−0.78
3A	3794	2.06	−0.87
3B	1786	2.45	−0.88
4	735	2.06	−0.84
5A	4347	6.45	−1.13
5B	9241	24.38	−1.81
6	3249	15.48	−0.85
7A	8304	17.15	−0.88
7B	15,908	13.28	−0.89

[a] *Estimated as the net agricultural return of an additional 1000 m^3 of water use in the region less delivery charges.*
[b] *1000 m^3 is 1 ML.*

Table 13.4 Cultivated Land, Share of Irrigated Land, Sown Area, Multiple Cropping Index, and Farm Household Income in the Ten Regions of the Yellow River Basin (YRB), 2002

Region	Cultivated Land (1000 ha)	Share of Irrigated Land (%)	Total Sown Area (1000 ha)	Multiple Cropping Index[a]	Farm Household Income (Dollars/Year)
1	79	4	59	0.75	216.9
2	913	26	1034	1.13	185.2
3A	1246	39	1350	1.08	229.1
3B	1910	51	1185	0.62	312.0
4	1315	11	1069	0.81	301.9
5A	1735	40	1773	1.02	374.4
5B	3160	34	4061	1.29	196.2
6	848	44	1277	1.51	282.3
7A	1411	82	2659	1.89	280.9
7B	3102	63	5370	1.73	355.9

[a] *Multiple cropping index implies the ratio of sown area over cultivated area.*
Data source: Authors' calculation based on county level data provided by Institute of Geographical Sciences and Natural Resources Research, Chinese Academy of Sciences.

ranges from more than 3 million hectares (such as regions 5 and 7B) to less than 1 million hectares (such as region 1, Map 13.1). The share of irrigated land varies considerably between regions with those in the lower reaches having the highest proportion of irrigated land. For example, in region 7A (Henan Province), more than 80% of cultivated land is irrigated. In contrast, only 4% of cultivated land is irrigated in region 1, in the upper reaches, and only 11% in region 4, in the middle reaches. This is because irrigators in the downstream reaches have access to ground water that is used conjunctively with surface water resources. Irrigation in upper and middle reaches depends mainly on surface water.

Land use intensity and cropping structures in these regions are also highly varied. Due to various irrigation conditions and other factors, land use intensity in downstream reaches is higher than that in the upper and middle stream reaches (Table 13.4). For example, the multiple cropping index in region 7A is near 2, while it is less than 1 in region 1.

The optimal reallocation of irrigation water was estimated by simulating trade between regions. With trade, the scarcity value of water is equated across regions, allowing for differential conveyance losses.

After the trade, water is reallocated from upstream to downstream and from low to higher returning uses, leading to an overall increase in irrigated agricultural production (Table 13.3 indicates the estimated scarcity value of water by region in the YRB). Under an optimal reallocation in the YRB, a large volume of water would be imported into regions 6 and 7A and, to a lesser extent, region 7B (Fig. 13.2). Much of this water is used for wheat production after the trade.

The increase in wheat production in the importing areas more than offsets the reductions in wheat production in the exporting areas and overall production in the basin is increased. Cotton production also increases after the trade.

The water is mainly exported from regions 3A and 5A, and to a lesser extent, from regions 2, 3B, and 4 where the scarcity value of water is very low. Much of this water was used for wheat and rice production, causing significant drops in production for all water exporting regions after reallocation. Regions where water is exported increase maize production as it is a high value crop used as animal feedstock. It is substituted for wheat in the cropping rotation.

The trade equalizes the scarcity value of water across all regions in the basin. After the trade, the price of water increases significantly in the upstream regions to region 5A and decreases in the downstream regions, region 6, 7A, and 7B.

There is an estimated welfare gain of almost $163 million per year in the importing regions after the water reallocation (Table 13.5). These benefits accrue as a result of significant increases in crop production in the importing regions. This is partially offset by reductions in agricultural production in the exporting regions, resulting in welfare loss of $36 million. The total benefit of water reallocation, comprising the increase in the value of agricultural production is around $129 million. This represents an increase in the value of agricultural production of around 1.8%.

The benefits that are generated as a result of water sales accrue to the holders of the property rights in the exporting regions. If, as is the case now, irrigators or irrigation districts do not own the water, the benefits of the water sales would go to the State. This would result in a loss in potential income from the sale of water to irrigators in the exporting regions of more than $64 million per year. That is, irrigators in the exporting regions are around $64 million worse off each year after the reallocation as they do not currently hold the property rights to the reallocated water. Alternatively, this is the amount that they would need to be compensated for the loss of access to water resources if the water was administratively reallocated. This is in addition to the lost agricultural production valued at around $36 million per year.

Table 13.5 Benefits of Water Reallocation in the YRB (Million Yuan)

	Water Importing Regions	Water Exporting Regions	Basin Wide
Change in value of agricultural production	172.2	−35.7	136.5
Value of the water transferred	−64.2	64.2	0
Total benefit of reallocation[a]	108.0	28.5	136.5

[a] *Including benefits of water sales accruing to irrigators in the exporting regions. For those farms that sold water, farm income after reallocation also includes annual lease value of water sold or the annualized value of the compensation received if the water was transferred administratively.*

Table 13.6 Impact on Farm Incomes of Water Reallocation (Dollars/Farm)

Region	Before Reallocation	Change in Farm Household Income Without Property Rights	Change in Farm Household Income With Property Rights[a]
1	217	−7.6	13.5
2	186	−2.6	5.2
3A	229	−8.3	7.4
3B	312	−6.3	4.9
4	302	−7.5	7.0
5A	378	−5.9	2.5
5B	196	0	0
6	282	22.7	14.6
7A	281	23.6	15.2
7B	356	3.5	1.9

[a] *For those farms that sold water, farm income after reallocation also includes annual lease value of water sold or the annualized value of the compensation received if the water was transferred administratively.*

The impact of water reallocation on farm household incomes is presented in Table 13.6. If exporting regions are not compensated for the reallocated water, reductions in farm household income in are greatest in the middle reaches of the basin, regions 5A, 3A, and 4—the regions with the greatest reductions in agricultural production. Conversely, with compensation all regions benefit from the reallocation with the largest benefits going to some regions that have the lowest farm household income in the basin, such as regions 2 and 3A. The increase in farm household income in the importing regions is considerably lower if farmers hold water property rights as they would need to pay for the water they received (in addition to delivery charges), reducing the benefits of water reallocation in that region.

ONGOING WATER RIGHT TRANSFER PROJECT IN THE YRB

Since 2000, YRCC has begun to promote the establishment of water right systems by conducting demonstration projects aimed at reducing water competition among sectors. With increasing water shortages, water becomes insufficient to support industrial development, especially energy industry development, in the upstream provinces such as Ningxia and Inner Mongolia. In order to solve this problem, in 2003, the YRCC established some water rights demonstration sites in the upstream reaches of the YRB. The purpose of these demonstration sites is to reallocate water from agriculture to industry through increasing irrigation efficiency (Wang et al., 2007; Li, 2007; Chen et al., 2007; Liu et al., 2007; Wang et al., 2006). In 2004, in order to promote the water transfer work in the YRB, the MWR issued the *Guidance on Water Rights Transfer Demonstration Works in Inner Mongolia and Ningxia*

Provinces. The YRCC also released two regulations titled *Management and Implementation Measures on Water Rights Transfer in the YRB* and *Management Regulation on Water-Saving Engineering*. These regulations have provided the legal foundation for water right transfers in the YRB.

According to the project design, there are three steps to conducting water right transfers. First, the local government needs to identify the components of water transfer fees paid by industry sector (Pei et al., 2007; He et al., 2007). The water transfer fees include construction fees for water conservation, operation, and maintenance fees for water-conserving investments, associated reconstruction and rehabilitation fees, and compensation fees for agriculture and ecology. Second, water rights need to be allocated to water users. This involves mechanisms for allocating water rights to villages and establishing WUAs that can allocate water rights to individual farmers. Finally, local governments need to identify the quantities of water that can be transferred without adversely impacting agricultural production and farmers' livelihoods.

One such water transfer pilot is operating in Ningxia Province (Chen et al., 2007). In Inner Mongolia, 16 projects signed transfer contracts, with a value of $100 million, and total water-saving investments, mostly canal lining, in the order of $31 million. Water right transfer projects involved seven types of industrial enterprises and two irrigation districts in Ningxia and Inner Mongolia. According to the contract and regulation of YRCC, the two IDs will transfer part of their water use rights to these industrial enterprises for a period of 25 years.

Despite some progress on water rights transfer, there are still considerable challenges facing both central and local governments regarding the implementation of water right transfers (Hu and Chen, 2004; He et al., 2007; Jiang et al., 2007; Yang et al., 2006). The first is the engineering problem. YRCC considers the construction of water-saving infrastructure to be very slow, thus constraining the progress of water right transfers. In addition, management of this new infrastructure remains a challenge. The second constraint relates to water rights. While some water right transfer projects have been established, a general water rights system has not been developed. Water users still have no clear understanding of their water rights. Water right transfers still depend on administrative power, not on developed water markets; in other words, they are a function of the central and local governments, and are not adjusted by market signals or economic measures. In fact, in China, establishing a real water right system and water markets still have a long way to go. The exact mechanism needed for effective promotion of the system of water rights is still hotly debated by many policymakers and researchers.

CONCLUSIONS AND POLICY IMPLICATION

Preliminary findings from the modeling work presented here suggest that there are considerable gains from water reallocation. In the scenario presented, water is moved downstream to higher value agricultural use under conditions of free trade.

The economic benefit, in terms of the increased value of agricultural production was around $129 million per year. If farmers in water exporting regions had the property rights to transferred water, income from water sales would more than offset the forgone income from reduced agricultural production. The income from water sales is estimated to be around $65 million per year. In the absence of property rights, the lost value of agricultural production lowers farm household incomes substantially. Conversely, with revenue from the sale of water, farm household incomes in the exporting regions would rise substantially. Importantly, without compensation the regions with the lowest incomes are likely to be affected the most by water transfers.

Water can be reallocated using by a number of means, two of which have been considered here—administrative reallocation and free water trade. While it may be theoretically possible to reach an economically efficient outcome by administering water reallocation, there are number of barriers that prevent this from happening in practice. For example, the information requirements are demanding. Information asymmetry between the administrative body and irrigators on the marginal value and opportunity costs of water mean that allocation decisions would be made based on imperfect information about where the greatest benefits can be generated.

Water markets, on the other hand, coordinate price signals and disperse information and preferences. Water markets would provide a mechanism to transfer water to higher value uses on a large scale and to the other productive uses, such as industry and the environment. For formal water markets to work efficiently, property rights to water must be private, exclusive and transferable. Secure ownership provides the incentive to invest in human or physical capital to improve the productivity of the resource. Transferability provides the flexibility to reallocate the rights according the changing demand and other conditions. The role of the state is to protect these property rights by enforcing contracts and reducing transactions costs and other barriers to exchange. However, legislation, institutions, and the necessary regulatory framework to support water reallocation do not currently exist in the YRB.

While the issues facing resource managers in China are unique in many ways, establishing and implementing water property rights structures to facilitate reallocation has been undertaken in many developed and developing countries. A body of literature exists drawing lessons from experience in these countries and assessing its relevance in others. Hu (1999), for example, explored the relevance of the Australian experience in the Murray Darling Basin to the Chinese context and concluded a similar legal and institutional framework would not be suitable because of the existing administrative framework in China and the incomplete and uncertain specification of resource access. Perhaps the most pervasive of reasons, however, is the small scale of farming in China and the consequent transactions costs of implementing water property rights at that scale. If, on the other hand, water rights are granted at a higher level, at the irrigation district level for example, there may not be sufficient incentive for farmers to engage in water saving practices unless they are adequately compensated.

REFERENCES

Chen, Y., Su, Q., Hu, Y., 2007. Do water transfer work well, promote the establishment of water saving society in the Yellow River Basin. China Water Resources 19, 51–53.

He, H., Xue, J., Fang, X., 2007. Research on compensation mechanism in the water rights transfer in the Yellow River Basin. China Water Resources 19, 59–61.

Hu, Z., 1999. Integrated catchment management in China: application of the Australian experience. Water International 24, 323–328.

Hu, Y., Chen, Y., 2004. Practice and consideration on water rights transfer in the Yellow River Basin. China Water Resources 15, 45–47.

Huang, Q., Rozelle, S., Dawe, D., Huang, J., Wang, J., 2005. Irrigation, poverty and inequality in rural China. Australian Journal of Agricultural and Resource Economics 49 (2), 159–176.

IWMI (International Water Management Institute) and YRCC (Yellow River Conservancy Commission), 2002. Yellow River Comprehensive Assessment: Basin Features and Issues. IWMI Working Paper 57, Colombo, Sri Lanka.

Jiang, B., Zhang, B., Li, E., 2007. Analysis of monitoring effects of water rights transfer in the Yellow River Basin. China Water Resources 19, 47–48.

Li, G., 2007. Exploring and practice of water rights transfer in the Yellow River Basin. China Water Resources 19, 30–31.

Liu, X., Wu, L., Wan, Z., 2007. Research on water rights transfer in the inner Mongolia in the Yellow river Basin. People's Yellow River 10, 16–17.

Ma, X., Han, J., Chang, Y., 2007. Research on evolution of water rights transfer in the Yellow River Basin. Research of China's Economic History 1, 41–47.

Ministry of Water Resources, 2000. China Water Resources Bulletin various issues (in Chinese).

National Statistical Bureau, 2003. China Statistic Yearbook (2003). National Statistical Bureau, Beijing, China.

Pei, Y., Ying, H., Wang, T., 2007. Research on water transfer price in the Yellow River Basin. China Water Resources 19, 56–58.

Wang, J., Huang, J., Rozelle, S., Huang, Q., Blanke, A., 2007. Agriculture and groundwater development in northern China: trends, institutional responses, and policy options. Water Policy 9 (S1), 61–74.

Wang, J., Xu, Z., Huang, J., Rozelle, S., 2003. Pro-poor Intervention Strategies in Irrigated Agriculture in China, Report Submitted to International Water Management Institute and Asian Development Bank.

Wang, J., Xu, Z., Huang, J., Rozelle, S., 2006. Incentives to managers and participation of farmers: which matters for water management reform in China? Agricultural Economics 34, 315–330.

World Bank, 1993. Water Resources Management: A World Bank Policy Paper. World Bank, Washington, DC.

Yang, X., Zheng, C., Chen, H., Yang, Z., 2006. Constraints Analysis of Water Rights Transfer in China. No 4, pp. 65–66.

YRCC (Yellow River Basin Conservancy Commission), 2005. Background Material for Yellow River Basin. http://www.lanl.gov/chinawater/documents/yellowriver.pdf (accessed 01.09.05.).

CHAPTER

Adoption of Water-Saving Technology

14

Over the last decade or more, official concern over impending water scarcity has increased as it has become apparent that China's water resources are becoming alarmingly scarce in some areas. Zuo (1997) notes that as of 1995, "The Party Central Committee and the State Council are much concerned with the problems arising from serious water shortages." Policy makers have begun to develop strategies. Some policies (eg, the requirement for receiving a permit before sinking a new well) have not been effective due to the vast number of villages in northern China and the problems involved with monitoring such a spatially dispersed economic activity. Others have not been implemented for political reasons. For example, water-pricing policies have not been implemented because the government has spent considerable policy effort in recent years to reduce taxes and fees.

China's government has begun in recent years to invest in research on water-saving agricultural techniques. Zuo (1997) reports that since "the beginning of the Seventh Five-Year Plan (1986–1990), water-saving and dry-land farming have been designated [as a] major scientific research [program] by the government, involving many specialists from different institutions, and more than 3000 practical achievements have been obtained in dry-land farming." International organizations and foreign governments have collaborated with China's government and research institutions on these projects. In addition to sponsoring research, government and nongovernmental organization–sponsored programs have promoted the adoption of specific water-saving technologies (WSTs), sometimes providing financial support for infrastructure.

Despite substantial investment in the development of WSTs and the potential impact of widespread adoption, there is little evidence that farmers have adopted the new techniques (Lohmar et al., 2003). The efficacy of current WST extension programs is a matter of debate (Deng et al., 2004). There has been little research on the extent of adoption in northern China, the conditions under which WST is adopted, or the impact of adoption on water use and rural welfare.

Our goal is to sketch a picture of the state of WST in northern China to increase awareness of past trends and its current status. We wish to establish a set of first-order facts about the role WST has been playing in China's agricultural sector. We pursue three specific objectives: (1) to illustrate the progress in adoption over the

Managing Water on China's Farms. http://dx.doi.org/10.1016/B978-0-12-805164-1.00014-2

past two decades; (2) to identify the characteristics of the most successful and unsuccessful technologies; and (3) to explain factors that might be promoting WST and factors that might be holding back its adoption.

WATER-SAVING TECHNOLOGY

During our survey of leaders and water managers in more than 400 villages, we discovered that there are many types of WSTs being used in northern China. For analytical convenience, we have divided WSTs into three groups: traditional, household-based, and community-based. We exclude discussion of a series of novel WSTs (such as drip, intermittent irrigation, and chemical-based sprays), because across our sample there were very low levels of adoption (ie, nearly zero).

Does WST save water? The answer to this question depends not only on the technical properties of each technology, but also on the hydrology of the system in which WST is used. In systems where irrigation water is being pumped from a shallow aquifer, water that is applied to a field but not evaporated from the soil surface or transpired by the growing crop recharges the aquifer and is not lost to the system. In cases like this (eg, the Luancheng county, Hebei study reported in Kendy et al. (2004)), real water savings come only from reduced evapotranspiration (ET) and adopting WST that reduces seepage (eg, underground pipe systems or lined canals) or applied water applications (furrow irrigation, level fields, or sprinklers for example) will not result in significant real water savings. Also, recharge in one area may impact the groundwater available for irrigation in another. In this case, reducing recharge by using WST could have a negative impact on groundwater availability elsewhere.

If, however, water that is not lost as ET is not available for irrigation elsewhere in the basin, adopting technologies that reduce seepage or applied water applications may result in real water savings. This is the case when water is being pumped from a confined aquifer, with no possibility of available recharge, or in surface irrigation systems where water lost through seepage is lost to the system.

The ultimate impact of WST adoption on water availability is also dependent on the effect that it has on other agricultural production decisions including crop choice and the demand for irrigation. If irrigated area expands in response to WST adoption (it becomes cheaper/more efficient to irrigate a larger area), the quantity of water applied as irrigation could actually increase. Kendy et al. (2004) concluded that crop change or reducing ET is the only effective water conservation measure.

Our use of the term "water saving" is limited to perceived, field-level applied irrigation savings. Our definition of water use efficiency is likewise limited to field-level measures of crop production per unit of water input. We understand that in some cases technology adoption may not save water when net water use is measured on an irrigation system- or basin-scale. The real water-saving properties of each technology depend not only on the technical features of the technology, but also on the hydrology of the system and the economic adjustments to production that are associated with adoption of the technology.

TRADITIONAL TECHNOLOGIES

Traditional technologies include border and furrow irrigation and field leveling. We have grouped these technologies because they are widely adopted and village leaders in a majority of villages report adopting these techniques well before the beginning of agricultural reform in the early 1980s. These irrigation methods have relatively low fixed costs and are divisible in the sense that one farm household can adopt the practice independent of the action of its neighbors. Both border and furrow irrigation have been practiced in China for many years. The definitions of these irrigation methods used during the survey were general and so we are not able to distinguish between traditional border/furrow irrigation practices and relatively new techniques that may even further increase field-level water savings. We assume that readers are familiar with border/furrow irrigation and the water-saving properties of these technologies, relative to flood irrigation.

A third traditional technology is targeted at the entire field plot. Field leveling includes any artificially flattening of the plot. Leveling a plot allows water to spread across the plot more evenly without designing bunds or channels to direct the water flow. It is reported to enhance water infiltration and reduce soil erosion, in addition to raising yields (Deng et al., 2004).

HOUSEHOLD-BASED TECHNOLOGIES

Household-based technologies include plastic sheeting, planting drought-resistant crop varieties, retaining stubble/low till, and surface level plastic irrigation pipe. We have grouped these technologies because they are adopted by households (rather than villages or groups of households), have relatively low fixed costs, and are highly divisible. Typically, adoption of these technologies is more recent than the adoption of the traditional technologies.

Plastic sheeting is a production technology rather than an irrigation technique. This term describes several more specific techniques that involve the use of plastic film to trap moisture between the ground and the sheeting. Plastic film is used to cover soil during or before the crop growing season. For example, one use of plastic sheeting is as a component of an agronomic system called Ground Cover Rice Production System (GCRPS—Abdulai et al., 2005). In experiments, GCRPS is reported to save 50–90% of applied irrigation water under experimental field conditions while requiring little training (Abdulai et al., 2005). In addition, farmers using GCRPS say that it increases soil temperature allowing earlier planting and harvesting. Plastic sheeting also increases soil temperatures under experimental field conditions (Li et al., 2003). A field experiment for wheat grown in Dingxi county in Gansu province found that using plastic sheeting in combination with presowing irrigation increased both yields and water use efficiency in addition to increasing soil temperature, but that plastic sheeting by itself did not increase yields (Li et al., 2004).

Drought-resistant varieties include any seed variety that is able to withstand relatively low water moisture conditions. China's wheat and maize breeding system has always prided itself on incorporating drought resistance into some of the highest yielding germplasm (Hu, 2000). Zuo (1997) also reports that drought resistant varieties of crops—including millet, sorghum, beans, tubers, buckwheat, and flax—have been developed and extended in China. In some cases, these varieties show yield increases of more than 10% over those varieties that are not drought resistant in years of below average rainfall.

Retaining stubble/low till is a technique in which the stubble from one crop is left on the field after this crop is harvested. Field studies in northern China show that low till methods can improve water use efficiency by reducing soil evaporation and increasing yields in comparison to traditional agronomic techniques including furrows (Deng et al., 2004; Pereira et al., 2003; Zuo, 1997). While in some sense this technology resembles no till practices that are used in many developed and developing countries, in most cases, the stubble is retained only after the wheat crop is harvested in the spring and before the maize crop is planted. Most farmers in northern China plow their fields after the maize crop is harvested during the fall (hence the name low till instead of no till).

Surface level plastic irrigation pipe refers to a coil of hose used to transport irrigation water to a farmer's field. Often white, surface level hose technology is made of soft, flexible plastic pipe. In China, due to its color and shape, farmers often call this technology a "white dragon." Field experiments have shown that surface water piping techniques, including low pressure pipes, can save up to 30% of water and small amounts of land (Zuo, 1997).

COMMUNITY-BASED TECHNOLOGIES

Community-based technologies include underground pipe systems, lined canals, and sprinkler systems. We have grouped these technologies because they tend to be adopted by communities or groups of households rather than by individual households. In most applications, they have large fixed costs and require collective action or ongoing coordination of many households. Sprinkler systems, for example, require substantial water pressure to operate. To attain sufficient pressure, some villages need to construct water towers and elaborate piping networks. In addition, the small size of plots and fragmented nature of most farm holdings in northern China mean that operating a sprinkler system requires coordination for use. It is difficult to use a sprinkler that irrigates in a large circular pattern on one plot without irrigating the plots of other households around it.

Despite the coordination problems, sprinkler systems can increase water use efficiency, given fixed plot areas and crop choice (eg, Peterson and Ding, 2005). Zuo (1997) also notes that sprinkler and drip systems save labor in addition to water, but have relatively high costs, which might limit the use of sprinkler technology to vegetable and fruit production.

Underground pipe systems include cement, metal, or plastic pipes used to transport water for irrigation. In China, almost all underground piping systems utilize PVC material. In many parts of northern China, installation requires digging trenches during the short period of time that elapses between the harvest of maize (or another summer crop) and the planting of winter wheat. Typically, underground piping systems have above-ground access fittings every 50–100 m. These techniques can save water (up to 30%) in addition to a small amount of land area, compared to unlined canal systems (Zuo, 1997).

Lining an irrigation canal with cement or other materials reduces seepage during conveyance. However, reducing seepage might not lead to water savings, particularly in situations where groundwater pumping relies on an aquifer recharged by canal seepage. In many villages lined canals have been installed or subsidized by a surface water irrigation district in conjunction with a local water resource bureau. Lined canals, like underground pipe systems, might increase water-use efficiency in some circumstances (Zuo, 1997).

FARMER PERCEPTIONS OF TECHNOLOGY TRAITS

Ultimately, the most important proximate determinant of technology adoption is the farmer's perception of the incremental benefits and costs to his own farm budget. Hence, we examine farm-level perceptions of the water-saving properties and other characteristics of each technology.

PERCEIVED WATER SAVINGS

Field level water savings and real, basin-wide water savings may differ due to several agronomic and hydrologic factors. WST adoption will increase in response to water shortage only if users (farmers and village leaders) perceive that adoption will lead to water savings or generate other benefits. Our survey captures these perceived water savings by asking respondents in villages where a particular technology was in use to estimate the water savings of that technology, relative to the status quo without use of the technology. Our data show that while the most commonly observed WSTs are perceived to save water, there are differences among the technologies (Table 14.1). For example, the highest perceived savings rate is for underground pipes (42%). The lowest perceived savings rates are for drought-resistant varieties (20%), plastic sheeting (28%), and retaining stubble/low till (8%). The estimated savings we report are higher than those of Yang et al. (2003, p. 147) who report that "officials and technicians interviewed in Henan, Ningxia and Hebei estimated that around 10–20% saving in water is attainable in their irrigation districts through application of conventional water-saving methods

Table 14.1 Village Leader Estimates of Water Savings, by Technology

Technology	Estimated Water Savings (%)
Traditional Technologies	
Border irrigation	38
Furrow irrigation	39
Level fields	33
Household-Based Technologies	
Plastic sheeting	23
Drought resistant varieties	20
Retaining stubble/low till	8
Surface pipe	35
Community-Based Technologies	
Underground pipe	42
Lined canal	30
Sprinkler	39

Note: These data from the authors' survey of village leaders and they include only observations from villages where the technology was adopted. Respondents were asked to estimate the average percent of water saved by the technology.
Data source: Authors' survey.

and better management." Our estimated savings rates may be a bit higher due to the way we asked the question, the status of our informant, and/or the nature of the sample.

OTHER BENEFICIAL TRAITS

One of the most striking findings of our research is the number of respondents who told us that, although farmers in their villages were adopting WSTs, they often were doing so for reasons other than water saving. In other words, technologies associated with water savings have other traits demanded by farmers. For example, in the case of plastic sheeting and retaining stubble/low till, water saving was not the primary motivation for adoption in more than half of the adopting villages (Table 14.2). In the case of plastic sheeting, although 46% of respondents report that water saving was the primary objective, in 84% of the remaining cases (ie, of the remaining 54%), the technology's main purpose was thought to be increasing the soil temperature around the crop in the early part of the growing season. In the case of retaining stubble/low till, saving water was cited as the primary motivation for adoption by only 19% of respondents. In 76% of the remaining adopting villages, saving fertilizer was the most frequently cited reason. These results are consistent with experimental findings about the effects of both plastic sheeting and retaining stubble/low till (Deng et al., 2004; Pereira et al., 2003; Li et al., 2004; Zuo, 1997; Abdulai et al., 2005). There were often secondary reasons for adoption, beyond water saving, even for technologies for which water saving was the primary objective.

Table 14.2 Was This Technology Adopted to Save Water? If Not, Why Was It Adopted?

Technology	Was This Technology Primarily Adopted to Save Water? Percent of Villages Responding "Yes"	Other Reasons for Adoption Only Listed for Technologies Which Less Than 2/3 of Villages Adopt to Save Water. Percent of Villages That Did Not Adopt to Save Water, in Parenthesis
Traditional Technologies		
Border irrigation	93	
Furrow irrigation	90	
Level fields	94	
Household Technologies		
Plastic sheeting	46	Moderate temperature (84%) Increase yield (35%)
Drought resistant varieties	74	
Retaining stubble/low till	19	Save fertilizer (76%) Increase yield (23%) Save labor (17%)
Surface pipe	83	
Community Technologies		
Underground pipe	93	
Lined canal	99	
Sprinkler	88	

Note: These data are from the authors' survey of village leaders and they include only observations from villages where the technology was adopted. If households in a village were using a technology, the respondent was asked whether or not the technology was primarily adopted to save water. If the technology was not primarily adopted to save water, the respondent was asked to list other reasons for adoption.

Data source: Authors' survey.

WATER-SAVING TECHNOLOGY ADOPTION

We track adoption with two sets of measures derived from our survey data. The first is a village measure in which a village is considered to have adopted a technology if at least one plot or farmer in the village uses the technology. While this does not mean that all, or even most, farmers in a village use the given technology, information on how many villages have at least one farmer using the technology provides an understanding of how spatially pervasive the practice has become. It also provides a convenient measure to track the diffusion of each technology over time. The second measure, the percentage of sown area using the technology, is a measure of the actual extent of adoption at the farm level.

VILLAGE ADOPTION

As the name implies, traditional WSTs have been used for many years (Fig. 14.1). The strongest distinguishing characteristic of traditional WSTs is that, even as of the early 1950s, they were being used in a relatively large share of China's villages. For example, in 1949 farmers in 55% of northern China villages were already leveling their land. Likewise, in the early years of the People's Republic, farm households in slightly less than half of northern China's villages were using border/furrow irrigation. Clearly, before the shortage of water across China began to elicit national and international attention, farmers in more than half of China were already using these traditional agronomic techniques. To the extent that they were doing so to save water, farmers have long been actively managing their water resources.

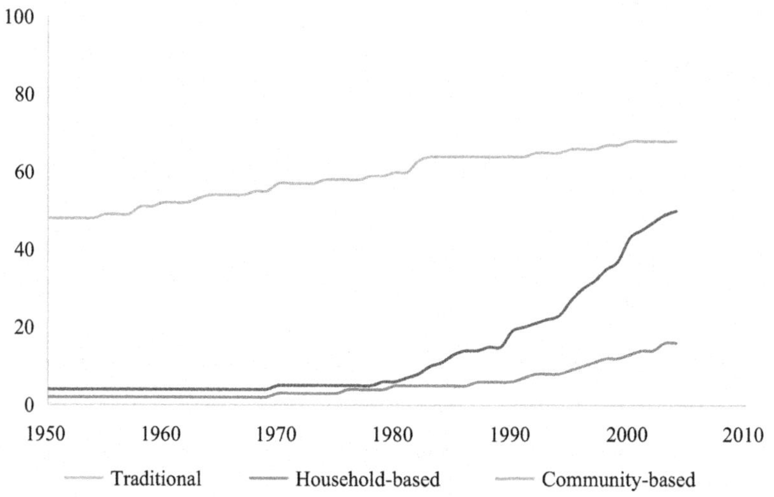

FIGURE 14.1

Adoption of water-saving technology.

During the reform period the adoption of traditional technologies grew slowly, in part because traditional technology adoption rates were already high in the prereform and early reform era (Fig. 14.1). Between the early 1980s and 2004, village level adoption rose from 68% to 77% for field leveling and from 60% to 68% for border/furrow irrigation. As traced in a typical S-shaped diffusion path, technology adoption growth rates are often relatively slow at the beginning of the adoption process, speeding up as public information and experience with the technology increases and then slowing down again as the pool of potential adopters dwindles (eg, Cabe, 1991). The high rates of early adoption and the recent slow growth rates of traditional technologies are consistent with a technology adoption (or diffusion) process that is near its maximum.

In contrast, household-based technologies have taken a different adoption path during the past 55 years (Fig. 14.1, middle set of lines). Although it is difficult to distinguish exact levels of adoption in 1949 from Fig. 14.1 (the paths are too tightly bunched), household-based WST adoption rates were all low initially, ranging from 1% (surface pipe) to 10% (retaining stubble/low till). Unsurprisingly, due to the relative abundance of water and the nature of farming at the time (collective-based with few incentives to maximize profits), household-based technology adoption rates at the village level remained low over the next 30–40 years. It was not until the early 1990s that their adoption rates accelerated. Between 1995 and 2004 village-level adoption of surface pipe more than doubled, from 23% to 48%. The emergence of villages in which farmers use retaining stubble/low till, plastic sheeting, and drought resistant varieties, was clear; the number of villages with at least one adopter for each technology rose by at least 17 percentage points. By 2004, farmers in at least 45% of villages were using each type of household-based WST. One explanation for the relatively rapid diffusion of household technologies is that at some time in the 1980s or early 1990s, some barrier(s) to adoption of these technologies loosened, and this initiated a surge of adoption activity. Likely, the increasing autonomy that producers were granted in the 1980s is at least partially responsible for the rising interest of households in WSTs.

Finally, although the basic pattern of community-based technology adoption follows the same fundamental paths as household-level technologies, these paths start lower and rise at a slower rate (Fig. 14.1, lower set of lines). Between the 1950 and 1980s, like household-level technologies, adoption rates are low. By the beginning of the reforms in the mid-1980s, the highest village-level adoption rate of a community technology (lined canals) is only 10%. Although, as in the case of household-level technologies, adoption rates begin to rise after the early 1990s, in 2004 the most commonly adopted community-based technology, lined canals, could be found in only 25% of northern China's villages. The average rate of increase of the three community-based technologies between 1995 and 2004 was only 9 percentage points.

Based on these adoption histories, it is unclear what is driving the adoption path of community-based technologies. Two sets of general economic forces might be at once encouraging and holding back adoption. Rising scarcity of water resources almost certainly is pushing up demand for community-based technologies, while the predominance of household farming in China (Rozelle and Swinnen, 2004) and the weakening of the collective's financial resources and management authority (Lin, 1991) have made it more difficult to gather the resources and coordinate the effort needed to adopt technologies that have high fixed costs and involve many households in the community. In contrast, household-based technologies may be more widely adopted due to relatively low fixed costs, divisibility, and minimal coordination requirements.

SOWN AREA EXTENT OF ADOPTION MEASURES

The most striking finding of our examination of the extent of adoption of WST is that, although it is growing rapidly, the extent of adoption is much lower than overall adoption rates (Table 14.3). The highest rates of adoption measured in terms of sown area are for traditional technologies (rows 1 and 2). Field leveling, for example, was adopted on 41% of sown area in 2004. Hence, farmers have yet to adopt even traditional technologies on most of northern China's sown area. Even the most basic, traditional WSTs are not used on at least 60% of sown area.

Table 14.3 Extent of Adoption: Proportion of Sown Area in Which Farm Households Use Water-Saving Technology in Northern China, 1995, 2004

Technology	1995 (%)	2004 (%)
Traditional Technologies		
Border/furrow irrigation	31	38
Level fields	39	41
Household-Based Technologies		
Plastic sheeting	5	11
Drought resistant varieties	11	18
Retaining stubble/low till	10	20
Surface pipe	7	17
Community-Based Technologies		
Underground pipe	4	13
Lined canal	5	9
Sprinkler	0	3

Note: These data are from the authors' survey of village leaders and they include the sown area of all villages, those that adopt and those that do not adopt. If households in a village were using a technology, the respondent was asked to estimate the amount of sown area on which each of the technologies was used. For convenience, we have combined border and furrow irrigation because they are not used simultaneously and are both plowing based, agronomic technologies. We have estimated percentages for the small number of observations for which the sown area in use is missing (0.04% in 2004 and 2.2% in 1995). Our estimates are predicted values based on regressions of sown area percent in the missing year on sown area percent in all nonmissing years (this includes 2001 data for the CWIM data set), total cash crop sown area, total staple crop sown area, surface water usage status, groundwater usage status, and dummy variables for each of the province-scarcity strata.
Data source: Authors' survey.

In the case of household and community-based technologies, the extent of adoption, as measured by percent of sown area, is generally growing, but is still quite low (Table 14.3, rows 3–9). For example, in the case of household-based technology, as in the case of village-level adoption figures, adoption rose substantially in relative terms. The extent of adoption of nearly all household-based technologies doubled or more than doubled in percentage terms (except for drought resistant varieties, which rose from 10% to 18%). Despite rapid growth rates after 1995, the adoption of household-based technologies was low, ranging from only 11% for plastic sheeting to 20% for retaining stubble/low till. In other words, as of 2004, averaging across the four most commonly observed household-based technologies, a typical household technology covered only 16% of sown area (the average of column 2, rows 3–6). The pattern of the extent of adoption of community-level technologies using sown area measures is similar, except that both the growth rates (in percentage terms between 1995 and 2004, only 5 percentage points, averaging across the technologies) and the final levels of adoption (in 2004, only 8%, on average) are lower.

WATER-SAVING TECHNOLOGY TRENDS: SUMMARY

Our data show a strong and consistent pattern of adoption of WST. Perhaps the most important single result is that the gains in WST adoption over the past decade or more have mostly come from household-based technologies. Traditional technologies are widely used but in fact are only slightly more widely adopted than in the past (thus deserving their name "traditional"). The typical community-based technology also has grown relatively slowly and in 2004 covered less than 10% of northern China's sown area. In contrast, household-based technologies have expanded at a relatively rapid pace. Almost half of all villages have farmers that use each of the household-based technologies.

Despite the growing use of all WSTs, the extent of WST use is still low in China, especially when using sown area coverage as a measure of adoption. No one type of technology covers more than 50% of sown area; no nontraditional technology covers more than 20% of sown area. In part, this may be due to the fact that not all areas of China are facing water shortages. In many areas, at least currently, there is no need for farmers to adopt WST (see Wang et al. (2006) for a discussion of the variability of water scarcity in China). However, the low levels of adoption in northern China might imply there are barriers to adoption. If policies can be created and incentives provided to farmers and groups of farmers to adopt new technologies, there is hope, at least at the field level, for large water-savings in the coming years.

Although the analysis to this point has relied almost exclusively on our data, when we compare our adoption rates with provincial level adoption rates (measured in percentage of sown area) the two sets of statistics are relatively consistent. The 2001 yearbook-based estimate for the adoption of sprinklers and drip irrigation is 3% (calculation based on EBCAY (2001)). This is within the range of our estimates of sprinkler and drip irrigation in 1995 (almost 0%) and 2004 (3%).

Likewise, the 2001 yearbook estimate of lined canals (3%) is close to our 1995 estimate (5%). For comparison with our statistics, province level yearbook data was aggregated using the same province level weights used in the authors' survey. The estimate for lined canals in 2001 is somewhat lower than our 2004 estimate (9%). The difference between our estimates and the figures generated by surveys run by the Ministry of Water Resources may be a difference in our samples and coverage, or it may also reflect differences in definitions. In our surveys, we included lined canals whether or not they were at the primary, secondary, tertiary, or field levels. Frequently, in national statistical reporting systems, the lowest levels of lined canals are not counted (since they are counted more as "ditches" rather than "canals").

Our findings and interpretations are also fairly consistent with those appearing in other literature. For example, in a survey of five irrigation districts reported by Yang et al. (2003), the research team concludes that canal lining, border irrigation, hose water conveyance, and plastic mulch are not widely used in Henan (Liuyuankou, and People's victory canal), Ningxia (Weining and Qingtongxia), and Hebei (Luancheng). With the exception of border irrigation, our results are in agreement. The partial nature of Yang's sample and the large areas of China that still do not have border irrigation (according to our data) suggests that even for border irrigation our results do not conflict.

THE DETERMINANTS OF WATER-SAVING TECHNOLOGY ADOPTION

We have seen that some types of technologies were popular before the 1980s; some have become increasingly common after 1990; and others have yet to be adopted. To begin explaining the pattern, we first examine the role of incentives as one of the key determinants of adoption. We also examine the role of the state in providing information, investment, and coordination.

ADOPTION AND WATER SCARCITY

Theory predicts that as a resource becomes more scarce, resource-conserving technologies are more likely to be adopted (Ruttan and Hayami, 1984). Irrigation costs that increase with water use should motivate farmers to reduce water use. For example, as the groundwater table falls, the cost of pumping increases, raising the average cost of irrigation for farmers using pumped groundwater. Farmers may respond to the rising cost of water by altering the quantity of water they apply to crops or by changing the mix of crops they produce. Foster et al. (2004) report that farmers in the North China Plain reduced the number of irrigations as groundwater levels declined.

Alternatively, consistent with a large amount of literature that shows the correlation between water scarcity and adoption of WSTs, farmers may respond by adopting new technologies. In China, Yang et al. (2003) demonstrate that farmers in groundwater irrigated areas adopt WSTs because they have control over the volume. When farmers bear the cost of the water they use, adoption rates for white dragons (surface piping) and other water-saving techniques are higher.

If farmers do not pay for water by volume, or if they otherwise do not have an incentive to save water, we should not expect them to adopt WSTs on their own. In northern China there are many situations in which farmers have little incentive to save water. In almost all irrigation districts, farmers using surface water rarely buy water by volume. Surface water irrigation fees are almost always based on sown area (Wang and Huang, 2004; Lohmar et al., 2003; Yang et al., 2003).

Our data show that the three variables we used to reflect water scarcity—share of irrigation water coming from groundwater sources, inadequate supply of surface water, and inadequate supply of groundwater—are all positively correlated with the three types of WSTs. For example, as the share of irrigation water coming from groundwater sources increased from 12% to 45%, the use of traditional WSTs increased from zero to 59%. Similarly, as the share of irrigation water increased from 24% to 43%, the use of household-based WSTs increased from 0 to 30% (see Table 14.3).

We find a negative relationship between the level of adoption of most WSTs and the use of surface water (Table 14.4). With the exception of lined canals and drought resistant varieties (which we do not include in Table 14.4), adoption rates are higher in groundwater villages for all technologies. Among all of the technologies, the

Table 14.4 Adoption Rates in Villages Using Groundwater and Surface Water, 2004

Technology	Groundwater-Using Villages (%)	Surface Water-Using Villages (%)
Traditional Technologies		
Border irrigation	73	61
Furrow irrigation	20	30
Level fields	83	81
Household Technologies		
Plastic sheeting	61	60
Drought resistant varieties	42	45
Retaining stubble/low till	62	57
Surface pipe	60	42
Community Technologies		
Underground pipe	34	22
Sprinkler	10	6

Note: These data are from the authors' survey of village leaders. We did not include lined canals since most of these are funded by surface water irrigation districts. In fact, our data bear this out: lined canals are found in 43% of surface water villages and in only 25% of groundwater villages.
Data source: Authors' survey.

differences are greatest for border/furrow irrigation and surface pipe. Interestingly the difference between plastic sheeting and retaining stubble/low till were not very large; however, as shown in Table 14.2, these technologies, in fact, were not primarily adopted to save water. Hence, this result is not surprising.

Inside groundwater villages, the incentives to adopt technology are much clearer; we expect groundwater villages with the lowest water levels to be most likely to adopt. Our data depict precisely this result when using either village-level or sown area-based measures (Table 14.5). With the exceptions of field leveling, retaining stubble/low till, and sprinkler technologies, farmers in villages with water from depths of 30–150 m are more frequently observed as using WSTs than farmers in villages with depths less than 10 m (columns 1 and 2). With the exception of field leveling, the proportion of sown area on which farmers use WSTs is greater in villages using deeper wells than those with shallow wells (columns 3 and 4). The differences are greatest for technologies designed to work with groundwater pumps. In villages that pump from deeper wells, farmers use surface pipe and underground piping systems in nearly double the number of villages and on nearly double

Table 14.5 Adoption Rates and Extent in Groundwater-Using Villages by Depth to Water, 2004

	Percent of Villages Adopting		**Percent of Sown Area Adopting**	
Technology	**Water Level 0–10 m**	**Water Level 30–150 m**	**Water Level 0–10 m**	**Water Level 30–150 m**
Traditional Technologies				
Border/furrow irrigation	86	96	42	62
Level fields	93	80	49	45
Household Technologies				
Plastic sheeting	52	62	9	15
Drought resistant varieties	34	57	11	22
Retaining stubble/ low till	68	62	21	23
Surface pipe	58	65	18	31
Community Technologies				
Underground pipe	17	63	13	33
Sprinkler	13	0	1	0

Note: These data from the authors' survey of village leaders. We did not included lined canals since most of these are funded by surface water irrigation districts. In fact, our data bear this out: lined canals are found in 43% of surface water villages and in only 25% of groundwater villages. The aggregated border and furrow irrigation adoption rates are estimated taking the covariance of adoption into account—only 34.6% of furrow adopters were not also adopters of border irrigation. The estimates of sown area for this category assume that the two technologies are exclusive.
Data source: Authors' survey.

the sown area (although there is more of a difference for underground piping). Similarly, the result that retaining stubble/no till is not related to the cost of water, is almost certainly related to the fact that the technology, in fact, was not primarily adopted to save water. It also is understandable that there were no villages that pump from deep wells in our sample that used sprinkler technology since sprinklers are only adopted in communities that receive large subsidies; apparently, the officials that make the decisions are not overly concerned with the cost of pumping. The field leveling may be a result of the fact that field leveling is correlated with a village's natural geography. A large share of China's shallowest wells are in areas that are naturally flat (making the cost of field leveling low and raising adoption).

ROLE OF THE GOVERNMENT

While there is considerable evidence that adoption of WST is associated with the cost of pumping and the need to pay for water volumetrically, perhaps a more surprising result is that it is not more correlated. Although there are explanations for certain technologies, the fact is that for a number of cases, farmers in villages with surface water and those pumping from shallow wells were adopting technologies at higher rates than those pumping deeper wells. In addition, there were many villages and considerable amounts of sown area in villages pumping from deep wells that were not adopting technologies that clearly provided savings in water (as well as energy in the form of electricity to drive the pumps). As a consequence, it would seem there must be other, nonpecuniary determinants explaining why some farmers adopt and others do not.

Adoption and Investment

Technologies with high fixed costs may be beyond the reach of farmers without outside assistance, posing a hurdle to adoption. Weakening of the collective's financial resources (Lin, 1991) indicates that the collective may have a declining ability to make such investments. Traditional and household-based technology investments come from farmers (Table 14.6). For community-based technologies, investment comes from three groups: farmers, villages, and upper levels of government. For sprinkler systems, the proportion of villages receiving upper level government investment is particularly large (51%). It could be that investments are low because there is a wealth constraint limiting the ability of farmers and village leaders to make such investments. Nevertheless, our data find that as government financial support increased, adoption rates of all three types of WST also rose.

Adoption and Extension Efforts

Extension may be an important factor in adoption. Abdulai et al. (2005) find that having access to advice from an extension agent is the "most important driving factor" in adoption of GCRPS in Hubei. They suggest this is because extension provides subsidized inputs and access to information. However, agents in the extension system face poor incentives and low budgets (Deng et al., 2004; CCICED, 2004).

Table 14.6 Primary Source of Investment in Water-Saving Technology (%)

Technology	Government	Village	Farmer	Water Manager	Others
Traditional Technologies					
Border irrigation	0	1	98	0	2
Furrow irrigation	2	0	95	0	3
Level fields	3	2	95	0	1
Household-Based Technologies					
Plastic sheeting	10	5	92	0	0
Drought resistant varieties	0	0	100	0	0
Retaining stubble/low till	2	0	96	0	3
Surface pipe	6	7	87	1	1
Community-Based Technologies					
Underground pipe	35	34	40	1	0
Lined canal	36	45	28	0	2
Sprinkler	51	13	48	0	1

Note: These data are calculated from the authors' survey of village leaders and includes only observations from villages where the technology was adopted. If households in a village were using a technology, the respondent was asked to name the primary source of investment.
Data source: Authors' survey.

Access to extension, a potential source of information, is more varied than source of financial investment, especially for household-based technologies, for which information comes from county governments, other governments, other farmers, and seed companies (Table 14.7). Traditional technology information comes primarily from other farmers. Community-based technology information comes from the village, the county government, and higher levels of government. Adoption of these technologies requires coordination and the source of information for adopters is concentrated in entities that can facilitate adoption (the village and upper levels of government). Hence, we expect adoption of community-based technologies to be responsive to government extension. Adopters of household-based technologies may also be responding to government extension, but to a more limited extent.

Data presented in Table 14.3 confirm previous findings. As the shares of villages receiving government extension support increased, so did the percent of areas adopting the three types of WSTs, suggesting a positive relationship.

The analyses we have presented thus far reflect only the simple correlation between variables but fail to consider other affecting factors that may also simultaneously affect WST adoption. Therefore, in the next section, we conduct more rigorous econometric analysis that controls for the impacts of other factors.

Table 14.7 Sources of Technology Extension: Percent of Adopting Villages by Source of Technology Extension (%)

Technology	Village	County	Other Government	Other Farmers	Traditional	Seed Co.	Outside Village
Traditional Technologies							
Border irrigation	5	8	4	19	65	0	1
Furrow irrigation	3	5	6	44	41	0	7
Level fields	2	7	7	26	55	1	3
Household Technologies							
Plastic sheeting	5	31	29	18	0	3	10
Drought resistant varieties	4	23	29	9	6	22	7
Retaining stubble/low till	9	23	26	23	11	0	7
Surface pipe	8	10	20	36	1	0	22
Community Technologies							
Underground pipe	25	14	43	11	0	0	3
Lined canal	28	23	32	12	3	0	0
Sprinkler	0	33	52	0	0	0	16

Note: These data are from the authors' survey of village leaders and includes only observations from villages where the technology was adopted. If households in a village were using a technology, the respondent was asked to name sources of extension or information about the technology.
Data source: Authors' survey.

ESTIMATION RESULTS OF ECONOMETRIC MODEL

In order to quantify the impacts of various factors on WST adoption, we develop an econometric model of WST adoption and estimate the model based on panel data from 538 villages in 1995 and 2005. The detail descriptions of econometric model are included in Methodological Appendix to Chapter 14.

The estimated results of WST are reported in Table 14.8, and the results show that overall performance as indicated by the Chi-squared test of the Tobit model is quite good and most of the estimated parameters are statistically significant and consistent with our expectations. We also tested the collinearity of the model. As the condition number between variables is very small, at only 15.1, the model does not have a collinearity problem.

Water scarcity is one of the major factors affecting farmers' decision to adopt various WST. Table 14.8 shows that all estimated parameters for the three variables reflecting water scarcity are positive and statistically significant in seven of nine parameters estimated in three types of WST (rows 1–3, Table 14.8), although different types of WST have different responses to water scarcity.

The parameter for the variable of share of irrigation water coming from groundwater sources is positive and statistically significant at 1% in all three types of WST adoptions. This demonstrates that villages with irrigation water from groundwater, in comparison with villages with irrigation water from surface water, are more likely to adopt WST.

The parameter for the variable of inadequate surface water also has a positive effect on all three types of WST, and is statistically significant at 1% in the traditional-based and household-based WST models. This indicates that the scarcer the surface water is, the more likely the adoption of traditional technologies and household-based WST. Moreover, the impact is twice as much in tradition-based (0.091, column 1, Table 14.8) than household-based (0.041, column 2, Table 14.8) WST. This might be because traditional technologies mainly rely on surface water.

Likewise, the parameter for the variable of inadequate groundwater is positive in all three WST models and statistically significant at 1% in household-based WST models and at percent in community-based WST models. This demonstrates that villages with scarce groundwater are more willing to adopt community-based and household-based WST. This might be explained by the fact that water used by community-based and household-based WST is mainly from groundwater and therefore any change of groundwater may greatly affect the adoption of these two kinds of WST.

Policy support is the other important factor that has facilitated adoption of WST. Of the three policies investigated, two—WST extension service and financial support—were found to have effects on WST (rows 4–6, Table 14.8).

The parameter estimated for government extension services on WST is positive and statistically significant at the 1% level in all three types of WST adoptions. This demonstrates that villages with government extension services,

Table 14.8 Tobit Estimated Results of Adoption Area of the Three Types of Water-Saving Technologies (%)

Explanatory variables	Area Share (%) of Adopting the Following Water-Saving Technology		
	Traditional	**Household-based**	**Community**
Water Scarcity			
Irrigation water completely from groundwater	9.313 (1.864)***	2.232 (1.092)**	5.134 (1.541)***
Inadequacy of surface water (%)	0.091 (0.021)***	0.041 (0.012)***	0.026 (0.018)
Inadequacy of groundwater (%)	0.030 (0.022)	0.036 (0.013)***	0.044 (0.018)**
Policy Supports			
Water-saving technology extension	6.383 (1.582)***	3.563 (0.922)***	3.489 (1.306)***
Financial support	0.883 (2.041)	1.582 (1.183)	10.006 (1.520)***
Demonstration village	3.080 (2.010)	0.366 (1.172)	−0.524 (1.561)
Cash crop area share (%)	−0.019 (0.040)	0.070 (0.023)***	0.088 (0.032)***
Type of Soil (Clay as a Comparison Base)			
Sand soil	−0.511 (2.188)	0.370 (1.283)	1.419 (1.825)
Loam soil	−1.352 (2.103)	−0.981 (1.228)	1.578 (1.751)
Villages Characteristic			
Cultivated land per capita (mu)	−2.047 (0.487)***	−0.107 (0.229)	−0.263 (0.358)
Share of irrigated area (%)	0.208 (0.023)***	0.043 (0.014)***	0.116 (0.020)***
Net income per capita (dollars)	0.000 (0.001)	0.001 (0.000)	0.003 (0.001)***
Nonagricultural employment share (%)	−0.076 (0.036)**	−0.006 (0.021)	−0.009 (0.029)
Share of villagers w/midschool education (%)	0.043 (0.032)	0.064 (0.019)***	0.070 (0.027)***
Distance from county capital (km)	−0.053 (0.030)*	−0.012 (0.018)	0.033 (0.024)
Provincial dummy variable	NA	NA	NA
Constant	−31.339 (6.040)***	−17.112 (3.375)***	−24.821 (4.383)***
Chi^2 checked value	489.8	332.61	360.65
Observations	1076	1076	1076
Degree of freedom	1049	1049	1049

*Note: In brackets are standard deviation; *, **, and *** respectively represent the significance level of statistical test is 10%, 5%, and 1%.*

in comparison with those without it, are more likely to adopt all three types of WST. This suggests that the government can effectively promote WST adoption in villages by providing better information and technology to farmers. Indeed, our decomposition analysis on the changes of WST from 1995 to 2005 indicates that government extension services alone could explain about 20% of total changes in WST.

The results also show that while the coefficient on government financial support is positive in all three types of WST adoption, it is only statistically significant in the community-based WST model (row 4 and column 3, Table 14.4). This might be because community-based technologies tend to require large investments and are thus more sensitive to government financial support policies. Our decomposition analysis on the changes of WST from 1995 to 2005 indicates that the explanation degree of this variable is as high as 25%. In other words, the variable of government's financial support can explain one-fourth of the changes in the adoption of community-based WSTs in our study areas.

It is interesting to note that our results do not provide any empirical impact of government demonstrations on any type of WST adoption. This may be because these demonstrations are often company sponsored and therefore feature more expensive technologies that may be difficult for individual households, or even villages, to adopt.

Finally, many of the control variables also have significant impacts on the adoption of one or more types of WST. For example, the share of irrigated area has a significant positive effect on all three types of WST, while the education variable (share of villagers with middle-school education) has a positive and significant impact on both household-based and community-based WST adoptions. Cash crop production and average net income in the village have both promoted adoption of community-based WST. This may be due to the higher return from cash crop production (making water more valuable), and higher investment potential from higher income villages. It is worth mentioning that the coefficients on arable land per capita and nonagricultural employment rate are negative in the traditional technology model, reaching significant level of 1%. This demonstrates that the two coefficients are negatively correlated with the adoption of traditional technologies, which may occur because the relatively simple traditional technologies often require more labor.

Our analysis is limited in several ways. Although typically knowledgeable about agricultural production and water management issues (and, thus, able to provide high quality information on many topics), we believe the quality of some variables is influenced by each village leader's knowledge of hydrology and water engineering. By turning to key informants in rural communities throughout northern China, however, we were able to amass a large volume of information, as seen from the farm and village points of view, and were able to ask questions that are both quantitative and qualitative.

CONCLUSIONS AND POLICY IMPLICATIONS

Our analysis in this chapter has shown that adoption of WSTs depends strongly on incentives. Farmers and villages alike appear to behave rationally in their decision to adopt WSTs, making a calculated cost-benefit analysis. Unsurprisingly, factors that make adoption cheaper or water more valuable have a significant positive impact on technology adoption. This includes perceived water scarcity, governmental financial support and extension services, average income, and the production of highly valued cash crops. Conversely, factors that make water less valuable have a significant negative impact on technology adoption. The availability of nonagricultural employment is an attractive alternative to investing in better irrigation technology. Likewise, when a village has a high rate of arable land per capita, it is less urgent that every piece of land is used as efficiently as possible.

Of the different types of WSTs, household-based technologies have grown most rapidly, and traditional technologies have the highest rates of adoption. The most successful technologies have been highly divisible and low cost ones that can be implemented without collective action or large fixed investments. Technologies that do not fit this description are adopted on a more limited scale, at least in part due to the failure of policy makers to overcome the constraints to adoption.

The adoption of WST in northern China has increased with increasing water scarcity, yet the rate of adoption is still quite low. Farmers in many parts of the region have not adopted even rudimentary WSTs. This suggests that the incentives are not in place to encourage efficient water use. The good news is that we have identified the conditions necessary to encourage technology adoption. Government policies seeking to encourage adoption can do so by adjusting these conditions. Increasing extension services and financial support is the most direct way, while appropriately directed policies—such as a water pricing policy—can also bring about large changes in the agricultural sector. Moreover, our research shows that one government strategy currently favored—village-level technology demonstrations—have no impact on technology adoption. Perhaps government resources might best be directed elsewhere.

As water becomes increasingly scarce, there is a fundamental question about how China is going to be able to develop a nation that is willing and able to be committed to save water. China's government is fast beginning to realize that water scarcity is not only going to affect the sustainable development of the nation's economy (as well as have an impact on social issues), it will also be able to influence China's political situation. As indicated by Moore (2013), "The water resource challenge to China's development is exceptionally complex, encompassing a blend of geographical, political, economic, and social dimensions." Because of this type of change of thinking, the 12th Five-Year Plan in China highlighted the importance of establishing a water-saving society by investing in WSTs and promoting policies that are targeted at saving water. In pursuit of this goal, China's Government has announced plans to spend 4 trillion RMB (over US$600 billion) on water conservation over the next 10 years.

The political pressure on China is demonstrated by its promotion of one of the strictest system of water resource management system in the world. This policy is called the Three Red Lines policy. The Three Red Lines policy was created to establish clear and binding limits on water quantity usage, efficiency, and quality. In early 2012, the State Council announced that the policy would limit total national water consumption to less than 700 billion cubic meters per year. This quantity is equal to approximately three-quarters of China's total annual exploitable freshwater resources. In addition, the policy attempts to increase irrigation use efficiency to 60% by 2030. Meeting this target is to be made one of the most important criterions in the annual official performance review. In January of 2013, the State Council issued a new document called the *Assessment Method for Implementing the Three Red Lines Policy* to further push the implementation of new policies. Obviously, the political importance of water is rising and likely will continue to rise as water becomes even scarcer in the future.

REFERENCES

Abdulai, A., Glauben, T., Herzfeld, T., Zhou, S., 2005. Water-saving Technology in Chinese Rice Production – Evidence From Survey Data. Working paper.

Cabe, R., 1991. Equilibrium diffusion of technological change through multiple processes. Technological Forecasting and Social Change 39, 265–290.

CCICED, 2004. Managing Rural China in the 21st Century: A Policy Prescription. Monograph, Center for Chinese Agricultural Policy. Chinese Academy of Agricultural Sciences, Beijing, China.

Deng, X., Shan, L., Zhang, H., Turner, N.C., 2004. Improving agricultural water use efficiency in arid and semiarid areas of China. In: New directions for a diverse planet. Proceedings of the 4th International Crop Science Congress, 26 September–1 October 2004, Brisbane, Australia. www.cropscience.org.au.

Editorial Board of China Agricultural Yearbook (EBCAY), 2001. China Agricultural Yearbook, 2001, English ed. China Agricultural Press, Beijing China.

Foster, S., Garduno, H., Evans, R., Olson, D., Tian, Y., Zhang, W., et al., 2004. Quaternary aquifer of the North China Plain – assessing and achieving groundwater resource sustainability. Hydrogeology Journal 12, 81–93.

Hu, R., 2000. A Documentation of China's Major Grain Varieties. Working paper. Chinese Center for Agricultural Policy.

Kendy, E., Zhang, Y., Liu, C., Wang, J., Steenhuis, T., 2004. Groundwater recharge from irrigated cropland in the North China Plain: case study of Luancheng County, Hebei Province, 1949–2000. Hydrological Processes 18, 2289–2302.

Li, F., Wang, P., Wang, J., Xu, J., 2004. Effects of irrigation before sowing and plastic film mulching on yield and water uptake of spring wheat in semiarid Loess Plateau of China. Agricultural Water Management 67, 77–88.

Li, Y., Shao, M., Wang, W., Wang, Q., Horton, R., 2003. Open-hole effects of perforated plastic mulches on soil water evaporation. Soil Science 168 (11), 751–758.

Lin, J.Y., 1991. Prohibitions of factor market exchanges and technological choice in Chinese agriculture. Journal of Development Studies 27 (4), 1–15.

Lohmar, B., Wang, J., Rozelle, S., Jikun, H., Dawe, D., 2003. China's Agricultural Water Policy Reforms: Increasing Investment, Resolving Conflicts, and Revising Incentives. Market and Trade Economics Division, Economic Research Service, U.S. Department of Agriculture. Agriculture Information Bulletin, No. 782.

Moore, S., 2013. Issue Brief: Water Resource Issues, Policy and Politics in China. http://www.brookings.edu/research/papers/2013/02/water-politics-china-moore (accessed 04.10.14.).

Pereira, L.S., Cai, L.G., Hann, M.J., 2003. Farm water and soil management for improved water use in the North China plain. Irrigation Drainage 52, 299–317.

Peterson, J.M., Ding, Y., 2005. Economic adjustments to groundwater depletion in the high plains: do water-saving irrigation systems save water? American Journal of Agricultural Economics 87 (1), 147–159.

Rozelle, S., Swinnen, J., 2004. Success and failure of reforms: insights from transition agriculture. Journal of Economic Literature 42 (2), 404–456.

Ruttan, V.W., Hayami, Y., 1984. Toward a theory of induced institutional innovation. The Journal of Development Studies 20 (4), 203–223.

Wang, J., Huang, J., Rozelle, S., Huang, Q., Blanke, A., 2006. In: Giordano, M., Shah, T. (Eds.), "Agriculture and Groundwater Development in Northern China: Trends, Institutional Responses and Policy Options". Groundwater in Developing World Agriculture: Past, Present and Options for a Sustainable Future. International Water Management Institute.

Wang, J., Huang, J., 2004. Water issues in the Fuyang River Basin. Journal of Natural Resources 19 (4), 424–429.

Yang, H., Zhang, X., Zehnder, A.J.B., 2003. Water scarcity, pricing mechanism and institutional reform in northern China irrigated agriculture. Agricultural Water Management 61, 143–161.

Zuo, M., 1997. Development of Water-saving Dry-land Farming. In: China Agriculture Yearbook 1996, English ed. China Agricultural Press, Beijing, China.

Methodological Appendices

Methodological Appendix to Chapter 2

We first decompose the total income Gini coefficient by income source. We begin by noting that if y_k is income from source k (eg, irrigated plots), then total household income, y_0, is

$$y_0 = \sum_{k=1}^{K} y_k, \quad K = 1, \ldots, K. \qquad [A.2.1]$$

Note the subscripts h and v are suppressed here. Following the method suggested by Stuart (1954) and Pyatt, Chen, and Fei (1980), and Lerman and Yitzhaki (1985), we can write the Gini coefficient for total household income per capita, G_0, as

$$G_0 = \sum_{k=1}^{K} S_k G_k R_k \qquad [A.2.2]$$

where S_k is the share of y_k in y_0; G_k is the Gini coefficient of y_k; and R_k is the Gini correlation between y_k and the distribution of y_0 and is defined as

$$R_k = \text{cov}(y_k, F(y_0))/\text{cov}(y_k, F(y_k)) \qquad [A.2.3]$$

where $F(y_0)$ and $F(y_k)$ are the cumulative distributions of total household income and income from source k respectively.

If income component j increases by a factor of e, such that $y_j(e) = (1 + e)y_j$ for all households, the marginal effect of this percentage change on total income inequality is

$$\partial G_0/\partial e_j = S_j(R_j G_j - G_0), \quad J = 1, 2, \ldots, K. \qquad [A.2.4]$$

where S_j, R_j, G_j, and G_0 are measured prior to the marginal income change. Dividing Eq. [A.2.4] by G_0, we obtain

$$(\partial G_0/\partial e_j)/G_0 = (S_j R_j G_j)/G_0 - S_j, \quad J = 1, 2, \ldots, K. \qquad [A.2.5]$$

The relative effect of a marginal percentage change in source-j income on the Gini coefficient for total income (elasticity of total income inequality with respect to income source j) equals the relative contribution of source j to overall income inequality minus the share of source j in total income.

Decomposing inequality by income sources, however, does not reflect the potential unequalizing effects of irrigation as manifested in income differences between groups that have different irrigation. Two distinct groups of farmers may exist: those who have access to irrigation and earn relatively high incomes and those who cannot gain access to irrigation facilities, perhaps due to high cost, and therefore earn relatively low incomes. In such a situation, irrigation may widen income differences between these two groups and therefore increase inequality. Thus, following Pyatt (1976), we divide our sample into an irrigated group and a nonirrigated group according to their irrigated land holdings and then decompose total income inequality into contributions from the within-group inequality and the between-group inequality. While the within-group inequality is the level of income inequality within irrigated group and within the nonirrigated group, the between-group inequality reflects partly the potentially unequalizing effect of irrigation. One limitation of this approach is that it does not separate the effect of irrigation from other factors that might be correlated with irrigation. For example, the quality of land and irrigation status are likely correlated since farmers are more likely to adopt irrigation for plots that have better quality.

The limitation of decomposing inequality by group can be overcome by using a regression-based approach to decompose total income inequality by income flows attributable to specific household characteristics. This approach follows the work of Taylor (1997) and Morduch and Sicular (2002). Using regression results in Table 2.9, total income can be written as:

$$y_{hv} = \widehat{\gamma} D_{hv} + \mathbf{X_{hv}} \widehat{\boldsymbol{\beta}} + \widehat{\alpha}_v + \widehat{\varepsilon}_{hv} \qquad \text{[A.2.6]}$$

In this approach, the estimated income flows contributed by characteristics such as area of irrigated land, level of education, and age are calculated using the estimated parameters ($\widehat{\gamma}$ and $\widehat{\boldsymbol{\beta}}$), and these flows constitute the various components of total income. The shares of income flows from the area of irrigated land per capita and other household characteristics take the form $\frac{\widehat{\gamma} D_{hv}}{y_{hv}}$ and $\frac{\mathbf{X_{hv}} \widehat{\boldsymbol{\beta}}}{y_{hv}}$, respectively. The decomposition by income flows uses the same approach as the decomposition by income sources except that each y_k is replaced by estimated income flows $\widehat{\gamma} D_{hv}$, $\mathbf{X_{hv}} \widehat{\boldsymbol{\beta}}$, $\widehat{\alpha}_v$, and $\widehat{\varepsilon}_{hv}$.

Methodological Appendix to Chapter 5

A.5.1 ECONOMETRIC MODEL ON THE DETERMINANTS OF TUBEWELL OWNERSHIP AND ITS EFFECTS ON GROUNDWATER TABLE, CROPPING PATTERN, AND INCOME

Based on the descriptive statistical analysis, we propose the following econometric model to analyze the determinants of tubewell ownership:

$$M_{jt} = \alpha + \beta W_{jt} + \gamma P_{jt} + \phi Z_{jt} + Dv_j + \varepsilon_{jt} \qquad \text{[A.5.1]}$$

In Eq. [A.5.1], M_{jt} represents the share of private tubewells in village j in year t. The variables on the right hand side of Eq. [A.5.1] are those that explain differences in tubewell ownership decisions among villages and over time. The variable W_{jt}, represents the degree of water scarcity, measured as the level of the groundwater table. We include a set of policy variables (Policy Interventions), P_{jt}, in order to assess the effects of policy on tubewell ownership patterns. Since we also use P_{jt} as instruments to identify tubewell ownership in the performance equations (see Eq. [A.5.3] below), a fuller discussion follows below.

In explaining tubewell ownership, we also control for a number other factors, including the degree of land scarcity measured as arable land per capita, and the village's ability to draw on its fiscal resources for investment measured as per capita village fiscal income. The rest of the control variables include the share of surface water irrigation, water quality (1 = good and 0 = bad), the share of labor force with higher than primary schooling, and the share of nonagricultural labor force. Finally, we also use village (Dv_j) dummy variable to control for unobserved village effects. The symbols α, β, γ, δ, ϕ, and η are parameters to be estimated and ε_{jt} is the error term.

Because we are concerned that tubewell ownership may be endogenous in the impact analysis (that is, in the second part of our chapter that uses multivariate analysis to measure the effect of tubewell ownership on sown area decisions, yields, and income), we need to include variables that will be able to identify the effect of tubewell ownership on agriculture decisions. To do so, in Eq. [A.5.1] we include the vector, P_{jt}, which is made up of the two policy intervention variables. The first variable equals one if the village received financial subsidies for investing in tubewells from county officials in the water bureau, and zero if not. The second policy variable equals one if the village received targeted bank loans for tubewell investment, and zero otherwise. We believe that the formulaic way in which upper-level officials allocate the grants and loans allow us to use these policy variables as instruments. In other words, officials used a predetermined formula as a basis on which they distributed the investment funds and loans. Assuming this is so, investment grants and targeted bank loans should have been expected to affect tubewell ownership but will have had no independent effect on sown area decisions or yields.

We also are interested that the possible effect of privatization of tubewell ownership on groundwater table. To do so, we specify the equation:

$$W_{jt} = \alpha + \beta M_{jt} + \gamma N_{jt} + \partial IV_{jt} + \phi Z'_{jt} + Dv_j + \varepsilon_{jt} \qquad \text{[A.5.2]}$$

In Eq. [A.5.2] the water table, W_{jt}, is specified as a function of tubewell ownership, M_{jt}, an interaction variable, N_{jt} (tubewell ownership multiply 2004 year dummy), a single instrumental variable, *IV*, other control variables, Z'_{jt} and village dummy variables (Dv_j). Besides being needed for our econometric estimation, the results of Eq. [A.5.2] should be of interest to policy-makers since they will be useful in assessing whether or not tubewell ownership reform accelerates the drawdown of the water table. We use the level of the groundwater table in 1990 as our instrumental variable in Eq. [A.5.2]. We assume that the instrumental variable can at least in part explain the level of the water table during the study period but has no direct or independent influence on tubewell ownership (except through its effect on water scarcity).

In order to analyze the impact of tubewell ownership on cropping patterns and income, we specify the equation:

$$y_{jt} = \alpha + \beta \widehat{M}_{jt} + \gamma \widehat{W}_{jt} + \delta N_{jt} + \phi Z''_{jt} + Dv_i + \varepsilon_{jt} \qquad \text{[A.5.3]}$$

where y_{jt} measures two types of performance indicators: either the share of crop area sown to one of the region's major crops (wheat, maize, cotton, and other cash crops) or real farmer income (1990 price). The variables on the right side of the Eq. [A.5.3] are those that explain the performance indicators. Using the identification strategy discussed above, we include the prediction of the tubewell ownership variable $(\widehat{M}_{jt})$ from Eq. [A.5.1] and the prediction of water table $(\widehat{W}_{jt})$ from Eq. [A.5.2]. The interaction variable, N_{jt}, is the same as that in Eq. [A.5.2]. We also include Z''_{jt} to control for other factors that might affect cropping patterns and income. In addition to those control variables in Eq. [A.5.1], Z'' also includes measures of water quality.

In order to analyze the impact of tubewell ownership on yields, we specify the equation:

$$y_{jt} = \alpha + \beta \widehat{M}_{jt} + \gamma \widehat{W}_{jt} + \delta N_{jt} + \phi Z''_{jt} + Dv_i + Dy_t + D\varepsilon_{jt} \qquad \text{[A.5.4]}$$

where y_{jt} measures the yields of major food grain crops (wheat and maize). Except for year dummy (Dy_t), the variables on the right side of Eq. [A.5.4] are the same as that in Eq. [A.5.3]. The year dummy is to control the impact of technological progress on yields.

Methodology Appendix to Chapter 6

A.6.1 ECONOMETRIC MODEL ON THE DETERMINANTS OF GROUNDWATER MARKETS

Based on the descriptive analysis and work on groundwater markets in other countries, the following econometric model is proposed to analyze the determinants of the breadth of groundwater markets:

$$T_{jt} = \alpha + \beta O_{jt} + \gamma W_{jt} + \delta L_{jt} + \phi Z_{jt} + \varepsilon_{jt}. \qquad [A.6.1]$$

In Eq. [A.6.1] T_{jt} represents the share of tubewells selling water in village j in year t. The variables on the right hand side of Eq. [A.6.1] are those that explain differences in the breadth of groundwater markets (or the share of tubewells that sell water) among villages and over time. The first variable, O_{jt}, represents the change of tubewell ownership and is measured as the share of private tubewells in village j. The two variables, W_{jt} and L_{jt}, measure the resource endowments of the village (both its water and land resources) and are included to measure if increasing resource scarcity (or the cost of using the resource) helps induce the development of groundwater markets. Specifically, the water resources variable (W_{jt}) is measured as the level of the groundwater table. The degree of land scarcity (L_{jt}) is measured as cultivated land per capita.

In Eq. [A.6.1] a set of control variables are included. The first set of control variables includes three policy variables which are included to assess the effects of policy on the development of groundwater markets. The first variable, fiscal subsidies for tubewells, is a dummy variable equal to one if there was a program of fiscal investment in the village that targeted tubewell construction (and zero otherwise). This government program, run by the local Bureau of Water Resources, is primarily targeted at individuals. A similar variable, bank loans for tubewells, is included to control for whether or not there was a loan program through China's banks that gives preferential access to low interest rate loans for investing in tubewells. Unlike the fiscal subsidy program most bank loan programs target local villages and leaders; the loans typically are supposed to be invested in collective wells. A final variable, well-drilling regulations, controls for the presence of local regulations that would, ceteris paribus, slow down the construction of tubewells. In our study areas, well-drilling regulations are the rules and directives produced by local officials that seek to influence the behavior of farmers before they drill new wells. When these are enforced, a farmer is supposed to apply for a permit for drilling a new well from the local government. The permit often says where and at what depth the well is to be drilled. Once it is granted, the farmer is legally allowed to sink a new well. Although there are such regulations in many, if not most, regions of China, these regulations are rarely implemented effectively. During our survey, we asked village leaders if farmers in their villages officially are supposed to apply for such permits before sinking new tubewells in their villages. If the answer was "Yes," the variable well drilling permit is coded as "1," otherwise it is coded as "0." Although there is no explicit government regulatory policy to encourage collective tubewells at the expense of private tubewells (or vice versa), without the support of government, it is likely that such regulations, if present (and enforced), would have a greater effect on slowing down investment in private tubewells. The hypotheses are that any government program that encourages (discourages) private tubewells relative to collective wells will encourage (discourage) the development of groundwater markets.

In explaining the development of the breadth of groundwater markets, the adoption of irrigation water conveyance technologies in the village is included as a way to control for the cost and efficiency of delivering water from the tubewell to the field. This variable is measured as a dummy variable, equaling one if the village had adopted conveyance-inducing technology, such as surface (white dragons) or underground pipe networks. It is thought that if the conveyance of water is easier (and more efficient), water markets will emerge more readily. Finally, several other factors are also controlled for. For example, village income per capita is included as a control for the village's socioeconomic conditions. The symbols α, β, γ, δ, and ϕ are parameters to be estimated and ε_{jt} is the error term.

In order to analyze the determinants of development of the depth of groundwater markets, the following econometric model has been specified:

$$M_j = \alpha + \beta O_j + \gamma W_j + \delta L_j + \phi Z_j + \varepsilon_j \qquad \text{[A.6.2]}$$

where M_j represents the share of water sold for tubewell j. While the basic structure of Eq. [A.6.2] is the same as Eq. [A.6.1], because of the nature of the dependent variable (and differences in the sample—the breadth of water markets analysis uses village-level data and the depth of water markets analysis uses tubewell-level data), the specification is slightly modified. The first variable, O_j, represents the ownership of tubewell j; and if the tubewell is owned by an individual (a single family), it equals to 1; otherwise, the tubewell is owned by a group of individuals equals to 0. Since the demand by the individual farm household for water from its own well is almost by definition less than the members of the shareholding group, a positive sign on the coefficient of the ownership variable would be expected (since there would be more of the excess capacity available for sale).

The relative scarcity of water and land might also be expected to affect the amount of water sold to other farmers. To control for water scarcity, the paper includes the variable W_j, which is measured by the depth of the groundwater table. Since the cost of pumping (and, hence, the price at which water can be sold to farmers) is directly related to the depth of the well, a negative coefficient on the depth of the groundwater table variable would be expected. However, it is possible that the sign is positive if as the groundwater table falls (or is lower in level), it is more difficult for farmers to sink their own wells and so groundwater markets emerge to meet the demand. The analyst also needs to be concerned about the endogeneity of such a variable since the development of groundwater markets may influence the level of the groundwater table. Consequently, in the analysis the chapter measures W_j as the groundwater table of the village in 1995, a time before the sample and a period before the takeoff of groundwater markets.

In the same spirit, the chapter includes a variable L_j in order to control for the degree of land scarcity (which measured as cultivated land per capita in the village in which tubewell j is located). The hypothesis here is that when land per capita is lower, the benefits of investing in one's own well falls and increases the demand for water markets.

In Eq. [A.6.2], as in Eq. [A.6.1], a set of three policy variables and a set of control variables are also included. The first variable equals one if the tubewell owner (or the shareholding group) received a fiscally subsidized rebate after investing in the well (and was zero if it was fully self-financed). The second policy variable equals one if the tubewell owner received a bank loan as part of the investment financing package of the well (and zero if not). Finally, a third policy variable equals one if the tubewell owner was issued a well-drilling permission certificate before the well was drilled, and zero otherwise. It would be expected that any policy that facilitates (discourages) the investment in tubewells would increase (decrease) the size and depth of the well and provide individuals with more (less) excess capacity from which they are able to sell water. The definitions (and expected signs) of the other control variables (village income per capita; dummy of adopting water delivery pipes) are the same as those in Eq. [A.6.1].

When estimating the determinants of the development of the breadth and depth of groundwater markets, a Tobit model is used. The Tobit model was developed by Tobin (1958) to treat data sets when the dependent variable is characterized as being censored. The Tobit model, in the form of $y = f(X) = \text{Max}(0, Xb + U)$, is a hybrid of the Probit model and the linear regression model (McDonald and Moffitt, 1980). The use of Tobit model has two advantages. First, it overcomes the information loss which would occur if a Probit model was used with censored data that was translated into binary data. Second, it overcomes the violation of the OLS assumptions that would occur if an OLS estimator was used with censored data. Recent social science research has shown a growth in applying the Tobit model (Helms and Jacobs, 2002; Gunderson, 1974; Langbein, 1986). This estimation strategy is needed since the dependent variables in both Eqs. [A.6.1] and [A.6.2] are in "share" form (that is, between 0 and 1). There are also a number of villages (tubewells) in which the value of the dependent variable is zero. Using the Ordinary Least Squares (OLS) approach might generate a biased set of estimated parameters.

In addition, we can address several potential statistical problems that might arise in the estimation of Eqs. [A.6.1] and [A.6.2]. Specifically, two possible multicollinearity problems come from using either the three policy intervention variables and/or the conveyance technology variable along with the actual share of private tubewells in the same regression (Eq. [A.6.1]). There are two ways to assess whether or not this is a problem. First, we use regression diagnostics to generate a condition number for the full model. Since the condition number is only 39, we can conclude that there is no evidence of multicollinearity in a statistical sense. We also can assess the impact of "including the variables" by reestimating Eq. [A.6.1], but without including the policy intervention variables (Table 6.5, column 2). As can be seen (comparing the results of all of the other variables in columns 1 and 2), the results vary little. The same is true when we reestimate the equation without using conveyance technology (Table 6.5, column 3) or without including conveyance technology and the policy intervention variables (Table 6.5, column 4—that is, there are few changes to the coefficients of interest). We adopt the same strategy for assessing the robustness of the results of Eq. [A.6.2]. While we can report that there is little change to the coefficients, due to space constraints, we do not include the detailed findings in a paper. Therefore, we believe that any potential multicollearity problem in Eqs. [A.6.1] and [A.6.2] is not serious.

Besides multicollinearity problems, we might also be concerned about the potential endogeneity problems in Eqs. [A.6.1] and [A.6.2]. It is possible that the estimated parameters of our variables of interest are biased since the share of tubewells selling water may be determined simultaneously with share of private tubewells. It also could be that the estimated coefficient is affected by unobserved heterogeneity. In such a situation, the estimate of the coefficient on the share of private tubewells variable could be biased. In order to control for any of this endogeneity, we use a fixed effects model. The results of the fixed effects model show that most coefficients do not change much. We take the same strategy for assessing the robustness of the results of Eq. [A.6.2]. While we can report that there is little change to the coefficients, due to space constraints, we do not include the findings in a paper. Therefore, we believe that any potential endogeneity problem in Eqs. [A.6.1] and [A.6.2] is not serious.

Methodology Appendix to Chapter 7

A.7.1 ECONOMETRIC MODEL ON THE EFFECTS OF GROUNDWATER MARKETS ON CROP WATER USE, CROP YIELDS, AND FARMER INCOME

In order to identify the impact of the various ways of accessing groundwater on crop water use, crop yields, and farmer income, we utilize a set of econometric models. The first econometric model, used to measure the effect on water use that results from the different ways of accessing groundwater, can be written as

$$w_{ijk} = \alpha + \beta B_{ijk} + \gamma C_{ijk} + \delta Z_{ijk} + \varepsilon_{ijk} \qquad \text{[A.7.1]}$$

where w_{ijk} represents water use per hectare for the ith wheat plot of household j in village k. The variables on the right hand side of Eq. [A.7.1] explain crop water use. B_{ijk} and C_{ijk}, our variables of interest, measure the ways in which farmers gain access to groundwater for irrigation. If farmers irrigate their plots by buying water from groundwater markets, B_{ijk} equals 1; otherwise, it equals 0. Similarly, C_{ijk} equals 1 if farmers irrigate their plots by pumping water from collective tubewells and equals 0 otherwise. If farmers pump groundwater from their own tubewells, both B_{ijk} and C_{ijk} equal 0. In other words, the plot on which a farmer uses his/her own tubewells for irrigation is the base case.

We also include Z_{ijk}, a set of control variables, to represent other factors that affect water use. Specifically, the first category of control variables includes two variables to assess the effects of the village's production environment on crop water use. We include variables measuring the share of irrigated area serviced by groundwater and the degree of water scarcity in the village measured as a dummy variable. The second category of variables controls for household characteristics, including age and education of the household head. Finally, our model also includes variables that control for plot characteristics, including plot area, the plot's soil type, and the distance of the plot from the home. The symbols α, β, γ, and δ are parameters to be estimated and ε_{ijk} is the error term.

While it is possible to estimate α, β, γ, and δ with OLS, in Eq. [A.7.1], there could be an endogeneity problem that biases the attempt to measure the true relationship between water use per hectare and access to groundwater. The fundamental problem is that there could still be other unobserved factors that affect both water use and access to groundwater. In order to estimate the parameters in Eq. [A.7.1] consistently when the explanatory variables B_{ijk} and C_{ijk} are endogenous, we use an instrumental variable (*IV*) approach to solve the problem. To do so, prior to estimating Eq. [A.7.1], we regress a set of variables measuring the access to irrigation, B_{ijk} and C_{ijk}:

$$B_{ijk} = \lambda_1 + \rho_1 IV + \varphi_1 Z_{ijk} + \mu_1 \quad [A.7.2]$$

$$C_{ijk} = \lambda_2 + \rho_2 IV + \varphi_2 Z_{ijk} + \mu_2 \quad [A.7.3]$$

where the predicted value of B_{ijk} and C_{ijk} from Eqs. [A.7.2] and [A.7.3], $\widehat{B}_{ijk}$ and $\widehat{C}_{ijk}$, would replace B_{ijk} and C_{ijk} in Eq. [A.7.1]. Eqs. [A.7.2] and [A.7.3] also include *Z*, which are measures of the other exogenous variables (which are the same as those in Eq. [A.7.1]—eg, measures of the village's production environment and the household's and plot's characteristics).

The *IV* approach is only valid if the variables in the *IV* matrix in Eqs. [A.7.2] and [A.7.3] have two properties: (1) the *IV* variables must be uncorrelated with the error term of Eq. [A.7.1]; and (2) they must be partly correlated (as a group) with the endogenous explanatory variable. The key instrumental variables in Eqs. [A.7.2] and [A.7.3] are two variables that measure the way in which policy makers have intervened into China's groundwater markets (in village *k*). The first variable, fiscal subsidies for tubewells, is a dummy variable that is equal to one if there was a program of fiscal investment in the village that targeted tubewell construction (and zero otherwise). This government program, run by the local Bureau of Water Resources, is primarily targeted at individuals. The second instrumental variable, bank loans for tubewells, is also a dummy variable to control for whether or not there was a program through banks that gives preferential access to low interest rate loans for investing in tubewells. Unlike the fiscal subsidy program, most bank loan programs targeted local villages and leaders, and loans were typically used for investment in collective wells.

Both our field work and regression results suggest that the choice of the instrumental variables (*IVs*) is satisfactory. First, officials in the local Water Resources Bureaus told us that these government programs were implemented on a fairly random basis; village leaders and farmers were rarely aware that they could influence these programs. Personal relationship (between officials governing over subsidy/loan programs and village leaders) was the most commonly cited reason for giving a grant or a loan to a villager or village leader (Luo and Kelly, 2004). In other words, our two instrumental variables, fiscal subsidies and bank loans for tubewells, are logically exogenous and should have no independent effect on water use, except through the influence on the way in which farmers gain access to groundwater.

Second, our *IVs* are partially correlated with the endogenous variables (B_{ijk} and C_{ijk}). The regression coefficients of our *IVs* are statistically significant in the regression results of Eqs. [A.7.2] and [A.7.3] (Appendix Table 7.1, columns 1 and 2, rows 1 and 2). In other words, our *IVs* are correlated with the decision of farmers to select how they obtain access to groundwater to irrigate. In summary, we can say we have basically solved the problem of endogeneity.

In order to answer the question of whether the emergence of groundwater markets affects crop yields, we use the following econometric model:

$$Q_{ijk} = a + bW_{ijk} + cX_{ijk} + dZ_{ijk} + e_{ijk} \qquad [A.7.4]$$

where Q_{ijk} represents the yield of wheat from the *i*th plot of household *j* in village *k* (which comes from our household survey). In Eq. [A.7.4], yields are explained by the variable of interest, W_{ijk}, which measures water use per hectare; X_{ijk}, which measures other inputs to the production process; and Z_{ijk}, which holds other factors constant, including characteristics of the production environment of the village, household and plot. Agricultural production inputs include measures of per hectare use of labor (measured in mandays), fertilizer (measured in the expenditure of fertilizer per hectare) and expenditures on other inputs, such as the level of the fees paid for custom services. The control variables for the village, household, and plot characteristics are the same as for Eq. [A.7.1]. We also added a variable that represents production shocks, measured as the farmer-estimated yield reductions percentage on a plot due to floods, droughts, or other "disasters." The symbols *a*, *b*, *c*, and *d* are parameters to be estimated and *e* is the error term.

The impact of the emergence of groundwater markets on crop yields is measured through the water use variable. For example, if the regression results from Eq. [A.7.1] show that buying water from groundwater markets will motivate farmers to reduce water use, and if production responds positively to water use (from the regression results of Eq. [A.7.4]), then we can deduce that buying water from irrigation service markets for groundwater will reduce yields.

In order to measure the effect of groundwater markets on income, we also specified the following econometric model:

$$y_{jk} = \pi + \sigma B_{jk} + \omega C_{jk} + \psi Z_{jk} + \xi_{jk} \qquad [A.7.5]$$

where y_{jk} represents either cropping or total income per capita for household *j*. Our interested variables, B_{jk} and C_{jk}, are the same as in Eq. [A.7.1]. Z_{jk} is a set of control variables affecting farmer income. Specifically, the first category of control variables measuring the village's production environment is the same as in Eqs. [A.7.1] and [A.7.4]. The second category of control variables represents household characteristics, including age and education of the household head and the size of the arable land of the household (measured on a per capita basis). The symbols π, σ, ω, and ψ are parameters to be estimated and ξ is the error term.

In order to estimate the parameters in Eq. [A.7.5] consistently when the explanatory variables B_{jk} and C_{jk} are endogenous, we use the same *IV* strategy as used for estimating Eq. [A.7.1].

Methodological Appendix to Chapter 9

A.9.1 TECHNICAL DETAILS OF THE MODEL OF CONTRACTUAL CHOICES

Since ultimate property rights reside with the leader as the curator of local assets in rural China, the leader will choose the contractual form that will further his interests. After the onset of the rural financial reforms, leaders have been encouraged to use village assets efficiently and are allowed to run them on a fee for service basis (Oi, 1999; Rozelle, 1994; Whiting, 2001). Description of policies and regulations that promote the efficient use of irrigation systems—both groundwater and surface water—are found in Lohmar et al. (2003) and Wang et al. (2005). Importantly, these policies implicitly allow the leader to claim profits from such service-oriented activities (eg, the water fee collected in the case of running irrigation systems) as the reward for building the village's treasury (Rozelle, 1994). As a consequence, we believe that it is reasonable to assume that the leader's interests frequently are consistent with profit-maximization (Oi, 1999). This situation is different from that described in Platteau and Gaspart (2003) in which the funding is entrusted to the leader by nongovernmental organizations for community development but is often used by the leader for his personal benefit. In this section, we develop a way to understand how different communities in China choose to manage their irrigation systems—as fixed-wage, fixed-rent, or profit-sharing contracts.

While we also study the conditions under which a certain type of resource managerial form arises, this chapter differs from the common property resource community management literature. Instead of examining the choices of users that self-organize for the purpose of resource management, in our case the leader is solving an optimization problem in order to decide how to manage the community's irrigation system. Like the water users in Ostrom (1990) who are searching for the institutional arrangement that minimizes transaction costs and internalizes externalities, the leader considers the characteristics of the environment and chooses a managerial form that maximizes the return to the irrigation assets. Such behavior is entirely consistent with the way in which China has implemented its reform era policies (Park and Rozelle, 1998).

Following Eswaran and Kotwal (1985), we outline a theory to explain the different ways in which villages manage their irrigation systems. The word outline is used since the authors borrow heavily from Eswaran and Kotwal's (1985) model of contractual choice in tenancy contracting. The authors take credit, not for producing a new analytical framework, but rather for applying an existing theory to a different question.

A.9.1.1 MODEL

In most surface-water irrigation areas in rural China, the distribution of water is organized at two levels: at the first level, the irrigation district (ID) is responsible for supplying water to irrigation units, which in China is almost always an administrative village; at the second level, the village committee (which is the governing body at the lowest level of quasigovernment in rural China). In our problem, the village committee decides to manage the village's water resources itself (traditional collective management) or contracts the canal system out to a manager (a profit-sharing contract or a fixed-rent contract), and it is this management unit that is responsible for delivering water directly into the fields of farmers. When water passes from the ID to the village (that is, at the first level), irrigation water is typically priced volumetrically because the volume of water can be measured fairly accurately as water passes out of the main, ID-controlled canal network into the village-controlled branch canal system. When purchasing water from the ID, the village committee (or the manager) usually pays a deposit to the ID before the irrigation season (as a down payment for the water to be used during the year); after the water is delivered to the village, the village committee (or the manager) needs to pay the balance.

Within the village, however, the situation is often fundamentally different. Households are obligated to pay a fixed water fee for irrigation services during an irrigation season or a year. Although the village leader or the manager would like to charge on a volumetric basis, due to the difficulty of measuring water at the plot level, water fees are often converted to a per unit of land basis. Although there is heterogeneity across IDs and even villages within the same ID, in most villages the water fee (which is fixed in terms of dollars per unit of irrigated land) was determined by the ID or the bureau that manages the ID (about 65% of the villages). In other villages (20% or so) the village committee set the household's water fee. In the rest of the villages, especially in those villages that contract out their canal network to a manager for a fixed-rent or profit-sharing contract, the water fee was determined together by the village leader and the canal manager (15% of all the villages). Importantly, in all of our sample villages, there was no difference in level of water charges among households within the same village; in fact, there was only a small degree of variation in the water fee between different irrigation seasons within the same year (summer/early cropping season and fall/late cropping season).

Hence, managing a canal in the village is similar to running a water company. The leader or the manager purchases water as an input from the ID and provides irrigation services to households. Output, in this case irrigation services (Q), generally includes the entire package of services, including the timing, frequency, and quantity of water delivered. Hence in our study, we measure Q as revenues from managing the village's canal system. Since, as discussed above, in villages that use surface water, households pay a fixed fee for each unit of irrigated land they have, the output, Q, is calculated as the total amount of fees collected from all the households within the village. The more irrigation services the leader/manager provides, and the less water they use, ceteris paribus, the higher is the profit.

In theory, output should be in units of irrigation services. Unfortunately, like most service sector firms, the output is difficult to measure in physical terms. Our use of revenue is reasonable under the assumption that irrigation service revenue in some sense is an aggregate measure of individual services (irrigation timing; frequency and quantity) that have been implicitly weighted by the prices of each service, the summation of which is the total revenue of the irrigation system. In our problem, this is measured as the sum of irrigation fees paid by all the households in the village. Such a strategy is common in the production economic literature when a firm is producing a number of varied products. For example, in estimating a household-level cropping production function of a farmer that produces a mix of grain and cash crops, it is common practice to add up the revenue of the farm's output and use total revenue as the dependent variable. Such a measure is still considered to be physical output: the prices are used as weights to aggregate across outputs that are measured in different physical units. Hence, in the same way that it is not correct to add the output of grain and cotton in physical terms, it is not correct to measure the output of the irrigation system in terms of the physical delivery of water (in cubic meters).

Since the irrigation services are not just the volume of water delivered, other inputs besides water are also required in providing irrigation services. One such input is the quality of the system's infrastructure (H, measured as the length of the canal that is lined). Most importantly, the right hand side of the production function also includes the two key inputs, time spent in providing collective activities (t, measured as the number of days spent by the manager in organizing the maintenance of the canal system, mitigating conflicts over water, etc.) and effort spent on operating the system. Effort, however, consists of two parts: the manual labor (L) required in running the irrigation system; and time spent on supervision (s) to monitor the labor and supervise the operation of the system (eg, operating sluice gates and collecting water fees). Both L and s are measured in days. The production function's parameters (δ_1 to δ_5) embody information about the irrigation system's technology, which will reflect the relative importance of the key inputs. We also believe that the inputs in providing irrigation services are more likely to be substitutes for each other. For example, when the price of water is higher, the village leader or the canal manager will use less water (W is lower). In addition, the leader or the canal manager will spend more time on supervisory activities to monitor more closely the opening and closing of the gates so that a higher percentage of the water purchased from the irrigation district is delivered to the field.

In the questionnaire, we had a section that collected information on the time spent on collective and supervisory activities in managing canals. In this section we created a disaggregated list of the activities that most leaders and/or managers had to perform (eg, operation of sluice gates, canal maintenance, water fee collection, mitigation of water conflicts and other administrative responsibilities). This list formed the basis for the block of the survey in the questionnaire that elicited time allocation of the leader/manager; we also left room for additional activities. For each activity, we asked who carried out these activities (the leader or the manager), whether they hired labor (or had their relatives/family members help them—that is, hire for a zero wage) and how much the hired labor was paid. We asked these questions on a month by month basis for all 12 months of the survey year 2001. In our analysis, the value of t (s) is calculated as the total amount of time spent on collective (supervisory) activities throughout year 2001. During the interviews, we asked the interviewees to convert the time into labor days using the criterion that one labor day was equal to 10 working hours. The descriptive statistics on t and s are in Table A.9.1.

Given these variables, we can relate the inputs to irrigation services using a Cobb–Douglas production function:

$$Q = \theta A t^{\delta_1} s^{\delta_2} L^{\delta_3} W^{\delta_4} H^{\delta_5}; \text{ and } \quad A > 0, \quad \delta_i > 0, i = 1, 2, 3, 4, 5. \qquad [A.9.1]$$

where θ is a positive random variable with an expected value of unity, intended to embody the effects of such stochastic factors as weather. For example, when there is abundant rainfall, irrigation services would be easily provided. On the other hand, in a dry season, water delivery is a difficult task. For the sake of brevity, in the theoretical model, we assume the leader and the canal manager are risk-neutral and thus the effect of the random variable, θ, is not discussed. In the results sections, we discuss the possible impacts of the risk attitude of either the leader or the manager on the contractual choice in a village.

The leader of each village considers the production function of irrigation services (Eq. [A.9.1]), the characteristics of his village, his own characteristics, and the characteristics of the pool of potential managers and solves three different problems for himself: how much profit would I make if I ran the irrigation system myself (as a fixed-wage contract); how much rent would I make if I leased it out (as a fixed-rent contract); and how much profits would be left for me if I split the duties and profits with a manager (under a profit-sharing agreement). He will then choose the contractual form that returns to him the highest level of profit.

In choosing the optimal contract, the leader must be able to provide the right level of incentives to the manager to operate the irrigation system—given its characteristics—most efficiently. To allow our model to reflect these decision-making criteria, we need to quantify the assumption that the leader has

Table A.9.1 Time Spent on Collective and Supervisory Activities per Unit Length of the Canal in Year 2001 (Unit: Labor Day/Meter)

Water Managerial Form	Fixed-Wage	Profit-Sharing	Fixed-Rent
Time spent on carrying out collective activities (t)	0.23 (0.37)	0.02 (0.03)	0.29 (0.48)
Time spent on carrying out supervisory activities (s)	0.04 (0.07)	0.04 (0.02)	0.05 (0.05)

Standard deviations in parentheses.

an absolute advantage in organizing collective activities and that the manager has an absolute advantage in carrying out supervisory activities. Hence, we define the time a leader spends on collective activities (supervisory activities) as t_1 (s_1—where 1 refers to the leader for the rest of the chapter) and define the time a manager spends on collective activities (supervisory activities) as t_2 (s_2—where 2 denotes the manager). The idea of differential ability is quantified by the introduction of two parameters, γ_1 and γ_2. It is assumed that 1 h of the leader's (manager's) time devoted to supervisory activities (collective activities) is equivalent to only a γ_1 (γ_2) fraction of 1 h devoted to supervisory activities (collective activities) by the manager (leader). Both parameters, γ_1 and γ_2, are between 0 and 1.

Defined in this way, the benefit of choosing a fixed-rent contract when an irrigation system requires relatively more tasks that are supervision-intensive is clear. The leader gives full incentives (that is, the manager becomes the residual claimant of profits after paying the contract fee) to the manager since the leader wants the manager to fully utilize the manager's own efficient supervisory skills. Unfortunately, the provision of incentives to the manager can only come at a cost; the manager also will be providing the collective activities, which he can do, but only inefficiently. Following this logic, it is easy to see why in other irrigation systems, for example those that require relatively more input of collective activities, the leader will decide to run the irrigation system himself and keep the profits, fully utilizing his skills in providing collective activities, albeit at the cost of less effective delivery of supervisory activity. Details of the model are available from the authors upon request.

A.9.1.2 OPERATIONALIZING THE MODEL

In creating a simulation model of the village's irrigation system that will allow us to study contractual choice, we use data (discussed below) to estimate the function coefficients (δ_1 to δ_5) of the irrigation services production function in Eq. [A.9.1]. In estimating Eq. [A.9.1], we recognize that our measures of t, time spent on managing collective activities, and s, time spent on carrying out supervisory activities, are measured with error. We assume that the time spent by the leader (manager) on supervisory (collective) activities is systematically less efficient than that of the manager (the leader). However, we could only collect data on the actual time they spent on these activities. In addition, the leader and the managers might have some incentive to overreport the time they spent on the activities they are responsible for. With such measurement error, there will be an attenuation bias in the estimated coefficients that measure the relationship between irrigation services and supervisory time (time spent managing collective activities). Although we have no way to correct for this (since we lack any valid instrument), in our simulations (presented below) we recognize the direction of the bias and use sensitivity analysis to show that our hypotheses (that use the parameters for simulation) are robust to the biased coefficients.

In our empirical specification, we used a fixed effects model at the county level. Using a fixed effects model allows us to account for unobservable differences in local characteristics that might affect the provision of irrigation services such as the level of precipitation. The estimation of irrigation production function performs well (Table A.9.2). Our data failed to reject the assumption that the production technology is characterized by constant returns to scale (CRS). The R^2 is relatively high. Each of the input variables also has the expected positive signs. Note that the estimated coefficient on the variable t is not significant. One explanation is that this is due to the attenuation bias caused by the measurement error associated with t. Unfortunately, we have no way to correct for this since we do not have any valid instrument. However, in the simulations t is still used because we believe that collective activities are essential in providing irrigation services. During our pretest, we asked the village leader and/or the canal manager to list all the major activities they have carried out in managing canals. Unanimously, collective activities such as cleaning canals and maintaining the canals were on the list for all the leaders and managers we interviewed. We were also told that when the village leader decided to offer a profit-sharing contract or a fixed-rent contract, one of the most important terms in the contract was to define the responsibilities of the leader and the manager regarding the collective activities. Hence, there is no doubt that collective activities are important in providing irrigation services and will definitely affect the contractual choice of the leaders. In our simulations we recognize the direction of the bias and use sensitivity analysis to show that our hypotheses are robust to the biased coefficients.

Using our estimates, we are able to parameterize our model in (1) with parameters that are consistent with our data. For example, our baseline parameters for the production coefficients are: $A = 3264.95$, $\delta_1 = 0.015$, $\delta_2 = 0.174$, $\delta_3 = 0.043$,

Table A.9.2 Estimates of Production Function Parameters for Use in Simulations $Q = At^{\delta_1} s^{\delta_2} L^{\delta_3} W^{\delta_4} H^{\delta_5}$, CRS: $\delta_1 + \delta_2 + \delta_3 + \delta_4 + \delta_5 = 1$

	With Fixed Effects at County Level
Time spent on collective activities (*t*)	0.015 (0.54)
Time spent on supervisory activities (*s*)	0.174 (2.31)**
Amount of labor (*L*)	0.043 (1.37)
Amount of water purchased (*W*)	0.725 (6.64)***
Length of the canal that is lined (*H*)	0.043 (1.88)*
Constant	8.091 (15.26)***
Observations	40
R-squared	0.9837
Number of counties	5

All variables are in log form. t, s, *and* L *are all measured using the unit of a labor day. Absolute value of t statistics in parentheses; * significant at 10%; ** significant at 5%; *** significant at 1%.*

$\delta_4 = 0.725$, and $\delta_5 = 0.043$. Since water is not priced volumetrically within the village, the revenue from providing irrigation services, which is calculated as the sum of water fees by all the households in the villages, is not perfectly correlated with the volume of irrigation water used in the village. However, since the water fee per unit of land is usually determined based upon an estimated average amount of water applied in irrigation (and a certain amount of service in some villages), the water fee is correlated with the volume of water. The degree of correlation depends on how close the water use of one household is to the average level. This may be part of the reason that the coefficient on the water input is closer to one than those of other inputs. We also set the opportunity cost parameters of the leaders and managers, $v = 1$ and $u = 0.8$, so that they are consistent with what we observe in the field; in our data the average daily wage of leaders exceeds that of managers by about 20%. We also set the level of the input prices, $w = 0.8$, $r = 0.004$, on the basis of our data; in our sample, the wage for labor is about \$2.42/day and the water price paid to the ID is around \$0.012/m^3 in many villages, so water price per cubic meter (r) is around 0.5% of the observed daily wage (w). Sensitivity analysis is performed throughout the analysis to ensure our choices of parameters are not driving the results.

Fig. A.9.1 illustrates the partitioning of the contract space based upon the relative efficiency parameters, γ_1 and γ_2. In Fig. A.9.1, γ_1^c and γ_2^c are the critical values of γ_1 and γ_2: the point at which the leader will choose to switch from a profit-sharing to a fixed-wage contract, or the point at which he will switch from a profit-sharing to

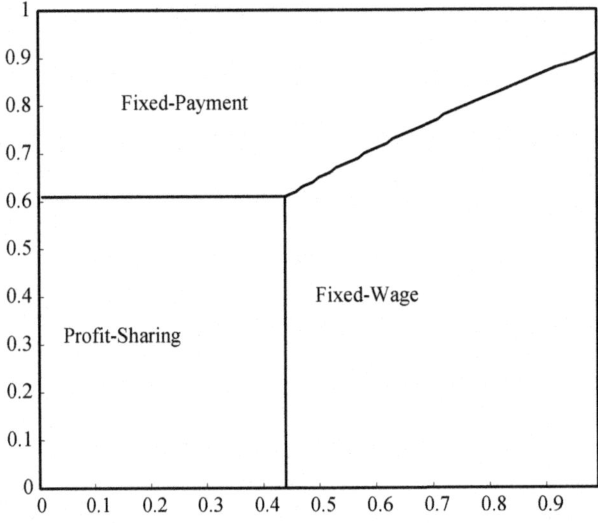

FIGURE A.9.1

Optimal contractual forms varying degrees of comparative advantage for village leader- and canal manager-supplied inputs.

fixed-rent contract. Values of γ_1^c and γ_2^c can be used to partition the contract space into those areas where each of the contracts is optimal. In the figure, we keep the values of the other parameters (A and δ_1–δ_5) constant. The only thing that varies is γ_1 and γ_2. The determination of the critical values can be found by simulating the model (solving the three problems of leader—fixed-wage, fixed-rent, and profit-sharing—and comparing the profit levels of each to find which generates the highest expected income) across a grid of γ_1 and γ_2 values (varying each from 0 to 1 by increments of 0.01). The result of such an exercise is denoted by the solid lines in Fig. A.9.1. If a characteristic of the village's irrigation system or the actors were changed, it is possible that γ_1^c and γ_2^c would change. Because it is possible that the values of δ_1 and δ_2 are underestimated (see discussion above), we repeated the simulations that underlie Fig. A.9.1 for different values of δ_1 and δ_2. Although the size of the area for profit-sharing and the other contracts change, the general configuration stayed the same; the same is true for other comparative static simulations using results from Fig. A.9.1.

A.9.1.3 PREDICTIONS

The basic results from Fig. A.9.1 can be used to gain some intuition about why the form of the contract between the leader and a manager varies across space. The relative efficiency of the leader (manager) to perform supervisory (collective) activities, $\gamma_1(\gamma_2)$, differs from village to village. If besides having a superior ability to organize collective activities, such as canal maintenance, a leader is willing and able to devote himself to supervisory activities, such as water allocation and fee collection (that is, the leader has a relatively high value of γ_1), the leader may choose to manage the village's canals by himself and at most only hire a foreman at a fixed wage to carry out certain routine irrigation tasks under his instruction. If on the other hand, a manager's ability to organize collective activities, γ_2, is high, the leader may prefer to lease out the canal to an individual manager for a fixed-rent contract. When the leader (manager) has little hope of increasing his ability to carry out supervisory (organize collective) activities and the values of γ_1 and γ_2 are both low, the leader may choose a profit-sharing arrangement whereby both the leader and manager share in the duties and also share the system's earnings.

From Fig. A.9.1, we can make the first prediction of the model:

Prediction 1: The dominant contractual form in a given village depends on the relative ability of the villager leader (manager) to perform supervisory (collective) activities. The more experienced the leader (manager) is at supervisory (organizing collective) activities, the more prevalent will be fixed-wage (fixed-rent) contracts.

Even if the ability of the leader (manager) to perform supervisory (collective) activities is constant (that is, γ_1 and γ_2 are constant), changes in other factors might lead to changes in γ_1^c and γ_2^c. Panels A to C in Fig. A.9.2 illustrate the results of three comparative static exercises.

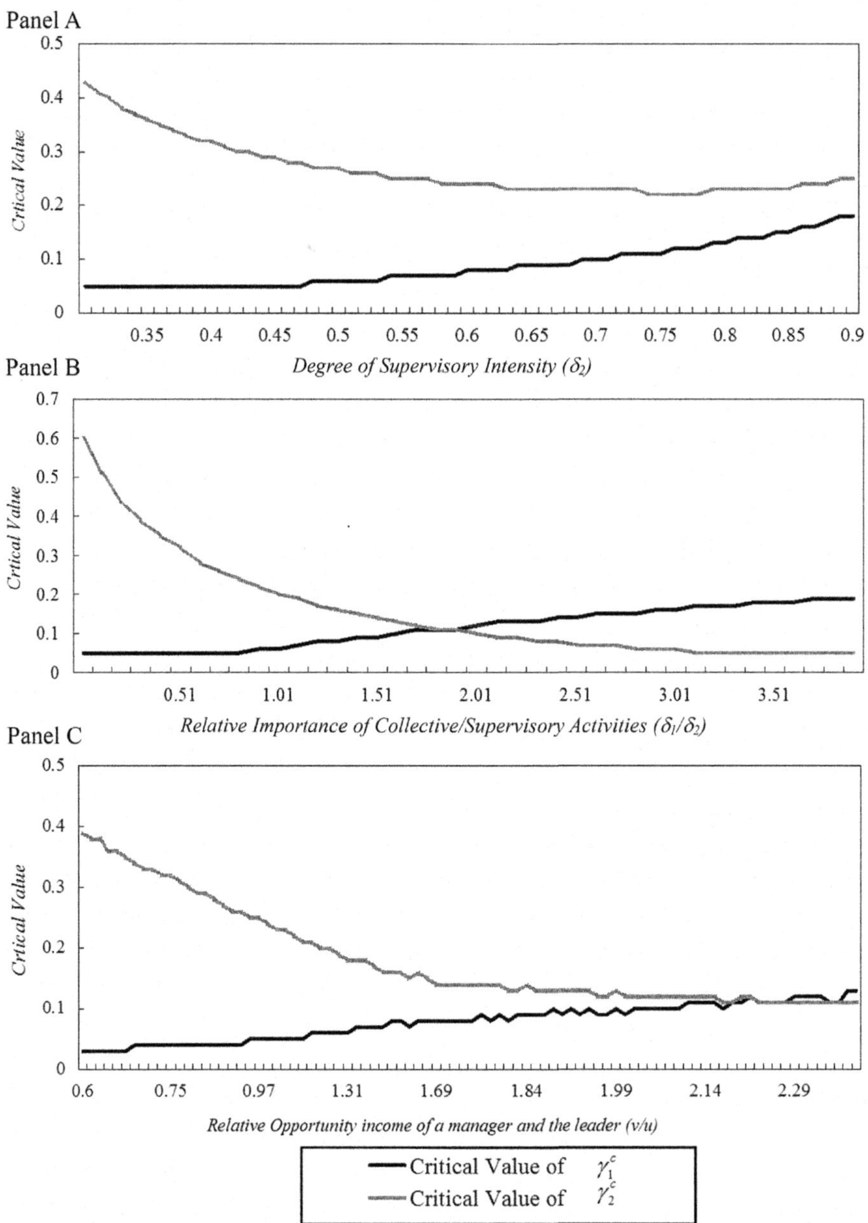

FIGURE A.9.2

Simulations of impacts of different factors on critical values of relative efficiency, Panel A. Changes in degree of supervisory intensity (δ_2), Panel B. Changes in relative importance of collective/supervisory activities (δ_1/δ_2), Panel C. Change in relative opportunity income of a manager and the leader (v/u).

According to the first exercise, the optimal managerial form depends on the nature of the village's cultivated land that can determine the relative importance of supervisory activities and collective activities (Fig. A.9.2, Panel A). In the notation of our model, when supervisory activities are highly valued, the value of the parameter δ_2 increases, and we can summarize with a second prediction:

Prediction 2: The optimal contractual form in a village depends on the nature of its cultivated land, such as the degree of land fragmentation. As land becomes more fragmented, supervisory activities become more valuable and the leader will be more likely to grant a fixed-rent contract to an individual manager.

Given the nature of village's cultivated land (and relative strengths of leaders and managers), the characteristics of the canal system also can determine the relative importance of the two activities. For example, lined canals typically require less maintenance (or are less maintenance intensive). In such a case, collective activities are less important. As a result, we might observe that δ_1 (the importance of collective activities in producing irrigation services) decreases relative to δ_2 (the importance of supervisory activities). Since the leader will not play a role in managing the day-to-day operation of canals, there is a relatively greater need for a motivated manager to operate the rest of the irrigation system. Under these circumstances, there should be a greater propensity for the profit-maximizing leader to move away from a fixed-wage contract into a fixed-rent contract (Fig. A.9.2, Panel B). Thus, we can make a third prediction:

Prediction 3: The optimal contractual form in a village depends on the design of the canal system that determines the degree to which the canal system is maintenance intensive. The less maintenance intensive the canal system, the less important are collective activities, and the more likely it is that the leader will select a fixed-rent contract.

Finally, our model can show that, holding other factors constant, if the opportunity to find an off-farm job is high for the pool of individuals that potentially could take on the role of manager, the opportunity cost (u) of forgoing other jobs to take on the tasks of managing the village's canals will increase (Fig. A.9.2, Panel C). In this chapter, we assume higher opportunities to find an off-farm job for villagers do not change the opportunity cost of the leader (at least during his current term). When becoming a leader, the individual commits himself to fulfilling a large set of duties that include tasks such as implementing family planning, managing the village's land, and adjudicating disputes among villagers. Burdened with these many duties, the leader almost never has time to spend intensively participating in an off-farm job during his tenure. In many places, rules prohibit him from physically leaving the village to take an off-farm job outside the village. In this case, γ_1^c will decrease (or γ_2^c will increase) relative to that of a village in which it is more difficult for managerial candidates to find other jobs. The change in the relative values of the opportunity cost parameter of the managerial candidate versus the leader will change the bargaining power of the managerial candidate; as u

rises, the leader will have to lower the fixed rent or increase the share (required by the managerial candidate) to attract the managerial candidate to take a fixed-rent or profit-sharing contract. As a result, under such circumstances, it would be less profitable for a leader to offer a fixed-rent or profit-sharing contract, and we can summarize this as follows:

Prediction 4: The optimal contractual form in a village depends on the opportunity cost of the leader and the pool of managerial candidates, which is associated with the social-economic environment of the village. The wealthier a village is (or the more opportunities there are to find an off-farm job), the more likely it will be that a fixed-wage contract is chosen.

A.9.2 MULTINOMIAL LOGIT REGRESSION TO EXPLAIN CONTRACTUAL CHOICE

Since the explanatory variables that we use are alternative invariant, a multinomial logit (MNL) model is used in our analysis. In equation form, the basic model can be written as

$$p_{ij} = \Pr[y_i = j] = \exp\left(\mathbf{x}_j'\boldsymbol{\beta}_j\right) \Big/ \sum_{l=1}^{3} \exp\left(\mathbf{x}_i'\boldsymbol{\beta}_l\right),\ j = 1, 2, 3. \qquad \text{[A.9.2]}$$

where j is the index for the alternatives and i is the index for our observations. The dependent variable y is defined as

$$y = \begin{cases} 1, & \text{if fixed-wage contract is chosen} \\ 2, & \text{if profit-sharing contract is chosen} \\ 3, & \text{if fixed-rent contact is chosen} \end{cases}$$

from Eq. [A.9.2], we know that

$$\ln \frac{p_{ij}}{p_{im}} = \mathbf{x}_i'\boldsymbol{\beta}_j \qquad \text{[A.9.3]}$$

The term, $\ln \frac{p_{ij}}{p_{im}}$, is the log of the relative probability that choice j will be chosen over choice m. In Greene (2004), this term is called the log-odds ratios. We will use this terminology in the rest of the chapter. Conveniently, the log-odds ratios are linear in the parameter vector $\boldsymbol{\beta}_j$. Suppose the choice m is the base category in the multinomial logit regression, the parameter vector $\boldsymbol{\beta}_m$ is normalized to be zeros. Now $\boldsymbol{\beta}_j$ represents the marginal effect of the independent variables on the log-odds ratio. For example, if in the regression results, $\widehat{\beta}_{jk}$ is positive and statistically significant, we can say that a marginal increase in the kth independent variable will make the village leader more likely to choose choice j as opposed to choice m.

A.9.2.1 ESTIMATION ISSUES

Although in our survey in 2001 we collected data on 80 villages in three provinces (Hebei, Henan, and Ningxia), we have only used the data collected in Ningxia province in this analysis for two reasons. First, in most areas in Ningxia province, surface water is the only source of irrigation water, which allows us to focus solely on the contractual forms that are used in managing canals. In contrast, in Henan province where both surface water and groundwater are used in irrigation; in Hebei province, the major source of irrigation water is groundwater and thus requires a separate study to analyze its managerial forms. Second, different villages in Ningxia province used different types of contracts to manage the canal systems. The variations in the way villages manage their canals offer a great opportunity to analyze the factors that have led to the differences in contractual choices. In contrast, in the surface water use areas in Henan province, most of the villages managed their canal systems in the same way (the fixed-wage contract) and so there was virtually no variation. Since we only have a small sample of data that we can use in our analysis ($n = 40$), there may be multicollinearity in the data. To check this, we conducted collinearity diagnostic to detect potential collinearity in the data (Belsley, 1991). Fortunately, we only detected a minor collinearity in the data (the condition index was less than 100).

We also tested various specifications of the model (that is, different combinations of explanatory variables) to examine the robustness of the estimation results. The results show that although the absolute magnitudes of the coefficients of the same variable vary with different combinations of explanatory variables, the relative magnitudes, the signs and the significances of the coefficients are largely consistent. Since the leaders choose from three different contractual choices, in the multinomial logit regression, there are two sets of parameters, one for each nonbase alternative. The relative magnitudes refer to the relative values of these two sets of parameters. Since the focus of our analysis is the directions of the impacts of the explanatory variables on the relative probabilities of different types of contracts being chosen by the village leaders, the variations in the magnitudes of the coefficients do not impair our analysis.

Given the differences in the estimation results generated by different model specifications, the natural question to ask is that which specification is the best. We compare the out-of-sample predictions of our models to choose the "best" model out of the different specifications for our analysis. In particular, we used a leave-one-out, cross-validation procedure (Allen, 1974). Using the results from the out-of-sample prediction, we then calculate the value of the Percent Correctly Predicted (PCP) measure, which is defined as the percentage of the sampled village leaders for whom the contractual form with the highest probability (as in the out-of-sample prediction) and the chosen contractual form (as observed in the data) are the same. It should be noted that the coefficients on the degree of fragmentation for the profit-sharing contract have the opposite sign in Model 1 and Model 2. This is the only major difference in the estimation results generated by Model 1 and Model 2. Hence, we did not include the impact of the degree of fragmentation on the choice of profit-sharing contract here. The model with the highest PCP should have the property of minimizing the impact of the small sample problems on the results of the analysis. Comparison of the values of PCP shows that models with fixed effects at the county level have higher values of PCP than models without fixed effects. Among the models with fixed effects at the county level, Model 2 in Table A.9.3 has the highest PCP. That is the basis of our choice of right-hand-side variables.

We also compare the values of PCP to the values in other studies. The PCP is 50% for our best model. This is comparable to the values in other studies.

Table A.9.3 Multinomial Logit Regressions Explaining Contractual Choice by the Village Leader in Ningxia Province (With Fixed Effects at County Level)—Marginal Effects in Model 2[a]

		(1) Fixed-Wage	(2) Profit-Sharing	(3) Fixed-Rent
	Condition of the natural environment			
1	Degree of fragmentation (number of plots/household)	−0.008 (2.67)**	−0.054 (16.69)**	0.065 (38.06)**
2	Water abundance (index—4 if abundant; 1 if scarce)	−0.107 (10.14)**	0.002 (0.38)	0.102 (14.34)**
	Characteristics of the canal system			
3	Canal lining (1 if lined, 0 otherwise)	−0.279 (44.92)**	0.220 (28.13)**	0.059 (5.70)**
4	Propensity to silt up (1 if yes, 0 otherwise)	0.076 (5.10)**	−0.348 (24.36)**	0.272 (44.52)**
	Opportunity cost of the pool of managerial candidates			
5	Share of off-farm income (%)	0.030 (36.03)**	−0.020 (22.98)**	−0.010 (18.35)**
	Human capital characteristics (ability proxies)			
6	Level of education of the leader (attainment in years)	0.020 (7.47)**	0.015 (7.30)**	−0.034 (21.15)**
7	Level of education of the managerial candidate (%, share of labor force with at least high school education)	−0.016 (21.23)**	0.001 (3.47)**	0.015 (28.35)**
8	Age of the leader (years)	−0.057 (48.64)**	0.040 (25.74)**	0.016 (17.51)**

*Robust z statistics in parentheses: * significant at 10%; ** significant at 5%; *** significant at 1%.*

[a] The marginal effects are calculated using two steps. In the first step, the changes in the quantities of interest are evaluated at each observation first. In the second step the sample averages of these changes are reported as the average marginal effects. The standard errors (and thus the z statistics) are calculated using the delta method.

For example, McFadden (1976) used a multinomial logit model to analyze the decisions of the California Division of Highways on the highway route. The values of his PCP measures ranged from 40.8% to 56.2%. The values of PCP ranged from 53.4% to 60.1% in the "best" models chosen by Matsukawa and Fujii (1994). It should be noted that both studies have larger sample sizes but have only used the in-sample prediction to generate the values of PCP. In summary, given the value of PCPs, our estimates are reasonably accurate and it is safe to make inferences based upon the estimates.

The dependent variable in our regression is the contractual form currently existing in the village in year 2001. Using the values of some of the key independent variables (water abundance; the share of off-farm income; proportion of migrants) from the same year may cause problems of endogeneity bias. To avoid the possible endogeneity problem, we chose to use 1995 values for these independent variables.

A potentially important drawback of the MNL model is the implicit assumption of independence of irrelevant alternatives property (IIA; Hausman and McFadden, 1984; Small and Hsiao, 1985). When we test whether the IIA property holds for our MNL regression, both the Small-Hsiao test and the Hausman test fail to reject the null hypothesis of IIA. Therefore, we believe that it is safe to assume that the IIA property holds for our data and that an MNL approach is appropriate.

For the sake of completeness of the analysis, we also report the direct marginal effect of the independent variables on the probability of a type of contract being chosen in Table A.9.3. The results of testing our predictions using Table A.9.2 are largely consistent with the results using Table A.9.3. For example, the results in Table A.9.2 show that when the land in villages is characterized by higher degrees of fragmentation, fixed-rent contracts are more likely to be chosen relative to fixed-wage contracts or profit-sharing contracts (row 1). The results in Table A.9.3 show that higher degrees of fragmentation increase the probability of a fixed-rent contract but decrease the probability of the other two contracts being chosen; interpreted together, these two results are consistent with the result that a fixed-rent contract is more likely to be chosen. Given that the results using either the log-odds ratio or the direct marginal effects are largely consistent, and the fact that it is much easier to use the log-odds ratio, we focus on the log-odds ratio in the chapter.

REFERENCES

Allen, D.M., 1974. The relationship between variable selection and data augmentation and a method for prediction. Technometrics 16, 125–127.

Belsley, D., 1991. Conditioning Diagnostics: Collinearity and Weak Data in Regression. John Wiley & Sons Inc., New York, USA.

Hausman, J., McFadden, D., 1984. Specification tests for the multinomial logit model. Econometrica 52, 1219–1240.

Lohmar, B., Wang, J., Rozelle, S., Huang, J., Dawe, D., 2003. China's Agricultural Water Policy Reforms: Increasing Investment, Resolving Conflicts, and Revising Incentives. Agriculture Information Bulletin No. 782. Market and Trade Economics Division, Economic Research Service, U.S. Department of Agriculture, Washington, DC.

Matsukawa, I., Fujii, Y., 1994. Customer preferences for reliable power supply: using data on actual choices of back-up equipment. The Review of Economics and Statistics 76, 434–446.

McDonald, J.F., Moffitt, R.A., 1980. The uses of tobit analysis. The Review of Economics and Statistics 62 (2), 318–321.

McFadden, D., 1976. The revealed preferences of a government bureaucracy: empirical evidence. The Bell Journal of Economics 7, 55–72.

Oi, J., 1999. Rural China Takes Off: Institutional Foundations of Economic Reform. University of California Press, Berkeley, USA.

Park, A., Rozelle, S., 1998. Reforming state-market relations in rural China. Economics of Transition 62, 461–480.

Platteau, J.-P., Gaspart, F., 2003. The risk of resource misappropriation in community-driven development. World Development 31, 1687–1703.

Rozelle, S., 1994. Decision making in China's rural economy: defining a framework for understanding the behavior of village leaders and farm households. China Quarterly 137, 99–124.

Small, K.A., Hsiao, C., 1985. Multinomial logit specification tests. International Economic Review 26, 619–627.

Wang, J., Huang, J., Rozelle, S., 2005. Evolution of tubewell ownership and production in the North China Plain. The Australian Journal of Agricultural and Resource Economics 49, 177–195.

Whiting, S., 2001. Power and Wealth in Rural China: The Political Economy of Institutional Change. Cambridge University Press, Cambridge, MA, USA.

Methodology Appendix to Chapter 10

A.10.1 ECONOMETRIC MODEL ON THE EFFECTS OF WATER MANAGEMENT ON CROP WATER USE

Based on the descriptive discussion, the link between water use per hectare and its determinants can be represented by the following equation:

$$w_{jk} = \alpha + \beta M_k + \gamma Z_{jk} + D_{jk} + \varepsilon_{jk} \qquad [A.10.1]$$

$$w_{jk} = \alpha + \beta I_k + \gamma F_k + \delta Z_{jk} + D_{jk} + \varepsilon_{jk} \qquad [A.10.2]$$

where w_{jk} represents average water use per hectare for household j in village k. The variable for household water use is created as follows: (1) the average water use per hectare per irrigation is a village average that varies by crop type; (2) average water use per hectare per irrigation was determined on the basis of information elicited from the village leader and water manager; (3) the number of irrigations for each crop is elicited by the household survey and so varies by household; (4) total water use for each farmer is created by multiplying the village average water use per hectare per irrigation times the number of irrigations per crop. Because, by construction (and in fact), there is correlation among the observations on water use within a village, our results are corrected for clustering at the village level.

The rest of the variables in Eqs. [A.10.1] and [A.10.2] are those that explain water use: M_k measures the type of the water management institution (such as water user associations (WUAs) and contracting); I_k and F_k, our variables of interest, separately measure the nature of incentives and participation of farmers; and Z_{jk}, a matrix of control variables, represents other village and household factors that affect water use. Specifically, we include a number of variables to hold constant the nature of the village's production environment and its cropping structure. We include variables measuring the source of water (either surface or ground), the degree of water scarcity, and the level of irrigation investment per hectare (a stock variable estimated as the sum of the investments made over the past 20 years). Cropping structure is measured as the proportion of the village's sown area that is in rice. The degree of water scarcity is an indicator variable developed from a question included in the village questionnaire. Enumerators asked village leaders to characterize the nature of water resources in their village. The leaders chose one of three precoded answers: 1 = water is very scarce; 2 = water is relatively scarce and frequently constrains agricultural production; and 3 = water is not short (at least currently). The indicator variable takes on the value of one if the leader responded either 1 or 2, and zero if he responded 3. Household characteristics include age and education of the household head and the household's land endowment. Finally, our model also includes D_{jk}, a dummy variable representing the ID that serves the household.

The symbols α, β, and γ are parameters to be estimated and ε_{jk} is the error term that is assumed to be uncorrelated with the other explanatory variables in our initial equations, an assumption that we subsequently relax.

In order to control for the potential endogeneity of water management types and incentives in the water use equation, we adopt an instrumental variable (*IV*) approach. To do so, prior to estimating Eqs. [A.10.1] and [A.10.2], we regress a set of variables on the water management institution variable, M_k, incentives for managers, I_k, and farmer participation, F_k:

$$M_k = \alpha + \beta \mathrm{IV}_k + \gamma Z_k + \varepsilon_k \qquad [A.10.3]$$

$$I_k = \alpha + \beta \mathrm{IV}_k + \gamma Z_k + \varepsilon_k \qquad [A.10.4]$$

$$F_k = \alpha + \beta IV_k + \gamma Z_k + \varepsilon_k \qquad [A.10.5]$$

where the predicted value of M_k from Eq. [A.10.3], $\widehat{M}_k$, replaces M_k in Eq. [A.10.1]; and the predicted values of I_k and F_k from Eqs. [A.10.4] and [A.10.5], $\widehat{I}_k$ and $\widehat{F}_k$, replace I_k and F_k in Eq. [A.10.2]. Eqs. [A.10.3]–[A.10.5] include Z_k, which are measures of the other village-level control variables. The control variables in Eqs. [A.10.3]–[A.10.5] are the same as those in Eq. [A.10.1] (eg, measures of the village's production environment and cropping structure).

This *IV* procedure, however, is only valid if the variables in the *IV* matrix in Eqs. [A.10.3]–[A.10.5] meet the definition of instruments. The key *IV* that we use in Eqs. [A.10.3]–[A.10.5] to address the endogeneity problem is a variable (IV_k) that measures the effect of the decision of regional policymakers to push water management reform in village *k*. Such a measure should function well as an instrument, especially in our setting, since the officials that were responsible for promoting water management reform believed that at least in the short run they were choosing villages on a fairly random basis. An official in one ID told us that, initially, he went to villages in which he personally knew the local officials. If the spectrum of the acquaintances of the typical water system officials are independent of the amount of water used in the village, our policy variable should meet the criteria of an instrumental variable: it is correlated with the decision of a village to participate in water management reform but does not have an effect on water use (or income or crop production) except through the influence of the reform. We also include the age and education of the village leader as *IV*s. We include village leader characteristics as *IV*s, following Li (1999). In the work of Li, the author claims that village leader characteristics may affect reform in the village, but the leader's characteristics would not have an independent effect on production decisions (in our case, water use).

A.10.2 ECONOMETRIC MODEL ON THE EFFECTS OF WATER MANAGEMENT ON PRODUCTION, INCOME, AND POVERTY

In addition to water management reform, other socioeconomic factors influence agricultural production, income, and poverty. In order to answer the question of whether water management reform affects outcomes, it is necessary to control for these other factors. To do so, we specify the link between agricultural production and its determinants as:

$$Q_{ijk} = \alpha + \beta W_{ijk} + \gamma X_{ijk} + \delta Z_{ijk} + D_{ijk} + \varepsilon_{ijk} \qquad [A.10.6]$$

where Q_{ijk} represents the yields of wheat, maize, or rice from the ith plot of household j in village k (which comes from our household survey). In Eq. [A.10.6], yields are explained by the variable of interest, W_{ijk}, which measures water use per hectare; X_{ijk}, which measures other inputs to the production process; Z_{ijk}, which holds other factors constant, including characteristics of the production environment of the village, household, and plot; and the irrigation district dummy, D_k. Agricultural production inputs include measures of per hectare use of labor (measured in mandays), fertilizer (measured in aggregated physical units), and expenditures on other inputs, such as fees paid for custom services. To measure fertilizer, we decomposed each type of fertilizer by nutrients, nitrogen (N), phosphorus (P), and potassium (K), and then summed across nutrients and fertilizer types. We also aggregated fertilizer by value, and our main results of interest do not change. The control variables for village and household characteristics are the same as for Eq. [A.10.1] except we do not use the village level cropping structure. We also add five plot characteristics, including measures of: two soil type variables; plot location (distance from the plot to the farmer's house); whether or not the crop on the plot is planted in rotation with another crop (single crop equals one, zero otherwise); and production shocks (measured as farmer-estimated yield reduction in percentage terms on a plot due to a flood, drought, or other "disaster").

We also establish the following equation to examine the relationship between income and other factors:

$$y_{jk} = \alpha + \beta I_{jk} + \gamma F_{jk} + \delta Z_{jk} + D_{jk} + \varepsilon_{jk}, \qquad [A.10.7]$$

where y_{jk} represents either total or cropping income per capita for household j, and the other variables are as defined above. In examining the effect of water management reform on poverty, we proceed in largely the same way. Because we are measuring poverty in terms of income, we use the same specification as in Eq. [A.10.7] and expect similar results, but with opposite signs.

Methodology Appendix to Chapter 11

A.11.1 ECONOMETRIC MODEL ON THE EFFECTS OF WATER USER ASSOCIATIONS ON CROP WATER USE AND CROP YIELDS

We have specified the following model for the impact of water management on water use in our sample villages in 2005:

$$\text{Water Use, crop (i)} = \text{a0} + \text{a11} * \text{Bank WUA} + \text{a12} * \text{non} - \text{Bank WUA} + \text{a2} * \text{Village characteristics} + \text{a3} * \text{Location dummies} + \text{e}. \quad [\text{A.11.1}]$$

where Water Use is measured as the amount of water applied to each crop i, where $i = 1$ for rice; and for $i = 2$ for an aggregate of rice + wheat + maize. In our analysis, we would have liked to run a separate regression for rice, wheat, and maize. However, there were not enough observations in the World Bank Study sites for wheat and maize to run them separately. Instead, we combined rice, wheat, and maize and included a set of interaction variables by interacting the wheat and maize dummies with the Bank WUA and nonbank WUA variables to control for the specific crop effects. The variable WUA is measured as a set of dummy variables, which measures whether or not a village is managed by a Bank WUA, a nonbank WUA (and in the regressions the omitted category are collective villages in the Bank survey site). We include a number of village characteristics as control variables, including Water Scarcity (which is measured as a dummy variable that equals 1 if the respondent said the village's water was scarce); Education of Labor Force (which is measured as the share of the labor force with an educational attainment above high school); Cultivated Land per capita (measured in mu); Share of Irrigated Area (which is measured as the percent of cultivated area that is irrigated); Distance to Township (which is measured in kilometers by the shortest route by road); Downstream (which is a dummy variable that indicates if a village is located in the lower reaches of the ID); age of the Party Secretary (in years), education of the Party Secretary (in years of educational attainment) and Water Management Experience of the Party Secretary (which is measured in the number of years during which the current party secretary at some point in his life was in charge of managing the village's canal system); and the Main Job of Party Secretary (which is a dummy variable that is equal to 1 if the party secretary relies primarily on farming for his income; and 0 if it is from a wage earning job or nonfarm self-employment). In addition and importantly, in the regression model we also include a set of county level dummies.

These will hold constant all nontime varying county-wide effects, including factors such as climate, soil, and varieties.

In addition, we also specify three other equations for analyzing the impact on yields (kg/ha.), income (measured in per capita terms in dollars), and cropping structure (share of total sown area in grain). As above, all of the explanatory variables are the same, including the variables used to measure the effect of WUAs (the two dummy variables) and the control variables. We also estimate this model for our village sample in 2005.

$$\text{Yields, crop (i)} = \text{a0} + \text{a11} * \text{Bank WUA} + \text{a12} * \text{non} - \text{Bank WUA} + \text{a2} * \text{Village characteristics} + \text{a3} * \text{Location dummies} + \text{e}. \quad \text{[A.11.2]}$$

Methodology Appendix to Chapter 12

A.12.1 HOUSEHOLD WATER DEMAND FRAMEWORK AND ESTIMATION APPROACH

In this section, we first introduce the household maximization model from which household water demand is derived. We then lay out the methods we use to estimate the production frontier and technical inefficiency parameters, the key elements in measuring the relationship between water and output.

A.12.1.1 HOUSEHOLD WATER DEMAND FRAMEWORK

Five inputs are used in production. Two inputs are variable inputs: capital (x_k) and fertilizer (x_f). Capital costs, which could also be called material costs, include expenditures on machinery, seed, plastic sheeting, herbicides, and pesticides. We assume that farmers can purchase fertilizer at unlimited quantities at the market price. Land (x_L), family labor (x_{fl}), and water (x_w) are treated as fixed allocable inputs. In rural China, the collective (or community) allocates land to each household based largely upon the size of the household. Except for rented plots, there is no cost for land. Only a small proportion of plots are rented (about 3% in 1995 and 7% in 2000; Brandt et al., 2004). Hence, we assume there is no cost for land, but that it is fixed. In addition to family labor, which is fixed by definition, labor input also may include hired labor (x_{hl}). Since in our data and in all of China, only a small percentage of farm labor is hired (Benjamin and Brandt, 2002), we believe that we can assume that labor is a fixed input. While there are no formal restrictions on pumping in the sample communities (Wang et al., 2005), our data show that in some communities the quantity of groundwater may be constrained (at least short run; ie, during the irrigation season).

Households are assumed to maximize the total profit from all three crops. The profit maximization assumption is supported by many previous studies on production behavior in China (eg, Huang and Rozelle, 1996; Lin, 1992). Given the small size of farms in rural China, it is reasonable to assume all households are price-takers. In our preliminary analysis, we assumed households were risk averse. The risk aversion parameter, however, is not statistically different from zero in our estimation. Therefore, without loss of generality, we assume households are risk neutral. Since there is no regulation on the rate of pumping by users in rural China, groundwater is an open-access common-property resource. In addition, there are on average 500 households per community and 30 households per well in our sample. Given so many uses, we can safely assume households are myopic users and only maximize profit from the current period. Based on these assumptions, the constrained profit maximization problem (Problem P1) can be expressed as:

$$\underset{x_{ij}}{\text{Max}} \sum_j p_j \theta f_j \left(x_{L_j}, \gamma x_{w_j}, x_{l_j}, x_{f_j}, x_{k_j}\right) - \sum_i c_i x_{ij}$$

$$\text{Subject to:} \quad \sum_j x_{ij} \le B_i \ \forall i = \text{Land, Labor, Water}$$

$$x_{ij} \ge 0$$

where the output price for crop j is p_j and the production frontier for crop j is f_j. The parameter, θ, is the technical inefficiency parameter. The cost for input i is c_i. B_i represents the vector of available quantities of the ith fixed allocable input. Although households pay for all the irrigation water they apply to their crops, only a proportion of that water is consumed by crop due to conveyance loss or return flow to the aquifer. The symbol, γ, denotes the proportion of the applied water that crops consume and thus contributes to crop growth. Kim and Schaible (2000) and Scheierling et al. (2004) have shown that if the amount of irrigation water applied, instead of the amount of water actually consumed by the crop, is considered as the amount of water that contributes to crop growth in crop production, the marginal benefit of water will be overestimated. The parameter γ is suppressed in the rest of the chapter for the sake of brevity.

Let W denote household level total water use. The total effect of a change in water price on W can be shown to be

$$\frac{dW}{dc_{\mathrm{w}}} = \sum_{j=1}^{J} \left(\frac{\partial x_{wj}}{\partial c_{\mathrm{w}}} + \frac{\partial x_{wj}}{\partial L_j} \frac{\partial L_j}{\partial c_{\mathrm{w}}} \right) \qquad [\text{A.12.1}]$$

The first term, $\frac{\partial x_{wj}}{\partial c_{\mathrm{w}}}$, represents the change in crop-level water use when the land allocation is held constant, that is, the intensive margin adjustment. Comparative statics show that $\frac{\partial x_{wj}}{\partial c_{\mathrm{w}}} < 0$. The second term, $\frac{\partial x_{wj}}{\partial L_j} \frac{\partial L_j}{\partial c_{\mathrm{w}}}$, represents the change in crop-level water use operated through land reallocation, that is, switching between irrigated area and rainfed area within the same crop, or changes in cultivated area across different crops. This is the extensive margin adjustment. Depending on the specific crop, $\frac{\partial x_{wj}}{\partial L_j} \frac{\partial L_j}{\partial c_{\mathrm{w}}}$ may be positive or negative. Summing over the intensive and extensive margins of all J crops, we get the household level response to changes in water prices. From Eq. [A.12.1], it is clear that to characterize the household level water demand, we need to consider both the intensive and extensive margin adjustments. To do so, we need to estimate two sets of parameters: the production frontier, f_j, and the technical inefficiency parameter, θ.

A.12.1.2 PRODUCTION FRONTIER AND GENERALIZED MAXIMUM ENTROPY

We estimate one set of crop specific production functions for each county, allowing production technology to vary by county but restricting it to be equal across communities within the same county. Since the price responsiveness of water demand depends on its own- and cross-price elasticities, a flexible functional form is used so that these relationships are not arbitrarily restricted by the choice of the functional form. We specify a quadratic function frontier:

$$Y_{jn} = \theta_n \left(\sum_i \alpha_{ij} x_{ijn} - \sum_i \sum_{i'} x_{ijn} z_{ii'j} x_{i'jn} \right) + e_{jn} \qquad [A.12.2]$$

The observed output and input use of household n for crop j are denoted by Y_{jn} and x_{ijn}, respectively. The symbol, θ_n, denotes the technical inefficiency of household n. The error term, e_{jn}, captures variation in outputs due to random events such as weather.

Estimating a flexible production frontier, however, would make the use of some estimation methods difficult or even infeasible. Since five inputs are used in production and a quadratic functional form is specified, there are 20 parameters to be estimated for each production frontier. If classical econometric methods (eg, maximum likelihood estimation) were used, the estimation problem would be ill-posed due to an insufficient number of data points. For example, when we estimate a production frontier for wheat in Tang and Ci County, there are only 18 and 19 data points respectively (Table 12.3, row 3 and 4, column 2). This means that there are fewer observations than the number of parameters.

As a solution, we choose to use the Generalized Maximum Entropy (GME) method that was developed by Golan et al. (1996). The GME estimator allows for estimation with any sample size. The GME estimator emphasizes both prediction and precision in its objective function and thus has the properties of being both subject to limited bias and to minimum variance (Golan et al., 1996). Under very general conditions, the GME estimator has desirable large sample properties, including both asymptotic efficiency and asymptotical normality. It should be noted that if the number of data points were sufficient, classical econometric methods and GME would produce the same estimation results.

In GME estimation, in addition to the data-consistent constraints, specified in Eq. [A.12.2], two additional sets of constraints are specified: the theoretical constraints and the numeric estimation constraints.

A.12.1.3 THEORETICAL CONSTRAINTS

Two sets of theoretical constraints are used so that the estimated production technology is consistent with the profit maximization behavior of households. The first set is the optimality condition constraints:

$$x_{ijn} = \frac{1}{2z_{ii}} \left[\alpha_{ij} - 2 \sum_{i' \neq i} x_{i'n} z_{ii'} - \frac{c_{in}}{p_j \theta_n} \right] + v_{ijn} \quad \forall i = \text{Fertilizer, Capital} \quad [A.12.3]$$

We only include the optimality conditions for variable inputs. The optimality conditions for fixed allocable inputs include the shadow values, which are not directly estimable.

The second set of theoretical constraints is the curvature constraint, which requires that **Z**, the matrix of all parameters on the squared and interaction terms of inputs ($z_{ii'j}$ s), be positive (semi)definite. The curvature constraint is imposed using the Cholesky decomposition (Paris and Howitt, 1998). The Cholesky decomposition is defined by $\mathbf{Z} = LL'$, where L is an $I \times I$ lower triangular matrix and I is the total number inputs. The positive (semi)definite property of **Z** is guaranteed by constraining the diagonal elements of L to be nonnegative ($L_{iij} \geq 0$). The Cholesky decomposition also ensures the symmetry of the matrix **Z**. In addition, we impose the monotonicity constraints: $p_j\theta_n\left[\alpha_{ij} - 2\sum_{i'}x_{i'n}z_{ii'}\right] - c_{in} > 0$. In our empirical estimation, the monotonicity constraints hold for all observations.

A.12.1.4 NUMERIC ESTIMATION CONSTRAINTS

When the GME method is used, instead of directly estimating the mean and variance of the coefficient, a probability distribution is estimated for each coefficient and the error term. Several possible values of a coefficient are chosen as the support values of the probability distribution and an unknown probability is assigned to each value. We follow Golan et al. (1996) and choose five support points for both coefficients and error terms. The coefficients and the error terms are then reparameterized in terms of unknown probabilities and support values. This set of reparameterization constraints are defined as $\alpha_{ij} = \sum_m p^m_{\alpha_{ij}} \overline{\alpha}^m_{ij}$, $z_{ii'j} = \sum_m p^m_{z_{ii'j}} \overline{z}^m_{ii'j}$, $v_{ijn} = \sum_m p^m_{v_{ijn}} \overline{v}^m_{ijn}$, and $e_{jn} = \sum_m p^m_{e_{jn}} \overline{e}^m_{jn}$, where m is the index of the support values and the ps are the unknown probabilities to be estimated. Symbols with upper bars denote the support values. The unknown probabilities are positive and all probabilities associated with the same coefficient or error term add up to one (the adding up constraints).

A.12.1.5 TECHNICAL EFFICIENCY PARAMETERS

The parameter, θ_n, is the technical inefficiency parameter that captures the degree of deviation of each household's actual production from the production frontier. More importantly, since technical inefficiency is often the result of a lack of managerial ability (Farrell, 1957; Leibenstein, 1966), θ_n reflects the interhousehold differences in managerial ability. Accounting for technical inefficiency is important since it can help us overcome a common problem associated with estimating a production function—the potential bias due to the existence of omitted variables. In particular, household managerial ability is often omitted because it is not directly observable to econometricians. Since the managerial ability affects both output level and the producer's choice of input, omitting it will bias the estimates of production function parameters (Griliches, 1957). The estimated technical inefficiency parameters can capture unobservable heterogeneity that is relevant in our analysis (that is, managerial ability); therefore, we believe our estimates will be less affected by omitted variable bias. We employ conventional panel data models, such as fixed-effects or random-effects models, to account for unobserved heterogeneity. Unfortunately, for most households in our sample, we do not have more than one observation for a single crop. Therefore, it is not possible to use a household fixed-effects approach.

We use the classic Farrell definition of the output distance function to measure θ_n. The output distance function for the nth household that produces a single output, $D_O(\mathbf{x}_n, y_n)$, is defined as: $\inf_{\theta_n}\{\theta_n > 0 : (y_n/\theta_n) \in P(\mathbf{x}_n)\}$. The parameter θ_n denotes the inverse of the factor by which the output y_n could be increased while still remaining within the feasible production set, $P(\mathbf{x}_n)$, for the given input level $\mathbf{x}_n$. The value of $D_O(\mathbf{x}_n, y_n)$ is less than or equal to one if y_n is an element of $P(\mathbf{x}_n)$. If $D_O(\mathbf{x}_n, y_n)$ is one, then the observation $(\mathbf{x}_n, y_n)$ is on the production frontier.

Data Envelopment Analysis (DEA) involves measurement of efficiency for a given observation in the sample data relative to the boundary (that is, the production frontier) of the convex hulls of the data intersected with the free-disposal hull. The distance function, $D_O(\mathbf{x}_n, y_n)$, is estimated by solving a Linear Programming (LP) problem (Charnes et al., 1978; Färe and Primont, 1995; Farrell, 1957). The problem (Problem P2) can be expressed as:

$$\underset{\theta_n, z_1, z_2, \ldots, z_H}{\text{Max}} \quad 1/\theta_n$$

$$\text{Subject to:} \begin{cases} \sum_{h=1}^{H} z_h y_{jh} \geq y_{jn}/\theta_n, j = 1, 2, \ldots, J \\ \sum_{h=1}^{H} z_h x_{ih} \leq x_{in}, i = 1, 2, \ldots, I \\ \sum_{h=1}^{H} z_h \leq 1 \\ z_h \geq 0, \; h = 1, \ldots, H \end{cases}$$

where the variable, z_h, represents the intensity level of the production activity of household h. The constraints in the LP problem specify disposability (which corresponds to the monotonicity of the production frontier) and the convexity of the feasible production set (which corresponds to the concavity of the frontier). Note these assumptions are consistent with the theoretical constraints in the GME estimation of the frontier parameters.

A.12.1.6 SUMMARY OF ESTIMATING DEMAND PARAMETERS WITH GENERALIZED MAXIMUM ENTROPY AND DATA ENVELOPMENT ANALYSIS

In GME estimation, the estimates of probabilities given their support values are obtained through maximizing the negative joint entropy of the distributions of the coefficients and the error terms conditional on the data-consistent constraints, the theoretical constraints, and other constraints. The objective function of Problem P2 is easily added since it is also an optimization problem. In summary, the estimation problem (Problem P3) can be expressed as:

$$\underset{p^m_{\alpha_i}, p^m_{z_{ii'}}, p^m_{e_n}, p^m_{v_{in}}}{\text{Max}} H\left(p^m_{\alpha_i}, p^m_{z_{ii'}}, p^m_{e_n}, p^m_{v_{in}}\right)$$

$$= -\sum_{j=1}^{J}\sum_{i}\sum_{m} p^m_{\alpha_{ij}} \ln p^m_{\alpha_{ij}} - \sum_{j=1}^{J}\sum_{i}\sum_{i'}\sum_{s} p^m_{z_{ii'j}} \ln p^m_{z_{ii'j}} - \sum_{n=1}^{N}\sum_{m} p^m_{e_n} \ln p^m_{e_n}$$

$$-\sum_{n=1}^{N}\sum_{i}\sum_{m} p^m_{v_{in}} \ln p^m_{v_{in}} + \sum_{n=1}^{N}(1/\theta_n)$$

Subject to

Data consistent constraints: $Y_{jn} = \theta_n\left(\sum_i \alpha_{ij}x_{ijn} - \sum_i\sum_k x_{ijn}z_{ikj}x_{kjn}\right) + e_{jn}$

Theoretical constraints

First order conditions: $x_{ijn} = \dfrac{1}{2z_{ii}}\left[\alpha_{ij} - 2\sum_{i'\neq i} x_{i'n}z_{ii'} - c_{in}/p_j\theta_n\right] + v_{ijn}$

Curvature conditions: $Z = LL'$; $L_{iij} \geq 0$

Monotonicity: $p_j\theta_n\left[\alpha_{ij} - 2\sum_{i'} x_{i'n}z_{ii'}\right] - c_{in} > 0$

Reparameterization: $\alpha_{ij} = \sum_m p^m_{\alpha_{ij}}\overline{\alpha}^m_{ij}$, $z_{ii'j} = \sum_m p^m_{z_{ii'j}}\overline{z}^m_{ii'j}$,

$v_{ijn} = \sum_m p^m_{v_{ijn}}\overline{v}^m_{ijn}$ and $e_{jn} = \sum_m p^m_{e_{jn}}\overline{e}^m_{jn}$

Adding up constraints: $\sum_m p^m_{\alpha_{ij}} = 1; \sum_m p^m_{z_{ii'j}} = 1; \sum_m p^m_{e_{jn}} = 1; \sum_m p^m_{v_{ijn}} = 1$

Output distance function $\sum_{h=1}^{H} z_h(Y_{jh} - e_{jh}) \geq (Y_{jn} - e_{jn})/\theta_n$; $h = 1, \ldots, H$

$$\sum_{h=1}^{H} z_h(x_{ijh} - v_{ijh}) \leq (x_{ijn} - v_{ijn})$$

$$z_h \geq 0; \quad \sum_{h=1}^{H} z_h \leq 1$$

Note that the production frontiers of all crops are estimated jointly. In the set of constraints imposed on the output distance function, we subtract e_{jn} from Y_{jn} and v_{ijn} from x_{ijn}. This subtraction avoids attributing any statistical noise to deviations from the frontier, a weakness of DEA noted by many researchers. We use a bootstrapping procedure that combines algorithms developed in Simar and Wilson (2007, 2000). Details of our bootstrapping are not presented here, but are available from authors upon request. The bootstrapping procedure allows us to simultaneously obtain standard errors of production frontier and technical inefficiency coefficients and also generate consistent estimates of technical efficiency parameters (Kneip et al., 1998; Simar and Wilson, 2007). Problem P3 is solved using the General Algebraic Modeling System (GAMS) software. STATA is used to generate bootstrapping samples.

For several reasons, we do not believe that our approach of estimating the production frontier suffers from the fundamental identification problems raised by Marschak and Andrews (1944). First, a system of equations is estimated including both the production frontier equation and the optimality condition equations. Second, in a separate set of analyses, we instrumented for inputs using a set of variables (input prices, whether there is a production shock or not, distance from house to plots, etc.), and the results do not differ much from the case using the raw input uses. Third, since levels of output used in estimation (after correcting for the impacts of production shocks) are close to the expected levels of outputs, we believe there is no simultaneous equation bias, a problem noted by Hoch (1958), which is associated with using actual output instead of expected output in estimation. Finally, the reasonable range out-of-prediction errors (not reported here) further shows that there is no serious bias in our estimates. This confirms what Golan et al. (1996) has stated: "this formulation (of GME estimation) may lead to parameter estimates that are slightly biased but have excellent precision."

The estimation approach produced reasonable estimates of the production frontier coefficients. In Table A.12.1, for the sake of brevity, we only report the results for wheat production frontiers. Bootstrapping results show that most estimates are statistically significant. The linear coefficients (the α_is) are all positive and statistically significant. The quadratic coefficients are also reasonable. For example, coefficients on the interaction term of capital and labor, z_{lc}, are negative and statistically significant for Xian and Tang County. We know from the optimality condition that the sign on the cross-price elasticity of input i and i' is the opposite of the sign on $z_{ii'}$, indicating that labor and capital are substitutes. This is consistent with findings of several studies on developing countries (eg, Garcia-Penalosa and Turnovsky, 2005; Khandker and Binswanger, 1992).

Table A.12.1 Estimation Results of the Wheat Production Frontier, $\sum_i \alpha_i x_i - \sum_i \sum_{i'} x_i z_{ii'} x_{i'}$

County	Input	α_i	$z_{ii'}$				
			Land	Water	Labor	Fertilizer	Capital
Xian county	Land	34.7379 (6.424)**	66.3825 (19.311)**				
	Water	1.7589 (0.136)**	−0.2111 0.074	0.0012 (0.000)**			
	Labor	0.9125 (0.330)**	−0.1297 0.198	0.0005 (0.00027)*	0.0039 (0.002)*		
	Fertilizer	1.3989 (0.411)**	−0.1046 0.151	−0.0004 0.001	−0.0001 −0.001	0.0028 (0.002)*	
	Capital	1.3232 (0.520)**	−0.2267 0.276	0 0.001	−0.0027 (0.0015)*	−0.0012 0.002	0.0101 (0.007)*
Tang county	Land	184.3451 (46.532)**	378.1395 (183.741)**				
	Water	0.4622 (0.100)**	−0.4763 0.355	0.0016 (0.001)**			
	Labor	0.9488 (0.218)**	−0.7962 0.615	0.0025 (0.001)**	0.009 (0.003)**		
	Fertilizer	0.5001 (0.134)**	0.0452 0.477	−0.0021 0.001	−0.0033 −0.002	0.0063 (0.003)**	
	Capital	1.571 (0.243)**	−1.7664 1.367	0.0008 0.003	−0.0037 (0.0021)*	−0.0045 0.005	0.0402 (0.020)**

Continued

Table A.12.1 Estimation Results of the Wheat Production Frontier, $\sum_i \alpha_i x_i - \sum_i \sum_{i'} x_i z_{ii'} x_{i'}$—cont'd

			$z_{ii'}$				
County	**Input**	α_i	**Land**	**Water**	**Labor**	**Fertilizer**	**Capital**
Ci county	Land	48.7264 (12.906)**	80.1194 (44.890)**				
	Water	2.0536 (0.169)**	−0.0748 0.129	0.0018 (0.001)**			
	Labor	0.6073 (0.136)**	−0.2187 0.281	0.0006 (0.000)*	0.0038 (0.002)*		
	Fertilizer	0.2492 (0.076)**	0.0085 0.197	−0.0012 0	−0.0013 0.001	0.003 (0.001)**	
	Capital	0.703 (0.293)**	−0.5294 0.504	−0.0014 0.002	−0.0005 0.003	−0.0023 0.002	0.0174 (0.010)**

For the sake of brevity, estimation results of the production frontier of maize and cotton as well as the set of technical inefficiency parameters are not reported here. Bootstrapped standard errors are reported in parentheses. When performing the GME estimation, we change the unit of land to be squared meter so that the magnitude of land is in range with that of other inputs. Because of the rescaling, coefficients on land are large in magnitudes. Asterisk (*), double asterisk (**), and triple asterisk (***) denote coefficients significant at 10%, 5%, and 1% levels, respectively.

REFERENCES

Brandt, L., Rozelle, S., Turner, M.-A., 2004. Local government behavior and property right formation in rural China. Journal of Institutional and Theoretical Economics 160 (4), 627–662.

Benjamin, D., Brandt, L., 2002. Property rights, labour markets, and efficiency in a transition economy: the case of rural China. Canadian Journal of Economics 35 (4), 689–716.

Charnes, A., Cooper, W.W., Rhodes, E., 1978. Measuring the efficiency of decision making units. European Journal of Operational Research 2 (6), 429–444.

Färe, R., Primont, D., 1995. Multi-output Production and Duality: Theory and Applications. Kluwer Academic Publisher, Boston/London/Dordrecht.

Farrell, 1957. The measurement of production efficiency. Journal of the Royal Statistics Society Series A 120, 253–281.

Garcia-Penalosa, C., Turnovsky, S.-J., 2005. Production risk and the functional distribution of income in a developing economy: tradeoffs and policy responses. Journal of Development Economics 76 (1), 175–208.

Golan, A., Judge, G.G., Miller, D., 1996. Maximum Entropy Econometrics: Robust Estimation with Limited Data. John Wiley & Sons, New York, USA.

Griliches, Z., 1957. Specification bias in estimation of production function. Journal of Farm Economics 39 (1), 8–20.

Huang, J., Rozelle, S., 1996. Technological change: rediscovering the engine of productivity growth in China's rural economy. Journal of Development Economics 49 (2), 337–369.

Hoch, I., 1958. Simultaneous equation bias in the context of the Cobb–Douglas production function. Econometrica 26 (4), 566–578.

Khandker, S., Binswanger, H., 1992. The Impact of Formal Finance on the Rural Economy of India. World Bank Policy Research Working Paper Series No. 949.

Kim, C.S., Schaible, G.D., 2000. Economic benefits resulting from irrigation water use: theory and an application to groundwater use. Environmental and Resource Economics 17 (1), 73–87.

Kneip, A., Park, B.U., Simar, L., 1998. A note on the convergence of nonparametric DEA estimators for production efficiency scores. Econometric Theory 14 (6), 783–793.

Leibenstein, H., 1966. Allocative efficiency vs. X-efficiency. American Economic Review 56, 392–415.

Lin, J.Y., 1992. Rural reforms and agricultural growth in China. American Economic Review 82 (1), 34–51.

Marschak, J., Andrews Jr., W.H., 1944. Random simultaneous equations and the theory of production. Econometrica 12 (3–4), 143–205.

Paris, Q., Howitt, R.E., 1998. An analysis of ill-posed production problems using maximum entropy. American Journal of Agricultural Economics 80 (1), 124–138.

Scheierling, S.M., Young, R.A., Cardon, G.E., 2004. Determining the price-responsiveness of demands for irrigation water deliveries versus consumptive use. Journal of Agricultural and Resource Economics 29 (2), 328–345.

Simar, L., Wilson, P.W., 2007. Estimation and inference in two-stage, semi-parametric models of production processes. Journal of Econometrics 136 (1), 31–64.

Simar, L., Wilson, P.W., 2000. Statistical inference in nonparametric frontier models: the state of the art. Journal of Productivity Analysis 13 (1), 49–78.

Wang, J., Xu, Z., Huang, J., Rozelle, S., 2005. Incentives in water management reform: assessing the effect on water use, production, and poverty in the Yellow River Basin. Environment and Development Economics 10 (6), 769–799.

Methodology Appendix to Chapter 13

A.13.1 ESTIMATING THE BENEFITS OF WATER REALLOCATION IN THE YELLOW RIVER BASIN

A.13.1.1 MODEL SPECIFICATION

The methodology used to model water reallocation in the Yellow River Basin involved four steps. First, a flexible production function was estimated to characterize the agricultural production technologies in the basin for each region and for seven agricultural crops: wheat, maize, rice, vegetables, cotton, soybean, and potatoes. Second, a nonlinear profit maximization problem was formulated for each region. The objective function embeds the flexible crop production functions and input costs. The constraints reflect regional resource limits on the availability of land and water as well as any policy restrictions. Third, the optimization model was solved over a range of water prices to estimate the parameters of a regional water demand function. Fourth, these water demand functions were incorporated into a spatial equilibrium model of regional water markets for the Yellow River Basin to estimate the potential gains from water transfers within the basin.

This methodology was chosen because data sources were limited and the majority of the data that was available is cross-sectional. Production and cost parameters are derived from empirical data. The second stage is to determine the resource costs of land and water, or any policy constraints, as these costs are not fully reflected in input prices. We derive these costs by calibrating the optimization to recreate the base year production and resource use data. The end result of this zero degree of freedom approach is the development of a positive rather than a normative model that can be used to evaluate policies designed to reallocate resources through trade or administrative means.

A.13.1.2 ESTIMATING PARAMETERS OF CROP PRODUCTION AND IMPLICIT LAND COST FUNCTIONS

For the purpose of this study the Yellow River Basin has been divided into 10 regions based on hydrologic, agro-climatic, and soil conditions. This is consistent with previous work undertaken by the World Bank (1993). The important administrative boundaries and provinces included in each region are provided in Table 13.2 and shown in Map 1. There are three basic principles underlying this classification: First, within regions, natural geographic conditions, water resource development and use, and water conservancy development and objectives are sufficiently similar or sufficiently dissimilar; second, distinctions important to different main-stem river sections or tributaries are maintained; third, if possible, the administrative boundaries and catchment areas corresponding to major works on the main stem or tributaries are preserved.

A.13.1.2.1 Assembling a Minimum Data Set

The data for the research comes from both field survey and secondary sources. As the field survey sources, we rely on data from the China Water Management Survey (CWMS) and the North China Water Resource Survey (NCWRS). The secondary sources include production cost data at the provincial level, and areas and yield of crops at the county level, collected by the Ministry of Agriculture.

The parameters of the production technology and regional resource use were estimated from a data set for the base year, 2002. The data set combined county-level agricultural statistics with village and farm-level survey data. The assembled data base included land area allocated, water use, labor and material inputs, product and input prices, and average yield for each region and crop. Water use was estimated by multiplying the area allocated by long term average of total crop evaporative requirement of water, which was estimated using Penman Evaporation data, crop coefficients, and the number of days of a year the crop occupies the land. Summing the land area allocated and the volume of water used over all crops gives an approximation of the regional land and water endowments that can be allocated between crop enterprises.

The minimum data set to estimate the parameters of flexible crop production functions and true costs also required, for each region, the scarcity or rental value of allocable land and water resources. As there are no market price data for land or water, a method suggested by Howitt (1995) was used and the resource scarcity values were estimated by fitting a linear program model for available data. These scarcity values are treated as implicit market prices. In each regional linear program model, profits from regional crop production are maximized subject to resource constraints levels, as estimated above, and a set of calibration constraints designed to exactly calibrate to base year land allocations. Depending on the relative endowment of land and water in the region, one of the two resources becomes scarcest with a positive scarcity value. The dual values or the shadow prices of the resource constraints of the linear program models are taken as their scarcity values.

A.13.1.2.2 Crop Production Functions

The specification of production technology that was used is more flexible than the commonly used fixed coefficient Leontief specification. This is preferred because the Leontief specification may produce misleading results from the impacts of policy changes. For each crop, the production technology is assumed to exhibit constant elasticity of substitution (CES) between inputs. For each region and crop a CES production function of the form given in Eq. [A.13.1] is to be estimated.

$$y_{ri} = \alpha_{ri}\left(\sum_j \beta_{rij} x_{rij}^{\gamma}\right)^{\frac{1}{\gamma}} \qquad \text{[A.13.1]}$$

where $\gamma = \frac{\sigma-1}{\sigma}$; σ = elasticity of substitution; β_{rij} = share of input j in the production of crop i in region r, with $\sum_j \beta_{rij} = 1$; α_{ri} = scale parameter for crop i, in region r.

For each region, r, and crop, i, production function Eq. [A.13.1], which uses j inputs, has j unknown parameters ($j-1$ share parameters of β_{rij} and α_{ri}). Just as in the case of calibration of computable general equilibrium (CGE) models, for each input, j, the share parameter, β_{rij}, is estimated using data on the use and unit cost of all individual inputs (Howitt, 1995). The unit factor cost used here includes scarcity value estimated as explained above on top of the nominal cost. In the case of land, the unit factor cost also includes a marginal crop-specific cost derived from the shadow price of calibration constraints to reflect the heterogeneity in land allocated to different crops. The unit factor costs used here represent true costs that exactly exhaust total revenue for each crop and thus ensure the model exactly calibrates to base year land allocation to alternative crops (Howitt, 1995). For each region, r, and crop, i, the remaining parameter, α_{ri}, is estimated by inverting equation Eq. [A.13.1] and then substituting estimated values of β_{rij} and base year y_{ri} and x_{rij} values.

A.13.1.2.3 Implicit Land Cost Functions

As the estimated share parameters, β_{rij}, include marginal crop specific costs due to land heterogeneity, and the parameters of the corresponding total land cost function should also be derived. This cost function accounts for an implicit cost that needs to be explicitly incorporated to exactly calibrate to base year land allocation data. For each region, r, and crop, i, the implicit cost, c_{ri}, is given by a quadratic function of the area allocated to that crop, $c_{ri} = \delta_{ri} x^2_{ri,j=\text{land}}$, and thus the marginal cost by $\partial c_{ri} / \partial x_{ri,j=\text{land}} = 2\delta_{ri} x_{ri,j=\text{land}}$. The parameters δ_{ri} are estimated by substituting the shadow prices of linear programming calibration constraints for the left-hand side, and the base year land allocation for $x_{ri,j=\text{land}}$ on the right-hand side of the latter (marginal land cost) expression.

A.13.1.3 REGIONAL AGRICULTURAL PRODUCTION PROBLEM

It is assumed that in each region, scarce land and water resources are allocated to alternative cropping enterprises to maximize aggregate profits. The corresponding profit maximization problem is given in Eqs. [A.13.2]−[A.13.4].

$$\text{Maximize} \sum_i p_{ri} y_{ri} - \sum_{i,j} \omega_{rij} x_{rij} - \sum_i \delta_{ri} x^2_{ri,j=\text{land}} \qquad \text{[A.13.2]}$$

subject to

$$y_{ri} = \alpha_{ri} \left(\sum_j \beta_{rij} x^{\gamma}_{rij} \right)^{\frac{1}{\gamma}}, \text{ and} \qquad \text{[A.13.3]}$$

$$\sum_i x_{rij} \leq \psi_{rj} \qquad \text{[A.13.4]}$$

where p_{ri} = price of product of crop i in region r (Y/ton); ω_{rij} = nominal price of input j used in crop i in region r (Y/'000 m^3 for water and Y/day for labor);

ψ_{rj} = endowment of resource j in region r ('000 m^3 for water, days for labor, and hectare for land).

For each region r, the production problem is to maximize the aggregate profit from crop production Eq. [A.13.1] subject to a CES production technology for each crop Eq. [A.13.3] and regional resource constraints Eq. [A.13.4], which state, for each input, j, its use by all crops cannot exceed the endowment, ψ_{rj}.

This model is specified and solved in GAMS as a nonlinear programming problem.

A.13.1.4 ESTIMATION OF WATER DEMAND ELASTICITIES

For each region, the model given in Eqs. [A.13.2]–[A.13.4] is run to estimate regional demand for water at 50 discrete scarcity values (or prices) of water, $\omega_{r,j=\text{water}}$, ranging from 0 to 36.6 USD/'000 m^3. For each region, the data set with 50 observations of water price and quantity demanded is used to estimate the elasticity of demand assuming a constant elasticity specification. The base year water demand, water scarcity value, and the estimated elasticities are given in Table 13.3. The derived market prices or water scarcity values differ substantially between regions and suggest that there can be significant gains from reallocating water between regions.

A.13.1.5 REGIONAL WATER MARKET MODEL

For each region, r, water use, d_r, is a declining function of the user price, p_r. Assuming a constant elasticity specification, the demand function can be given as $d_r = \phi_r p_r^{\eta_r}$, where ϕ_r is the scale parameter and η_r is the elasticity of demand. For each region the value of the scale parameter, ϕ_r, is estimated by substituting the estimated demand elasticity and the base year water use and price given in Table 13.1. For each region, water supply is assumed to be equal to base year water use. The relatively high scarcity value of water (or implicit price) in region 5B is due, in part, to it drawing the majority of water used for irrigation from the Wei River. This region cannot be an importer of water after the reallocation as it is not possible to substitute Wei River water for Yellow River water for environmental reasons. It can, however, be an exporter of water.

A.13.1.5.1 Interregional Water Trade

Fig. 13.1 illustrates price quantity combinations before and after trade and measures of welfare changes for the case of two regions. Demand and supply schedules are denoted by D and S, respectively, and quantity and price by Q and p, respectively. Subscript 1 and 2 are used to denote the region. Price quantity combinations for region 1 and 2, prior to trade are given by (Q_1, p_1) and (Q_2, p_2), respectively and after trade by (Q_1', p_1') and (Q_2', p_2'), respectively. The unit cost of transport losses of water between the two regions is given by $p_2' t$. When trade between the two regions is allowed, region 1 sells q volume of water to region 2, which receives $q(1-t)$ after netting out the transport losses.

Water users in region 1 lose consumer surplus measured by the trapezoidal area $p_1p_1'ac$ while they gain from sales of water measured by the rectangle $Q_1'abQ_1$. The gains from trade for region 1 are thus given by $Q_1'abQ_1$ less $p_1p_1'ac$. Water users in region 2 gain consumer surplus measured by the trapezoidal area $p_2p_2'ed$. Total net gain from trade is given by $p_2p_2'ed$ plus $Q_1'abQ_1$ less $p_1p_1'ae$. If water is transferred administratively rather than traded through the market, assuming the same price and quantity combinations given in Fig. 13.1, the area $Q_1'abQ_1$ represents a fair compensation payment for region 1 water users.

A.13.1.5.2 Trade Flows

$$\phi_s p_s^{\eta_s} + \sum_r q_{sr} \leq s_s + \sum_r q_{rs}(1-t_{rs}), p_s\left[\sum_r q_{sr} - s_s + \sum_r q_{rs}(1-t_{rs})\right] = 0,$$
$$\text{for} \quad s = 1, \ldots, 10$$

[A.13.5]

For each region, s, water use in the region plus water exported to downstream regions, q_{sr}, cannot exceed the regional water supply, s, plus water imported from upstream regions, q_{rs}. If the regional water use plus water exports is less than the supply plus imports from other regions after adjusting for transport losses, then the price of water is zero, and if the price is positive then the first expression is satisfied with strict equality.

A.13.1.5.3 Price Arbitrage Conditions

$$p_r \geq p_s(1-t_{rs}), \quad q_{rs}[p_r - p_s(1-t_{rs})] = 0, \quad \text{for} \quad r = 1, \ldots, 10 \qquad [A.13.6]$$

At each region, r, there is a common price or scarcity value of water traded from the region to the adjacent downstream region, p_r, that cannot be lower than the price of water in the adjacent downstream region net of the cost of transport losses involved in transferring water to it. If the price in region r exceeds the price in the downstream region s net of the cost of transport losses then there are no sales of water to the downstream region. If water trade between the two adjacent regions is positive, then the price in these two regions differ only by the cost of transport losses between them.

This model is also specified and solved in GAMS but as a Mixed Complementarity Problem (MCP).

Methodology Appendix to Chapter 14

In order to quantify the impacts of various factors on the adoption of water saving technology (WST), we have developed the following econometric model:

$$\text{WST adoption} = F(\text{Water Scarcity}, \text{Policy Supports}, \text{Control Variables})$$

The dependent variable, WST adoption, is measured as the percentage of areas that adopted WST. Because there are three kinds of WST studied in this chapter, we have three measures of WST: traditional WST, household-based WST, and community-based WST.

We use three variables—complete reliance on groundwater, reliability of surface water, and reliability of groundwater—to measure overall water scarcity at village level. The dummy variable representing complete reliance on groundwater equals one if the village's irrigation water comes completely from groundwater, and zero otherwise. The reliability of surface water is a lagged variable and is measured as percentage of years when water channels in the village were short of water in 1993–1995 (for the first period of panel data, 1995) and in 2002–2005 (for the second period of panel data, 2005). The reliability of groundwater is also a lagged variable and measured as percentage of years when water wells in the village were short of water in 1993–1995 and in 2002–2005.

In terms of direct government support, we choose three variables related to policies which might affect WST. To minimize problems of endogeneity, these three variables are lagged 5 years. They are: (1) Whether the government had carried out activities to extend WST in the past 5 years, (1 = yes; 0 = no); (2) Whether the government has provided financial support to the village for adopting WSTs in the past 5 years (1 = yes; 0 = no); (3) Whether the county had set up demonstration villages or experimental bases in the village for adopting WSTs in the past 5 years (1 = yes; 0 = no).

In order to control for the impacts of other factors, we include several control variables in the model. They are share of cash crop in total crop area, to control for the impact of the cropping system; two dummy variables on soil type; six variables related to the village's basic characteristics, including arable land per capita (mu/person), share of irrigated area (percent), net income per capita (dollars/person), share of nonagricultural employment (percent), share of villagers with middle-school or high-school education (percent), and the distance from the village committee to the county capital (km). Net income per capita means the net income per capita after adjustment for consumer price index of rural residents taking 2005 as basic term. The nonfarming employment proportion here means [number of villagers earning incomes in their own village number of villagers earning incomes from outside of their village (those out in the morning and back in the evening) + number of villagers numbers of villagers working outside]/total labor force. Finally, we also include provincial dummies to control for the differences among provinces.

As the dependent variables are limited dependent variables with many observed zero values, the method of OLS may result in biased and inefficient estimates. Instead, we use a Tobit estimation method, which is more appropriate for these kinds of limited dependent variables.

Index

'*Note*: Page numbers followed by "f" indicate figures and "t" indicate tables.'

Y

Printed in the United States
By Bookmasters